SOTHIS PRESS

Sothis Press aims to completely reimagine the public domain book with an emphasis on universal access to captivatingly designed print editions. We believe that technology will enable every single book ever published to be perpetually accessible in beautiful print form. Sothis is beauty in ubiquity.

www.sothis.press
1 Waterhouse Square, London, ECIN 2ST

Coal Mine Surveying

A. T. Shurick

SOTHIS PRESS

ISBN 978-1-0368-0000-0

This edition is published by
Sothis Press © 2023. The company
is registered in England and Wales.
The Company Number is 14900389.

MINING WITHOUT TIMBER

BY

ROBERT BRUCE BRINSMADE, B. S., E. M.

FORMERLY PROF. OF MINING ENGINEERING AT WEST VIRGINIA UNIVERSITY, MEMBER OF AMERICAN INSTITUTE OF MINING ENGINEERS, COAL MINING INSTITUTE OF AMERICA, SOCIETY FOR PROMOTION OF ENGINEERING EDUCATION, ETC.

McGRAW-HILL BOOK COMPANY
239 WEST 39TH STREET, NEW YORK
6 BOUVERIE STREET, LONDON, E. C.
1911

*Printed and Electrotyped
by The Maple Pres
York, Pa.*

To

MY PARENTS

THIS BOOK IS GRATEFULLY

DEDICATED.

PREFACE

The rapid depletion of the primitive forests of America by fires and axe in recent years has raised the price of wood so rapidly that the mining industry is becoming alarmed as to its future supply of the big timber of which it has hitherto been such a prodigal consumer. Economy in the use of timber has been essential to commercial success in European mining for several generations and it was to the Old World that our operators went for their first systems of timberless mining.

The use of steel and masonry for the support of mine shafts and tunnels has long been practised in Europe not only because of dear timber but because most mines there are considered to be long-term investments rather than temporary speculations. As this replacement of timber supports by other material involves no new mining system and has been thoroughly covered by other writers, it will not be described in this treatise. Though most of the mining methods considered consume some timber, their economy in that respect is so marked as to justify the use of a title "Mining without Timber."

The work is not intended for a complete treatise on mining, but is meant to deal only with the various systems of excavation with such additional matter regarding exploring, blasting, explosives, and the control of ground as is necessary for the elucidation of the main theme. The emphasis is placed on the peculiar problems of the miner and little space is given to those mining topics which fall chiefly within the provinces of the mechanical, constructing or electrical engineer. Aqueous excavation by hydraulicing, by dredging and by solution for such deposits as those of placer-gold, salt and sulphur has been omitted because that subject can be treated best in special treatises of which there are already several on the market.

The examples of practice have been taken mostly from North America, supplemented by a few from Australia and South Africa. European practice has not been cited not only because its valuable features, modified to meet American conditions, will all be found in the examples given, but because the subject has recently been specially elaborated in Mayer's "Mining Methods in Europe." Where timbering is involved in the examples its details have been condensed since framing diagrams for all purposes are available in such books as Storm's "Timbering and Mining." The costs of the mining are mentioned in many of the examples and in the final chapter is given an outline of the manner of collecting and calculating the data for mine evolution. But no attempt has been made to treat the financial side of mining in detail for that has been lately comprehensively done in modern works like Ingalls' "Economics of Mining," Hoover's "Principles of Mining" and Finlay's "Cost of Mining."

As timberless mining systems are now in the course of development, this book is necessarily somewhat fragmentary and incomplete. It is merely an attempt to chronicle the generally accepted theories and the leading examples of practice so that the student may learn their present status, and the practising engineer may have access to the record of others' experience as a basis for the solution of his own problems. It aims to cover mining systems broadly, rather than particularly, in order to be equally useful to both coal and metal miners.

The examples of practice where they have not been drawn from the author's own articles, work and observations in the mines concerned, have been adapted, as acknowledged in the Appendix, from articles published recently in the technical press. Thanks are due the editors of "Engineering and Mining Journal," "Mines and Minerals," "Mining and Engineering World," "Mines and Methods," "Mining Science," "Mining and Scientific Press," "Transactions of the American Institute of Mining Engineers" etc., for permission to republish much valuable material and use many plates. The author also takes this opportunity to express his gratitude to the numerous mining men including mine-owners, managers, engineers, accountants, foremen, and miners, whose unfailing courtesy to him, on his visits of investigation to their mines, alone has made this book possible.

ROBERT BRUCE BRINSMADE.

MORGANTOWN, W. VA.
November, 1911.

CONTENTS

CHAPTER IX.

CHAPTER X.

CHAPTER XI.

CHAPTER XII.

CHAPTER XIII.

CHAPTER XIV.

CHAPTER XV.

CHAPTER XVI.

CHAPTER XVII.

CHAPTER XVIII.

CHAPTER XIX.

CHAPTER XX.

CHAPTER XXI.

CHAPTER XXII.

CHAPTER XXIII.

MINING WITHOUT TIMBER

CHAPTER I

EXPLOSIVES AND THEIR USE IN MINING

An explosion may be defined as a sudden expansion of gas. The substances which we call explosives are so unstable when exposed to a suitable flame or shock that they suddenly change into many times their original volume of gas with the evolution of heat. If the change to a gas takes place in the open, there is a flame and a whiff or a report. It is only, however, when explosives are set off in confined spaces like drill-holes that they do their chief work in mining. Consequently a blast or explosion may be said to be a rapid combustion in a confined space.

Explosives have two essential constituents, namely, combustibles and oxidizers. They may be broadly divided into three classes according to the relation which the combustibles bear to the oxidizers. Class I includes the mechanical explosives, or those in which the ingredients constitute a mechanical mixture; class II includes the chemical explosives or those in which the ingredients are in chemical combination; class III inc udes the mechanico-chemical explosives which are formed of a mixture of class II and an absorber.

Methods of Firing Explosives

Explosives are set off by two means—ignition and detonation. Because through ignition the combustion is transmitted by heat alone, it gives a slower explosion than one started by detonation which transmits the reaction by the rapidity of vibrant motion. By their nature class I is adapted to ignition, and classes II and III to detonation.

Ignition is commonly performed by squibs, fuse or electric igniters. A squib is really a self-impelling slow match, made by filling one-half of a thin roll of paper with black powder and the other half with sulphur. For their use in blasting, a drill-hole *ab*, Fig. 1, is loaded with an explosive *bc* and before filling the hole with the tamping *cd*, a needle *ac* is inserted into the explosive so that when it is withdrawn, a hole of a larger diameter than the squib is left through the tamping from *a* to *c*.

The squib is then inserted in this hole with the sulphur end out, and when lit the slow-burning sulphur allows time for the miner to escape before the powder of the squib takes fire and its reaction forces the squib along the holes to ignite the powder at *c*.

A fuse is merely a thread of black powder wrapped with one or more thicknesses of tape. In loading the hole, Fig. 1, the fuse would be inserted in place of the needle *ac*. A fuse burns commonly at the rate of 2 ft. a minute. Therefore a sufficient length should be used in the hole to allow the miner to retire in safety, after splitting and lighting the outer end, before the flame reaches the explosive at *c*.

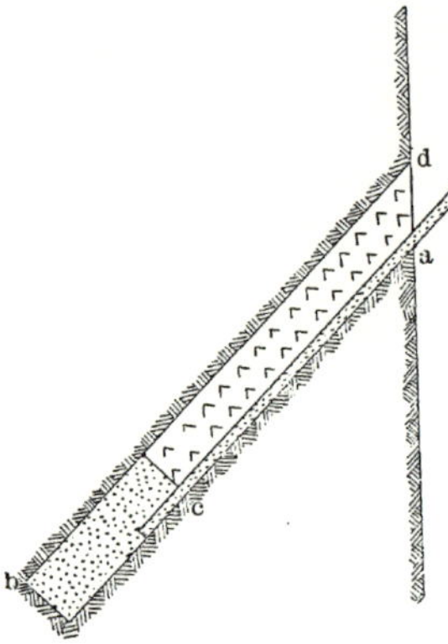

FIG. 1.—Drill-hole section.

The electric igniter consists of a shell *a*, Fig. 2, enclosing a charge of fulminate mixture in *b* and of sulphur cement in *e*. The copper wires *c* pass through *f* and enter *b* where they are connected by a platinum bridge at *d*. For ignition, the shell *a* is made of pasteboard and the igniter is placed within the explosive while the wires extend outside the hole to a blasting machine. The last is simply a small armature revolving between its poles and sending a current through the igniters in the circuit when its handle is shoved down. All the common electric igniters on one circuit are exploded simultaneously, but a recent invention is a delay-action igniter which permits electric firing in sequence.

Detonation is performed by fuse and cap or by electric caps. A blasting cap is simply a cylindrical copper cup with a small charge of fulminate mixture in its bottom, the fuse being inserted into the cup and fastened to it by crimping pincers. The cap is then inserted into one cartridge of the explosive and its attached fuse tied firmly to it by a string, in order to make a primer which is placed near or on the top of the explosive. The loaded hole will then resemble Fig. 1, the explosive being in *bc*, the cap and primer at *c*, and the fuse along *ca*. Lighting the fuse is the same as for ignition, only the fuse now fires the cap whose explosion detonates the explosive.

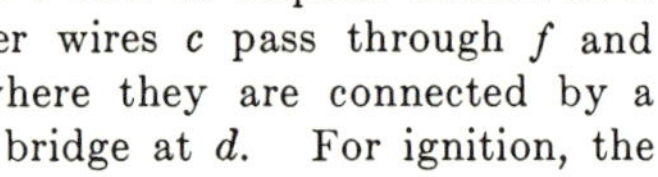

FIG. 2.—Electric exploder.

The electric cap resembles the electric igniter, Fig. 2, but has a copper instead of a pasteboard case *a* and the quantity of charge of fulminate mixture at *b* is increased as the sensitiveness of the explosive diminishes. The electric cap is inserted in and fastened to a primer-cartridge like fuse and cap, the electric cap being fired by a blasting battery in the same way as the electric igniter.

Loading and Tamping

A mechanical explosive like black powder usually comes in bulk. For loading it is poured into a cartridge (the size of the hole) which is made by rolling a piece of paper around a pick handle. For damp holes the cartridge must be oiled or soaped on the outside. This paper cartridge is pressed down into the hole by a soft iron tamping bar whose tip should be an expanding copper cone grooved on the edge for the purpose of allowing the copper loading needle or fuse to pass. Tamping bars with iron tips or iron needles are highly dangerous in formations containing pyrite or other hard minerals, on which the iron might strike a spark, and their use is therefore prohibited by law in many places.

A mining explosive of class II or III is handled in paper cartridges which can be ordered of a diameter to fit the hole. Before loading they are slit around lengthwise to permit of the explosive taking the shape of the hole when it is pressed down by a tamping bar which should be of wood for these explosives, instead of copper-tipped iron, on account of their being more sensitive to any shock than black powder.

In coal mines, coal dust is commonly used for tamping black powder, but this is a very unsafe practice in dangerous mines, for a windy or blown-out shot will have its normal flame increased, both in length and duration, by the ignition of the tamping. The best materials for tamping are a fine plastic clay or loam and ground brick or shale, and although sand is too porous to do well for black powder, it answers for higher explosives but must be confined in paper cartridges for use in uppers.

Water is used as tamping for nitro-glycerine and high explosives in wet down-holes, but it is little better than nothing. The fact that higher explosives will break rock without any tamping has caused many miners to abandon tamping them altogether on account of the ease of recapping untamped charges in case of a misfire. Mechanical explosives must be tightly tamped, nearly to the collar of the hole, or they will blow out instead of breaking the rock, and although the tamping may be shortened with detonating explosives, as they become quicker and stronger, a short length of tamping adds to the efficiency of the highest explosives.

Where only quick-acting explosives of classes II or III are at hand and it is desired to blast with the slow action of class I, the object can be partially obtained by special methods of loading. These methods provide an air cushion between the explosive and the rock and tamping by either having the stick of explosive of considerably smaller diameter than the drill hole or by having a very porous cellular tamping to separate the tight tamping from the explosive.

Before examining the various mine explosives in detail, let us consider an illustration of the method of calculating, from the chemical equation of an explosive, its calorific power, its temperature, and the number of expansions and its consequent exploding pressure. Let us assume the simplest case of a mechanical mixture of hydrogen and oxygen at a temperature of 0° C. and at sea-level pressure of 760 mm. of mercury. Then the chemical equation for complete combustion is

$$2H_2+O_2=2H_2O. \qquad (1)$$

the molecular weights being $4+32=36$. (2)

If t=thermometer temperature in degrees centigrade of the explosion;

T=absolute temperature in degrees centigrade of the explosion;

Σ=sign for summation;

WW_1W_2, etc.=weights in grams of various combustibles of the explosive;

CC_1C_2, etc=calorific power in calories of various products of combustion of the explosive;

ww_1w_2, etc.=weights in grams of various products of combustion of the explosive;

ss_1s_2, etc.=specific heat in calories of various products of combustion of the explosive;

V=volume of explosive originally;

V_1=volume of explosive due to chemical reaction alone;

V_2=volume of explosive due to chemical reaction and resulting temperature, t;

P=pressure of explosive originally;

P_2=pressure of explosive finally;

then we have from thermo-chemistry,

$$T=\frac{WC+W_1C_1+W_2C_2, \text{ etc.}}{ws+w_1s_1+w_2s_2, \text{ etc.}}=\frac{\Sigma WC}{\Sigma ws} \qquad (3)$$

For the given problem we have from equation (2),

W=4 grams of H gas;

w=36 grams of H_2O vapor.

From thermo-chemistry we have,

C=28,780 cal. for H;

s=0.4805 cal. for H_2O vapor;

substitute in (3) and

$$T=\frac{4\times 28,780}{36\times 0.4805}=6660° \text{ C.}$$

Then, from Avogardro's law, that the molecules of equal volumes of all gases under like conditions occupy the same volume, we have from (1),

$$2 \text{ vols. H}+1 \text{ vol. O}=2 \text{ vols. } H_2O,$$

or

$$V_1=2/3V. \qquad (4)$$

From Charles' law, the volumes of gases vary directly as their absolute temperature we have thus

$$\frac{V_2}{V_1}=\frac{T}{0+273}$$

or

$$V_2=\frac{6660V_1}{273};$$

substitute from (4) and we have

$$V_2=\frac{6660\times 2V}{273\times 3}=16.2\ V \tag{5}$$

From Boyle's law, if the gas of volume V_2 is prevented from expanding beyond volume V, we have for the final pressure P_2 in the explosive chamber P,

$$\frac{P_2}{P}=\frac{V_2}{V}$$

or

$$P_2=\frac{V_2}{V}P \tag{6}$$

Substitute in (6) from (5) and, as $P=1$ atmosphere $=14.7$ lbs. per sq. in., we have

$$P_2=\frac{16.2VP}{V}=16.2 \text{ atmospheres.}$$

or 238 lbs. per sq. in.

From physics, $T=t+273$,
hence

$$t=T—273=6660—273=6387° \text{ C.}$$

In practice, this theoretical pressure and temperature, resulting from the explosion, would have to be multiplied by a fractional factor of efficiency to allow for imperfect combustion and loss of heat through radiation and leakage. In large charges, these losses are proportionally less than in the case of small charges. This fact, coupled with the greater likelihood of their meeting weak places in the blast's burden, accounts for the higher efficiency of the former. These theoretical calculations are especially useful in comparing the relative strength of different explosives of the same type. In France, they are used extensively in the inspection of permissible explosives to determine if their final temperature is sufficiently low for use in dangerous coal mines

The practical usefulness of explosives depends upon (1) their cost of manufacture; (2) their safety and convenience as regards transportation and storage; (3) method necessary for their loading and exploding; (4) their exploding pressure; (5) the rapidity with which they explode; (6) the length and temperature of the flame. These six factors will now be discussed seriatim. Factor (1), or the cost, is often the most important factor in commercial operations like mining, although for purposes of war it is often little considered. Factor (2) or safety, affects the desirability for all purposes, the more sensitive the explosive, the higher the freight rate by rail or boat, and if sensitive beyond a certain point, it cannot be shipped thus at all. Those explosives which, like dynamite, freeze at ordinary winter temperatures are at a disadvantage as are also those which, like black powder, are handled loose and can be easily ignited by a spark struck by a hob-nailed shoe on a floor spike. Some explosives, like imperfectly washed guncotton, are liable to explode by spontaneously generated heat, while others become dangerously sensitive if exposed to the sun during shipment. The desirability of explosives belonging to either of these last two mentioned classes is plainly discounted because of these attributes. The next factor (3) or loading and exploding, is important in connection with conditions such as prevail in dangerous coal mines (where an open light is prohibited), in subaqueous blasting (where both explosive and exploder must be unaffected by water), or where misfires could not be corrected. Factor (4), or the pressure, is what determined the real effective breaking force of the explosion, but it is modified in practice by (5), or the rapidity of the explosion. Slow and fast explosives are comparable to presses and hammers for forging steel. The former exerts its pressure gradually until the strain exceeds the tensile strength of the material and the rock gives way along a surface of fracture. The latter gives a sharp quick blow which will shatter the surface of rock exposed to the explosive before any fracturing action is exerted on the blast's burden of rock.

The slow explosive will detach the rock in large masses while the fast type may crush it to bits. Black powder is an example of the first and nitro-glycerine of the second. Explosives with all graduations of rapidity between these extremes are on the market. The fastest explosives are applicable where the rock is very hard to drill as, for example, in the case of certain Lake Superior hematites, or where a tremendous force must be exerted from confined spaces as in breaking the cut for development passages; also where a shattering rather than a fracturing action is needed, as in chambering the bottom of drill holes or in shooting oil wells. The slowest explosives are used in quarrying, for the purpose of detaching monoliths, or in consolidated or soft rock which can be fractured by a slow, pressing movement but only dented by a quick hammer blow.

Factor (6), or the flame and temperature, is an important consideration for blasting in gassy or dusty coal mines. The so-called "permissibles" are explosives made to fall below a minimum legal requirement as regards length and temperature of flame. When one considers that a permissible like carbonite gives, in practice, a flame height of 15.8 in. and a flame duration of 0.0003 seconds, as compared with 50.2 in. and 0.1500 seconds respectively, for black powder, we can see how much safer the permissible is to use.

We will now consider the properties of the three classes of explosives:

Class I, or Mechanical Explosives

The common representatives of this class are black powder and mechanical permissible explosives. Black powder was discovered before 600 A. *D.* by the Chinese, and by Roger Bacon in 1270, but it was not used for mining until Martin Weigel introduced it at Freiberg in 1613. It can be made from a single combustible, charcoal, mixed with an alkaline-nitrate oxidizer, but in order to lower its ignition temperature for blasting to about 275° C., part of the charcoal is replaced by sulphur. For the cheaper blasting powders, the oxidizer is sodium nitrate which, being easily affected by dampness, is replaced in the higher grade powders by potassium nitrate. The ingredients are first ground then mixed thoroughly while moist and finally pressed in cakes, dried, broken and sized. Assuming the equation for the complete combustion of black powder to be.

$$3C+S+2KNO_8=3CO_2+N+K_2S. \qquad (7)$$

We have by calculation for its percentage composition,

$$\begin{aligned} \text{carbon} &= 13.4 \\ \text{sulphur} &= 11.8 \\ \text{sodium nitrate} &= 74.8 \\ & \overline{100.0} \end{aligned}$$

and for the percentage composition by volume of its resulting gas,

$$\begin{aligned} CO_2 &= 75 \\ N &= 25 \\ & \overline{100} \end{aligned}$$

The theoretical exploding temperature is 4560° C. and the pressure is 5820 atmospheres. In practice the composition is varied according to the experience of each maker. As the combustion is imperfect, poisonous and combustible gases like carbon monoxide, hydrogen sulphide and hydrogen and unpleasant vapors, like the sulphide, sulphate, hyposulphite, nitrate and carbonate of potassium, are given off by the explosion and sometimes render breathing or the carrying of open lights in the fumes a dangerous procedure. In fact, Bunsen's experiments proved that only one-third of the ignited gunpowder really followed the reaction of equation (7).

Black powder is sold in grains which vary in size from the fine sporting gunpowder to the 2-in. balls of artillery powder. For blasting, the grains vary in diameter from one-eighth to one-half of an inch, and the rapidity of the explosion decreases with an increased diameter of grain. The grains should be of uniform size, quite dry and thoroughly tamped in the hole in order to get good results. The specific gravity of lightly shaken black powder is about the same as water. Its cheapness, non-freezing, comparative safety for shipping and handling, easy explosion by ignition and slow action are the favorable qualities of black powder which cause its wide use. For coal mines free from dangerous gases and dust, it is a better explosive than detonating permissibles whose quicker action breaks up the coal and injures the roof more. Black powder is rendered inefficient for many other purposes, however, because of its necessitating much tamping, its low power, the readiness with which it is spoiled by moisture and its long flame.

Of the mechanical permissibles bobbinite has been extensively used in England. Its percentage composition is,

$$
\begin{aligned}
\text{Potassium nitrate} &= 65.0\\
\text{Charcoal} &= 20.0\\
\text{Sulphur} &= 2.0\\
\text{Paraffin wax} &= 2.5\\
\text{Starch} &= 8.0\\
\text{Water} &= 2.5\\
\hline
&\quad 100.0
\end{aligned}
$$

It is thus chemically very close to black powder excepting that it contains more charcoal and less sulphur and makes up that discrepancy by the addition of wax, starch and water. The lack of sulphur raises its ignition temperature while the wax forms a waterproof coating for the grains of powder. The starch and water absorb heat, shorten the flame and decrease the exploding temperature to under 1500° C. It is handled in compressed cartridges with wax coverings. It has a central hole to admit the fuse, for ignition by squib is not allowed in dangerous coal mines.

Class II, or Chemical Explosives

The five common explosives of this class are guncotton, nitro-glycerine, nitro-gelatin, fulminates and picrates. They all contain nitryl (NO_2) and their detonation is made possible by the unstable quality of nitryl compounds.

Guncotton.—This was discovered by Schönbein in 1846, but it was little used until it was found that its dangerous instability was not inherent but due solely to the surplus acid left in its tissue by imperfect washing methods during its manufacture. The equation for making it is,

$$C_6H_{10}O_5 + 3HNO_3 = C_6H_7O_5(NO_2)_3 + 3H_2O. \quad (8)$$

cotton + nitric acid = guncotton + water.

The ingredients are allowed to stand in a cold place for some time before the washing out of the free acid is begun.

The reaction on exploding is,

$$2C_6H_7O_5(NO_2)_3 = 3CO_2 + 9CO_2 + 3N_2 + 7H_2O. \quad (9)$$

Equation (9) shows that the explosion gives no solid product like the K_2S of equation (7) and that the percentage composition by volume of the resulting gas is,

$$CO_2 = 13.7$$
$$CO = 40.8$$
$$N = 13.7$$
$$H_2O = 31.8$$

100.0

By the method of calculation already explained, it is found that guncotton theoretically has an exploding temperature of 5340° C. and a pressure of 20,344 atmospheres.

The combustible qualities of the large percentage of carbon monoxide resulting from its explosion render guncotton unfit for use in coal mines, and its poisonous qualities make it unsuitable for any underground use. For surface work, it is very powerful, smokeless, does not freeze and is not volatilized or decomposed by atmospheric temperature. It ignites between 270 and 400° F. and if unconfined will then burn quietly. When dry, it is sensitive to percussion and friction, but under water it is insensible to ordinary shocks. Immersed, it absorbs from 10 to 15 per cent. of water, but even then it can be exploded without drying by the use of an extraordinarily strong detonator. Its chief disadvantage above ground is its high cost and the fact that it comes in hard compressed cartridges (specific gravity about 1.2) which fit drill holes only imperfectly and therefore lose in efficiency. For any destructive work without the use of drill holes, like demolishing walls, dams and the like, the sharp, sledge-hammer blow of its explosion renders it very efficacious.

Nitro-glycerine or "Oil."—This was discovered by Sabrero in 1847, but did not become commercially valuable until 1863 under the direction of Alfred Nobel. The equation for its making is,

$$C_3H_8O_3 + 3HNO_3 = C_3H_5O_3(NO_2)_3 + 3H_2O. \quad (10)$$

glycerine + nitric acid = nitro-glycerine + water.

Strong sulphuric acid is an ingredient of the mixture, but it does not take part in the reaction, which must take place at a moderate temperature to be safe. The resulting "oil" is much easier to wash than gun-cotton and consequently is cheaper. It is a yellow, sweetish liquid poisonous both to the blood and the stomach. Its specific gravity is 1.6. Its freezing-point is about 45° F. and to insure against freezing the temperature must be above 52° F. When frozen, it is insensible to ordinary shocks, as is also the case when it is dissolved in alcohol or ether. It is, therefore, commonly shipped either in tin cans, packed in ice, or in so ution in wood alcohol. It can be precipitated from the latter before use by an excess of water.

Nitro-glycerine does not evolve nitrous fumes until 230° F. As it begins to vaporize at about 100° F., it is important in thawing it not to exceed this temperature. Thawing, therefore, is only safely done by heating the explosive over a water bath at less than 90° F., or by leaving it in a room of the same temperature for some time. The explosive ignites at only 356° F. and if then pure and free from all pressure, jar or vibration, it will burn quietly. These safe-igniting conditions, however, are difficult to obtain, for a small depth of liquid causes sufficient pressure to explode it when ignited. Thus a film of it, heated on a tin plate, burned without an explosion only if under one-fourth inch thick. The exploding temperature is 380° F. This 24° margin above the igniting temperature accounts for the numerous cases of conflagration without explosion. The reaction of the explosion is,

$$4C_3H_5O_3(NO_2)_3 = 12CO_2 + O_2 + 3N_2 + 10H_2O. \qquad (11)$$

From equation (11) the explosive product is gaseous and its percentage composition by volume is

$$\begin{aligned} CO_2 &= 46.0 \\ O &= 3.8 \\ N &= 11.8 \\ H_2O &= 38.4 \\ & \overline{100.0} \end{aligned}$$

By the previous calculating method, it is found that theoretically the exploding temperature is 6730° C. and the pressure is 29,107 atmospheres. From the fact that its explosive product contains no carbon monoxide, "oil" can be used underground, but only when mixed with an absorber. Alone, it is too sensitive to be safe, while being liquid, if unconfined, it would leak from holes in porous rock, and if confined in canisters it will not fill the drill hole. With its great speed and strength it also tends to shatter locally any enclosing rock, except the toughest, rather than detach it. These characteristics render it inefficient for most mining work.

For shooting oil wells, however, its shattering quality renders it peculiarly suitable. For this purpose, a cylindrical canister of a diameter to fit the well and containing from 100 to 200 lbs. of nitro-glycerine, is carried to the well swung from the body of a spring buggy. After filling the well with water, the canister is topped with a cap and lowered to the proper depths by a rope, along which a weight, called a "go-devil," is dropped onto the cap to cause the explosion.

Nitro-gelatin.—This was discovered by Nobel in 1875 and is a yellowish jelly of considerable toughness, but easily cut with a knife. It is made by dissolving guncotton in nitro-glycerine. Authorities differ in the proportion of guncotton, some recommending only 7 per cent. To balance all the free oxygen of the nitro-glycerine by the excess carbon of the guncotton alone, takes 87.3 per cent. of the former to 12.7 per cent. of the latter and gives the following equation:

$$9C_3H_5O_3(NO_2)_3+C_6H_7O_2(NO_2)_3=33CO_2+15N_2+26H_2O. \quad (12)$$

From equation (12) the percentage composition of the solely gaseous product is,

$$CO_2=44.6$$
$$N=20.2$$
$$H_2O=35.2$$

By the theoretical calculation, the exploding temperature is 7080° C. and the pressure is 27,100 atmospheres. The last figure shows nitro-gelatin to be only 7 per cent. weaker by weight than nitro-glycerine, while its somewhat higher cost is due to its guncotton ingredient. When used alone for military purposes, about 4 per cent. of camphor is dissolved in the nitro-glycerine along with the guncotton to make a product called military gelatin. The last explosive is so insensitive that it can be punctured without effect by a rifle bullet. The common nitro-gelatin is much less sensitive than No. 1 dynamite, to shock or friction, and unaffected by a short immersion in water at 158° F. and by an 8-day immersion at 113° F.

It will not exude nitro-glycerine under a high pressure or any atmospheric temperature. Its specific gravity is 1.6 and it can be set off only by a strong detonation. It ignites at 399° F. and will then only burn when unconfined. When it freezes, which is between 35 and 40° F., it becomes more sensitive than normally owing probably to the partial freeing of the nitro-glycerine ingredient.

Nitro-gelatin is now used for mining wherever the highest power explosive is needed and is especially adapted to wet or subaqueous blasting, either alone or as "gelatin" dynamite.

Fulminates.—Mercuric fulminate is the common commercial salt. It is made as follows from mercuric nitrate and alcohol:

$$Hg(NO_3)_2+C_2H_3O=Hg(CNO)_2+3H_2O+2O. \quad (13)$$

The explosive reaction is

$$Hg(CNO)_2=HgO+CO+C+2N. \quad (14)$$

Equation (14) shows that mercuric fulminate is a poor explosive because it produces the poisonous fumes of HgO and CO as well as unburned carbon. If a little damp, it explodes very feebly and if quite wet, not at all. However, its non-freezing quality, its quick hammer-like vibrant explosion and its uniform sensitiveness to ignition or shock cause its use as the chief ingredient of percussion-cap mixtures for detonating other explosives. Its exploding temperature is 305° F.

Picrates.—These salts are founded on picric acid, which is made by mixing carbolic and nitric acid according to the equation,

$$C_6H_6O + 3HNO_3 = C_6H_3(NO_2)_3O + 3H_2O. \qquad (15)$$

Its explosive reaction is

$$C_6H_3(NO_2)_3O = H_2O + H + 6CO + 3N. \qquad (16)$$

Picric acid comes in yellow crystals which are soluble in hot water or alcohol, and melt at 230° F. It is used very largely in dyeing. It is expensive to make and difficult to explode. Equation (16) indicates that it produces much of the poisonous carbon monoxide which shows incomplete combustion and consequently a decreased power. Picrates are the basis of the military explosive lyddite, but the recent commercial failure of the excellent mining picrate jôveite may discourage future attempts to adapt them to blasting.

Class III, Mechanico-Chemical Explosives

This class will be considered under five groups: (1) guncotton; (2) nitro-glycerine; (3) nitro-gelatin; (4) fulminate; (5) nitro-benzol. Detonating permissibles for coal mining fall mainly under groups (2) and (5) and will be considered last.

Guncotton Group.—The evaporating of guncotton, after it has been dissolved in a suitable solvent such as alcohol or acetone, produces a hard, horny material which is the basis of most modern smokeless gunpowder. Its chief blasting powder, however, is tonite which is formed by adding enough barium nitrate to guncotton to just completely oxidize the gases caused by the explosion as follows:

$$4C_6H_7O_5(NO_2)_3 + 9BaNO_3 = 24CO_2 + 21N + 14H_2O + 9BaO. \qquad (17)$$

The percentage composition, by volume, of the gaseous product of equation (15) is,

$$\begin{aligned} CO_2 &= 45.7 \\ N &= 20.0 \\ H_2O &= 34.3 \\ &\overline{100.0} \end{aligned}$$

By calculation, the exploding temperature is 3590° C. and the pressure is 10,300 atmospheres, which are two-thirds and one-half, respectively, of the corresponding figures for guncotton. As an offset to lessened power tonite is plastic, cheaper than guncotton and 50 per cent. denser. Its harmless fumes adapt it to underground use and, like dynamite, it is packed in paper cartridges. It has been extensively used in England, where it is shipped under the same safety regulations as black powder. It is hard to ignite and when alight, it normally burns slowly without explosion. Tonite, like guncotton, is non-freezable and is detonated only by a strong cap. Potassium nitrate has been used, instead of bar um nitrate, as the oxidizer, in another guncotton mixture of similar properties which is called potentite.

Nitro-glycerine Group.—These mixtures are called dynamites. They were introduced by Nobel to lessen the sensitiveness of nitro-glycerine and at the same time retain its other good qualities. The absorber of the "oil" is called the "dope," which may be selected to be either *inert* or *active* in the explosion.

The freezing temperature of all dynamite is that of nitro-glycerine, as is also its behavior when frozen and its method for being safety thawed. Dynamite that does not leak nitro-glycerine under the conditions under which it is to be used is one of the safest explosives known. It should not be shipped, however, in rigid metallic cases, which accentuate shocks and vibrations, but in wooden boxes in paper cartridges packed in saw-dust. Thus packed, it has failed to explode when dropped on the rocks from a considerable height or when struck by heavy weights.

Dynamite can be heated with less danger than nitro-glycerine. If set on fire, it will usually burn quietly unless unfavorable conditions are present. If the dynamite is in a closed box, its smoke cannot escape and consequently the pressure may be raised enough to cause an explosion. If caps or gunpowder are present, the fire will explode them and the resultant shock will detonate the dynamite, If the heat from the fire causes the "oil" to exude from the cartilages, this "oil," if under a static head, will explode when ignited, as explained above. Again, the heat from the burning dynamite may heat the adjoining unlighted cartridges to the exploding temperature of 380° F. before they get sufficiently exposed to the air to ignite. Heated gradually in the open so much of the "oil" may be evaporated that a mere whiff ensues when the exploding temperature is finally reached.

In spite of all these dangerous contingencies, several instances are on record where several tons of dynamite have burned in conflagrations without exploding. If afire in cartridges, it burns slowly like sulphur, but if loose it will burn quickly like chaff.

The dope first used was inert infusorial earth or kieselguhr, which will safely absorb three times its weight of nitro-glycerine. The resulting kieselguhr dynamite when strongest contains 75 per cent. "oil." It is a pasty, plastic, unctuous, odorless mass of a yellowish color with a specific gravity of 1.4. The effect of the dope is to cushion the "oil" so that the shock to explode it must be stronger as the percentage of dope becomes greater. It is not possible to explode kieselguhr dynamites which contain under 40 per cent. of "oil" and even with 60 per cent. it takes a strong cap.

The disadvantage of 75 per cent. dynamite is the exudation of "oil" on a warm day or under water so that dangers may arise from having to deal with the sensitive "oil" before suspecting its presence. It is thus ordinarily unsafe to ship or use and the 60 per cent. strength is now commonly sold as No. 1. The strength of kieselguhr dynamite is almost equal to that of its contained "oil."

The active-dope dynamites have no such narrow limitations as the inert types and not only may numerous absorbers be used, but the percentage of nitro-glycerine may vary from 4 to 70 per cent. These explosives go under various names. The common active absorbents are such combustibles as wood meal or fiber, rosin, pitch, sugar, coal, charcoal, or sulphur, and such oxidizers as the alkaline nitrates or chlorates. The chemical composition of the oil-dope mixture should be such as to give only completely oxidized products on combustion. The strength of this type is equal to that of the "oil" plus that of the explosive dope when *completely* burned. In other words, black powder mixed with enough "oil" to detonate it would *all* burn as shown by the reaction of equation (7), thus giving several times more power than when ignited alone. The density and appearance, as well as the necessary strength, varies with the dope and the percentage of "oil." The commercial method of rating dynamite, by its percentage of "oil," is misleading as no account is taken of the varying strength of the explosive dopes.

Nitro-gelatin Group.—A mixture of this group is called a *gelatin* dynamite. Somewhat more expensive than nitro-glycerine, it is preferable wherever the highest power is desired and, being unaffected by water, it is the best powder for subaqueous use. It is more plastic and less sensitive than common dynamite and therefore easier to load and safer to transport, but it requires a stronger cap for exploding. The military powder gelignite, a favorite in England and Japan, and forcite come under this group.

Fulminate Group.—For percussion-cap filling, mercuric fulminate is mixed with a sufficient amount of some oxidizer to insure complete combustion on exploding. Alkaline-nitrate oxidizers may be used but potassium chlorate is the favorite. The latter gives the following exploding reaction:

$$Hg(CNO)_2 + KClO_3 = HgO + KCl + 2CO_2 + 2N. \qquad (18)$$

Equation (18) shows that potassium chlorate should form 30 per cent. by weight of the mixture, which also contains a little gum to give coherence. Caps are designated by numbers or letters according to the amount of fulminate contained. The common series is.

Cap No.	$Hg(CNO)_2$ Grains.
1	4.5
2	6.0
3	8.0
4	10.0
5	12.0
6	15.0
6.5	19.0
7	23.0
8	30.9

The larger the cap, the more expensive, but if the cap selected is too small to insure perfect detonation of the explosive, incomplete combustion will ensue with noxious fumes and loss of power. In general, dynamite requires stronger caps as the percentage of nitro-glycerine or the temperature decreases.

Nitro-benzol Group.—Although nitro-benzol contains nitryl it does not contain sufficient oxygen to be an explosive and, when unmixed with its oxidizer, it can be shipped as an ordinary chemical. On this account, the nitro-benzol or Sprengel group is especially adapted for use in isolated places far from dynamite factories. The favorite Sprengel explosive is rackarock, which is a mixture of nitro-benzol with the chlorate or nitrate of potassium or with sodium nitrate, as an oxidizer. By mixing 77.6 per cent. of mononitro-benzol with 22.4 per cent. of sodium nitrate, we can get the following reaction on detonation:

$$2C_6H_5(NO_2) + 10NaNO_3 = 12CO_2 + 6N_2 + 5H_2O + 5Na_2O. \qquad (19)$$

From equation (19) the percentage composition, by volume, of the gaseous product is,

$$\begin{aligned} CO_2 &= 52.2 \\ N &= 26.1 \\ H_2O &= 21.7 \\ & 100.0 \end{aligned}$$

By calculation, the theoretical temperature is 5300° C. and the pressure is 13,800 atmospheres. Unlike ignited black powder, rackarock, when properly detonated, follows closely its theoretical reaction which shows harmless gases and a temperature of 79 per cent. and a pressure of 47 per cent. of the figures for nitro-glycerine. For practical use, the oxidizer of rackarock is handled alone in wax-paper cartridges and the required quantity of nitro-benzol is not poured into a cartridge until just before charging the drill hole.

Detonating Permissibles.—These explosives practically all contain either nitro-glycerine, nitro-gelatin, nnitro-bezol or ammonium nitrate as the detonated ingredient and some contain two or more of them. Their exact composition is usually kept secret by the manufacturers, but they must pass the government tests for temperature and flame. These explosives are made of various strengths and require stronger caps than common dynamites. Detonation means a quick generation of a small quantity of hot gas while the ignition of black powder means the slow production of a large quantity of impure gases and vapors. A large quantity of fine, unstable salt like magnesium carbonate, of a steam-generating salt like ammonium nitrate, or of a substance with much hygroscopic moisture like wood meal, are the ingredients relied upon to cool the quick small flame of permissibles. The compositions of a few typical permissibles are as follows:

Name.	Nitro-benzol.	NH_4NO_2	Ground Wood.	Water.
Amvis	4.50	90.0	5.0	0.50
Ammonite	12.00	88.0	...	....
Electronite ..... 19.0 (BaNo),	......	73.0	7.5	0.50
Westfalit, No. 1. 4.5 (rosin)[3]	......	95.0	...	0.50
Bellite, No. 3	5.25	94.0	...	0.75
Carbonite ....25.0(" oil ")	0.50(soda)	34.0 ($NaNO_3$)	40.5	....

Misfires

The cause of misfire depends upon both explosive and the manner of firing. The three classes of explosives with their methods of firing will now be considered.

Mechanical Powders of Class I.—In breaking coal with igniting powders, it is inadvisable to attempt to use a missed hole if the tamping must first be dug out, therefore a new hole is bored, charged and fired alongside the first. In rock breaking, where boring holes is expensive, the tamping may be dug out safely if only copper tools are used when approaching the powder. However, if the explosives are well selected, and kept dry, and care is taken in locating and loading the holes, misfires will seldom occur.

With *squib-ignition* misfires may be caused by (*a*) wetness of powder; (*b*) dampness of squib; (*c*) loss of powder from squib; (*d*) squib-hole clogged by dirt; (*e*) hole too long for squib to recoil and reach powder.

With *fuse-ignition* misfires may be due to (*a*) damp powder; (*b*) cutting of fuse in tamping; (*c*) imperfect fuse; (*d*) damp fuse; (*e*) loss of powder from end of fuse.

With *ignition by electric igniter* misfires may occur from (*a*) imperfect igniter; (*b*) damp igniter; (*c*) wire broken in tamping; (*d*) circuit imperfectly wired; (*e*) current leakage from poor insulation; (*f*) current deficiency from imperfect or overloaded blasting machine. The completeness of the circuit can be tested before the exploding by passing a feeble current through a galvanometer placed in the circuit.

Detonating Powders of Classes II and III.—In breaking coal with these powders it is better, as with igniting powders, to bore and load a new hole than to dig out the tamping from a missed hole. In rock work, it is good practice to dig out the tamping from a missed hole to within only half an inch of the powder and then insert a new primer cartridge with detonator and retamp. The excavation of tamping should be cautiously done when approaching the powder and care be taken not to strike the cap.

Dynamite should not be allowed to remain long before firing in water holes, for the water may displace the "oil" and perhaps cause a misfire or the escape of "oil" into adjoining crevices when it may later be struck by a pick or drill and explode. Powder should never be used when even partly frozen, for the thawed portion may explode alone and leave the frozen residue in the hole or blow it out into the muck to become in either case a source of danger for the next shift of miners.

In firing a round of holes in sequence, the explosion of one hole may blow off the primer of an adjoining hole whose remaining charge is therefore left unexploded in the hole-stump. Except for the last contingency, and that of two holes exploding simultaneously, the counting of the exploding reports gives a check on detonating in sequence which is lacking in simultaneously firing by electricity. An electric cap may be damp and conduct the current through the circuit, without exploding itself, and a missed hole will thus result. A fuse may have a broken thread of powder whose wrapping may catch fire and smoulder some time before igniting the powder beyond the break For all these reasons the stumps of blasted holes should be carefully examined before resuming work, and where misfires are suspected a half-hour interval should elapse before revising the broken face.

Fuse and cap detonation has the last four causes of misfires already given for fuse ignition and, in addition, is liable to failure of the cap, either from dampness, imperfection, or insufficient strength for the given explosive.

The causes of misfires already given for electric ignition can be made to read correctly as the causes with *electric detonation* by simply substituting the word *cap* for *igniter* and adding the requirement that the cap must be of adequate strength.

CHAPTER II

PRINCIPLES OF BLASTING GROUND

It is only in recent years that engineers have had much to do with the details of underground excavation, as it was thought that all the schooling necessary for the successful miner could be gained by practice with a drill and shovel. It is evident, however, that where rock breaking forms such an important item of expense as it does in most mines, it will well repay study to ascertain if science cannot duplicate here the same success it has gained over empiricism in other departments.

After an explosion of powder in the bore hole, Fig. 3, the sudden expansion of the resulting gases will exert its force equally in all directions on the bore hole, until either the enclosing rock or the tamping yields

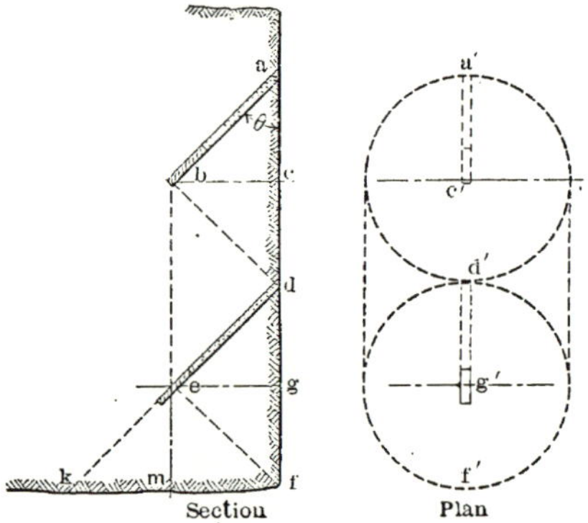

FIG. 3.—Cones of blasting rock.

and the gases escape. The rock will yield along what is called the line of least resistance, which would be *bc* in the assumed homogenous rock of Fig. 3. It is evident that the angle θ, which the hole *ab* makes with the exposed surface or the free face of the rock, can vary from nothing to 90 deg. At $\theta=0$ deg., there would be no hole and at 90 deg. the hole would be in the position *bc*, the line of least resistance, and would give a blown-out shot. The quantity of rock thrown out by the explosion would have the volume of a cone with an altidue *bc* or h, and a base with a radius *ac*, whose volume $v=1/3\ h\pi(ac)^2$ and where $\theta=45$ deg. (the usual condition for the maximum volume) $ac=h$ and we have $v=\dfrac{\pi h^3}{3}=$ (nearly) h^3, or if m is a constant, depending on rock, then $v=mh^3$.

For a case with two free rock faces if the powder charge be placed at *e*, with the lines of least resistance *eg* and *em* of equal length, the explosion will break out two cones *def* and *fek*, or nearly double the volume for one free face, so that $v = 2mh^3$. It is similar for three or more free faces, so that as a general equation we have, if n = the number of free faces, $v = nmh^3$.

From this formula it can be seen that a system of mining should be adopted which utilizes as many free faces as possible in breaking. In development work for vertical, horizontal or inclined drives or passages, we start each round of holes with one free face and with our cut holes break out either a cone or a wedge whose surface forms another free face for the benefit of the other holes of the round. In stoping work, which must be started from a drive, we can always manage to maintain two and often three free faces in homogeneous rock, and in stratified formations sometimes four or more faces, as a bedding plane is often nearly the equivalent of a free face.

In stratified formations the correct principles of breaking are especially important for economy's sake. The simpliest case is that of beds 2 to 4 ft. thick. Here the holes should be drilled in a plane parallel to the beds because it is evident that we can more easily separate two wet coins on a table by sliding one sideways than by trying to lift it off vertically. Also these parallel holes do not weaken the blast by allowing the powder gases to escape through the bedding seam. Where the beds are thin, say under 8 in., we encounter the possibility, with holes parallel to the bedding, of having only the small bed blown out that contains the hole. For this reason it is advisable to first make a cut by driving holes across the bedding planes and then break to the cut with the balance of the holes drilled parallel to the bedding plane, but which now exert their maximum force perpendicular instead of parallel to the beds.

The method of firing also affects the pointing and the necessary number of holes to drill for breaking. There is a great advantage in simultaneous or electric firing wherever a weak roof or the greater danger from misfires w th unskilled miners do not militate against it. In Fig. 3 it is evident that only the cone *abd* and the double cone *dek* would be broken out by the charges at *b* and *e* fired separately, but if *b* and *e* are not too far apart and are fired together the line of detachment will be along the lines *abek* instead of *abdek* and the extra volume *bde* will be broken with no extra powder or drilling. In any case of breaking, the pressure p produced by the explosive multiplied by the area of its section a (taken along the axis of the hole) must equal the ultimate tensile or shearing strength T of the rock multiplied by the area of its surface of fracture S or $pa = TS$.

If pa is greater than TS it means an excess of explosive over that required for detaching the burden. This excess causes a "windy" shot, resulting in a greater air blast, a louder report and a longer, hotter flame than from a normal shot. A normal charge leaves traces of the drill hole, but an insufficient charge leaves "candlesticks" in the rock and loose pieces of the burden have to be blasted off.

Underground Development

In illustrating we will take the case of driving horizontal headings or drifts as the same principles of breaking apply equally well for inclined and vertical shafts and raises. The practical difference in the latter arises from the setting of the drills and the handling of the muck and the water, and the fact that the length of the section in shafts generally makes the central cut advisable. We will also assume, to simplify the illustrations, a heading small and soft enough to allow its breakage by rounds of nine holes in three rows of three holes each, although often nine holes are more effective in four rows, one of three and the balance of two holes each. For longer headings with harder rock, the same principles would apply, but more holes must he added for breaking the round. On this basis we will now consider the following six cases of formation.

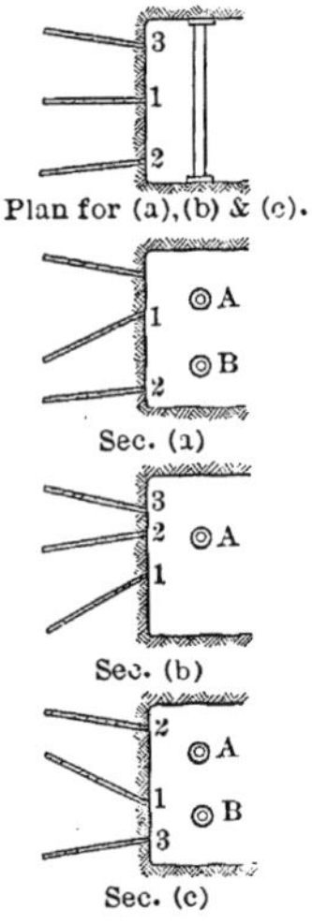

FIG. 4.—Holes for headings with horizontal bar.

Case I. Homogeneous Rock Free from Bedding Planes or Joints in the Face of the Heading.—Since this formation breaks equally well in any direction, the holes should be placed for the most convenient drilling and mucking. For setting the bar horizontally, as is usual where it is desired to begin drilling before the muck from the last round is cleaned up, the placing of Fig. 4 (*a*) is a favorite. Here the adjustable arm is unnecessary and the first setting of the bar is at *A* to drill the upper and middle rows with the machine above the bar. The second setting of the bar is at *B* and the machine is turned under it for drilling the bottom row 3 of lifters. The horizontal rows of holes are usually fired in the order 1, 2, 3, Fig. 4 (*a*).

The side instead of the bottom cut is handiest if we wish to set the bar vertically. We first set up at *A*, Fig. 5, and drill row 1, then at *B* with the machine on one side to drill row 2 and on the other to drill row 3. Here the vertical rows of holes are fired in the order 1, 2, 3, Fig. 5 plan. In other to keep the passage straight, the cut holes of row 1 will be put for the next round on the opposite side to what is shown, so that the finished sides have a zig-zag appearance, alternately right and left as shown in the plan of Fig. 5. The middle hole of vertical row 1 points downward, like hole *c*, instead of flat-wise, like the balance of horizontal row 1, so as to throw out a bottom cut and avoid a horizontal inclination to the face too acute for rapid progress in a narrow heading.

For large tunnel headings, 8 ft. square, in hard homogeneous rock, the cone or "Leyner" center-cut system has recently permitted of very

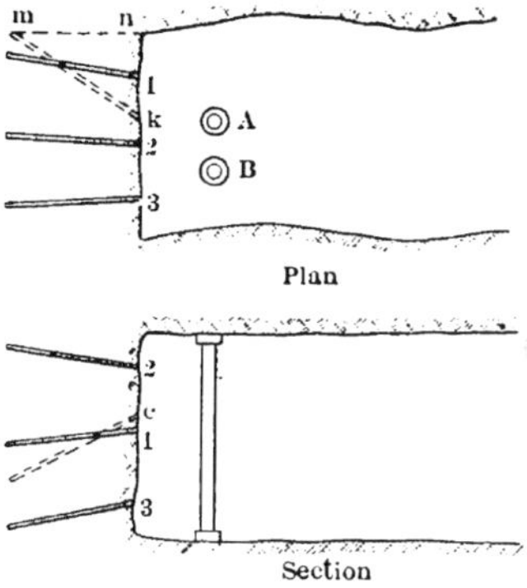

FIG. 5.—Holes for headings with vertical bar.

fast driving in western metal mines. It is especially adapted to the water Leyner drill on account of the many upper holes used and the fact that th's drill is short enough to allow the sharp pointing of the holes with two settings of the bar. For hard steel ore and jasper in a Michigan iron mine, this system was thus applied.

In Fig. 6, *A* is the bar in first position for two machines and from its

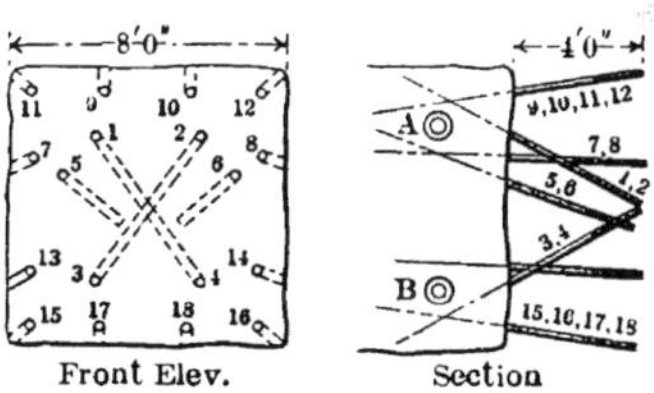

FIG. 6.—Holes for Leyner tunnel cut.

top the four back holes, Nos. 9, 10, 11 and 12, are drilled. The machines are then tipped forward until the crank can just turn and clear the back or top of the drift for drilling the top center cut holes Nos. 1 and 2, while finally they are turned under the bar for side holes Nos. 5, 6, 7 and 8. The bar is then changed to position *B*, the machines are set up on top and side holes Nos. 13 and 14 are drilled. Then, after turning the machines under the bar, they are tipped up in front so the crank just clears the bottom of the drift and holes Nos. 3 and 4 are drilled about to meet Nos. 1 and 2 in the center of the heading. The four lifters, Nos. 15, 16, 17 and 18, are the final holes. In softer and better-breaking ground, cut holes Nos. 5 and 6, one lifter and one back hole can be left out, but the four cut-holes, Nos. 1, 2, 3 and 4, are nearly always used and are pitched up and down and in, to meet about in the center.

The five remaining cases are given for regularly stratified rock, but the joints or cracks of massive rock may, like bedding planes, often be utilized for breaking.

Case II. Rock in Horizontal Beds; (a) Medium Thick Beds.—Here the best results from the powder can be got by two settings of the bar vertically and following the drilling and firing directions given above for the method illustrated by Fig. 5. In the disseminated lead mines of southeastern Missouri (Example 9, Chapter VIII), this method is modified as follows:

For a drift 10 ft. wide by 6 1/2 to 7 ft. high, 12 to 13 holes are needed, placed in three rows horizontally by four rows vertically. The bar is set up once to drill each vertical row of holes, four set-ups being necessary to complete a round. Each vertical row is fired separately by fuse and dynamite and as only three or four holes are fired at a time, not enough smoke or broken rock is produced to prevent the drillers from setting up again very soon after blasting. This method with three shifts of two drill men each allows an advance of 5 to 7 ft. in 24 hours with 2 3/4-in. drills. By the former center-cut system, two drills and four men were able to advance only 10 to 15 per cent. faster than by the one drill and the side-cut method just described, all loading and tramming, in each case, having been done by muckers.

(b) Thin Beds.—Here, as already explained, the cut-holes must cross the bedding planes. A bottom cut is advisable. The bar is set horizontally at *A*, Fig. 4 (*b*). Often all three rows can be drilled direct although sometimes the use of the adjustable arm on the bar is necessary to get the correct pointing of the holes. The holes of row 1 are fired first and break out the cut to the bedding plane on the floor of the heading. Before loading the row of cut-holes, it is often helpful to stop up their bedding planes, around the powder, with clay but this precaution is unnecessary in the two upper rows where the holes are parallel to the beds.

Case III.—Rocks in Vertical Beds Parallel to Heading; (a) Medium Thick Beds.—This case requires the bottom cut of Fig. 4 (*a*) which has already been described under *Case I.* The use of this method in the vertical copper veins of Butte, Mont., is as follows: The placing of holes is shown in Fig. 7 for the 12-hole system, although for most rock nine holes are ample, the center holes of rows 2, 3, and 4 being omitted. For this arrangement the drill bar (with adjustable arm) need only be set

up once vertically, as shown. The round of holes is usually loaded and fired at one time and goes of in the order of 1, 2, 3, 4. Some of the miners regulate the explosion by cutting the fuse of different lengths and spitting them simultan o ly while held together in the hand, and others by cutting all the fu of the same length and spitting them separately in the required ord .

(*b*) *Thin Beds.*—The sclu)n of this case follows Fig. 5 and also resembles *Case II* (*b*) except t .t here the side instead of the bottom cut is used. With one setting of e bar, the three vertical rows *K*, 2 and 3 may be drilled and shot in th same order, row *K* breaking out the cut, along a side bedding plane, *n* , and rows 2 and 3 breaking to the cut.

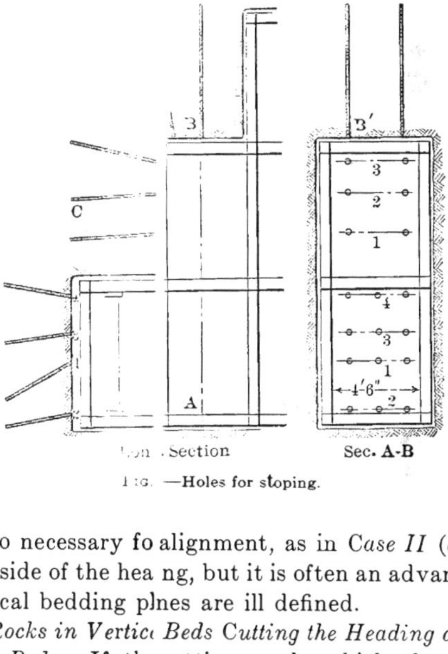

.on . Section Sec. A-B

I :G. —Holes for stoping.

Here it is not so necessary fo alignment, as in *Case II* (*a*), to alternate the cut on each side of the hea ng, but it is often an advantage especially where the vertical bedding plnes are ill defined.

Case IV.—Rocks in Vertic Beds Cutting the Heading at an Angle; (*a*) *Medium Thick Beds.*—If the cutting angle which the bedding plane makes with the side of the ading is 45 deg. or less, the method of Fig. 4 (*a*) is usually preferab. If the cutting angle is more than 45 deg., the choice between the methods of Fig. 4 (*a*) and of Fig. 5 will often be merely a question of onvenience in setting the bar horizontally or vertically, respectively.

(*b*) *Thin Beds.*—With a ctting angle of 45 deg. or less the method of Fig. 5 is the best. Where the cutting angle is more than 45 deg., the choice between the methcs of Fig. 4 (*a*) and Fig. 5 depends on setting the bar as in *Case IV* ().

Case V.—Rocks in Inclined Beds Dipping Toward the Floor of the Heading.—For either medium thick or thin beds the method of Fig. 4 (*a*) is the best. Care must be taken, however, in the ease of beds dipping over 45 deg. to stop the holes of the horizontal row 1 at the last bedding plane which intersects the face of the heading above the floor.

Case VI.—Rocks in Inclined Beds Dipping Away from the Floor of the Heading.—For either medium thick or thin beds the method of Fig. 4 (*c*) should be used. The bar is set up at *A* for row 2 and at *B* for rows 1 and 3. The order of firing the horizontal rows of holes is 1, 2 and finally 3. The end of the holes in row 1 should be stopped beneath the last bedding plane intersecting the face of the tunnel under the roof in order to utilize this plane as a free face in breaking.

Surface Excavation and Underground Stoping

In some kinds of deposits, especially the huge copper-bearing porphyry lenses and the Lake Superior iron mines, much time is often saved by drilling all the holes possible in the periphery of a heading in the ore from the same set-ups that are used in drilling the face. These peripheral holes can then be left untouched until the stoping of that section begins, when they can be easily loaded and fired.

Holes for stoping may be placed according to the direction in three groups, (1) down holes (2) flat-holes, and (3) uppers. A dip of about 45 deg. downward and upward can be assumed to make the limit between groups (1) and (2) and of (2) and (3) respectively, although the division between (2) and (3) is really marked by the angle of repose of the cuttings, that is, when the hole becomes self-cleaning, which may often mean a steeper dip than 45 deg. The speed of cutting with reciprocating drills depends on the removal of cuttings after each stroke to expose a fresh face. Therefore with these drills down holes drill easiest, then uppers, and lastly flats. Using the hammer drills with hollow bits cleaned by water or air-jets, there is less difference in drilling speed for different directions of pointing.

Down Holes; Underground.—Down holes are used underground in the underhand benches of tunnels or metal mines. To start this system, a heading *ah*, Fig. 8, is run at the top of the tunnel or stope and the down holes put in its floor for the first bench. The depth of this bench is limited by the length of the bit which can be inserted in the hole and that depends on the height of the heading which is usually around 7 ft. so that the ordinary railroad tunnel, 20 to 25 ft. high, requires two benches and two settings of the tripod at *a* and *b*, Fig. 8 to reach the bottom. These bench holes point downward anyhow but often an advantage may be taken of the structure. Thus with horizontal beds, the holes of the first bench can be terminated at a bedding plane which the gases from the explosion will enter and thus exert a lifting action on the mass to be broken off.

Where there is a choice of plans, a heading can often be given in a direction that will take the maximum advantage of the bedding and joint planes for breaking, both in driving and stoping. On this principle, the rooms of coal mines are usually laid out perpendicular to the line of the main joint planes of the coal seam or to the "face cleat."

Down Holes; at Surface.—Above ground the only limit to the depth of the hole is the capacity of the drill. In considering breaking from deep holes we have a choice of two methods (*a*) multi-charging, (*b*) chambering.

In the drill hole *ab* of Fig. 9, it is evident that a charge of explosive at any point *b* will only break out a cone like *cbd* where *eb* is the line of

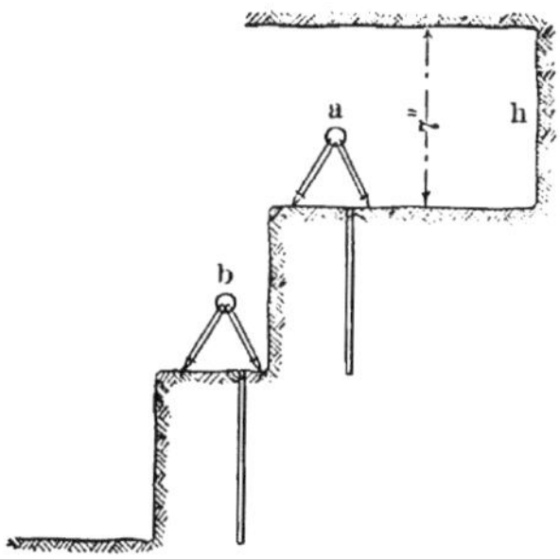

FIG. 8.—Holes for underhand stope.

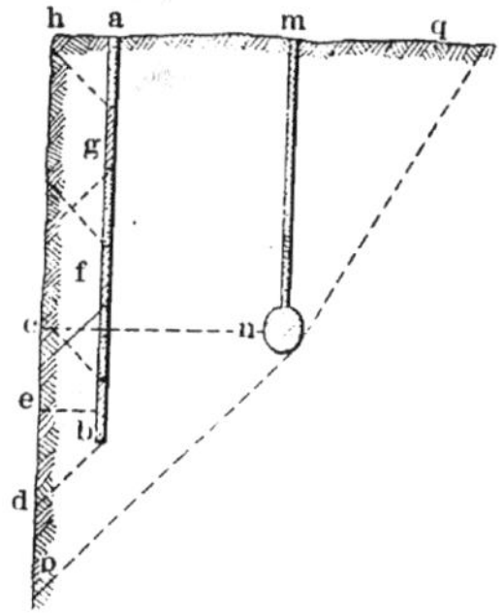

FIG. 9.—Holes for high bench.

least resistance. In order to break the whole length *ab* by multi-charging, other charges of explosives as *f* and *g* would be placed along the hole, with tamping between, and all be set off by simultaneous firing. In this way the whole mass *abd* would be detached.

By chambering, the breaking from a long hole would be achieved differently. Instead of the hole being placed near the face *hd* of the bench as is the hole *ab* (because of its small section for developing explosive pressure), the hole *mn* would be placed back from the face so that *nc*, the line of least resistance in homogeneous rock, would be only a little shorter than the length of the hole above the chamber at *n*. The chambering is effected by shattering the bottom of the hole with high-power dynamite so that the final shape of the chamber approaches a sphere. In France this chambering, in limestone, is performed with hydrochloric acip, each dose of neutralized acid being washed out and a new one poured in until the chamber is of the required size. When the chamber is filled with gunpowder or low-power dynamite and exploded, it will exert nearly as much force upward as horizontally and will break out a mass along the surface of fracture *qnp*.

The choice between multi-charging and chambering depends on the varying conditions of formation, drilling and exploding. In a fissured formation, chambering has often an advantage because the explosive may be localized in a solid portion of the rock, although it often needs the use of two kinds of explosives, one for chambering and the other for breaking. Where it is desired to break off only a thin slice like *hab*, Fig. 4, from the cliff, it is evident that multi-charging should be resorted to. When an even topography will allow the handling of the portable steam or electric churn drill for a 3-in. to 12-in. hole (instead of the reciprocating drill for a 1 1/2-in. hole), the multi-charging method will permit the drilling and breaking of a much longer hole than would be feasible by chambering.

Flat Holes; Underground.—Of the three groups, flat holes are the most difficult to drill, especially those which are pointed above the horizontal for the reason that they neither hold water or discharge their cuttings by gravity. This group is much used in the overhead stoping system with piston drills as the drill tripod can be set on the broken rock. Thus flat water holes which are easier to load than uppers and free from their dust can be drilled at a fair speed. In overhand stoping with a weak back, as in the vertical veins of Butte, Mont., flat holes have also an advantage over uppers as the timber sets can be carried next to the back and the drilling can proceed under the lagging. Thus in Fig. 7 at *C* rows 1 and 2 are water holes and only row 3 need be drilled dry.

In the zinc district of Joplin, Mo., flat holes are used instead of the usual down holes to break the benches below the heading of the underhand stoping system as described under Example 10 of Chapter VIII.

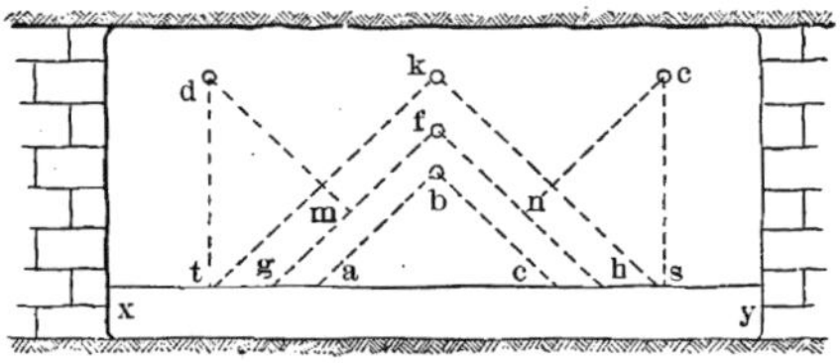

FIG. 10.—Holes for seam.

In driving coal headings or rooms by "blasting off the solid," flat holes bored by augers are generally used and are placed similarly to those shown for headings in flatly bedded rock in Fig. 5. In the location of the horizontal rows of holes, the character of the bedding planes between the coal seam and its roof and floor must be considered. If the roof is "tight," the shot must exert a strong shearing force to separate it. This is achieved by slanting the row of holes sharply upward and terminating them at the tight plane. A similar remedy is applied to a tight floor. In many seams the coal is cut up into cubes by two sets of joint-planes perpendicular to the bedding planes, called the "face" and "end" cleats, which condition makes breaking easy.

The shearing of a coal face, before shooting, takes the place of the cut holes in blasting off the solid and the smaller charges allowable for the former method not only save explosive but prevent the shattering of the roof. With coal sheared vertically along one rib of a heading, the holes for breaking would be placed like vertical rows 2 and 3, Fig. 5. Where the shear is made horizontally as in the undercut *xy*, Fig. 10, it is customary in a thick seam of coal to place the first or "buster" shot at *b* in order to break out the triangular prism of coal *abc*. Then when the shattered strip *gfh* has been removed by the pick, we have *dm* and *cn* instead of *dt* and *cs* for the line of least resistance from the corner holes *d* and *e*, by which last the balance of the undercut coal can now be easily shot down. For a thin vein of coal, the "buster" shot would be located at *K* on a level with the corner holes and it would break out the triangular prism *tKs* as thick as the seam.

The undercut shown in Fig. 10 is that made by a hand or power pick. Being a height of 12 in. or so in front with a downward slope to 4 in. in the back, its shape allows the "buster" shot to throw much of the coal out of the undercut, so that the strip *gfh* can be easily extracted by the pick to prepare for the corner shots. When the undercut, however, is made by a chain machine, it is of uniform height of only about 4 in., and the "buster" shot may not throw the coal outward. It is then often advisable to place an extra "snubbing" shot at *f* to flatten down the detached prism *abc* so that the shots *d* and *c* can be made effective without first cleaning out the broken coal underneath.

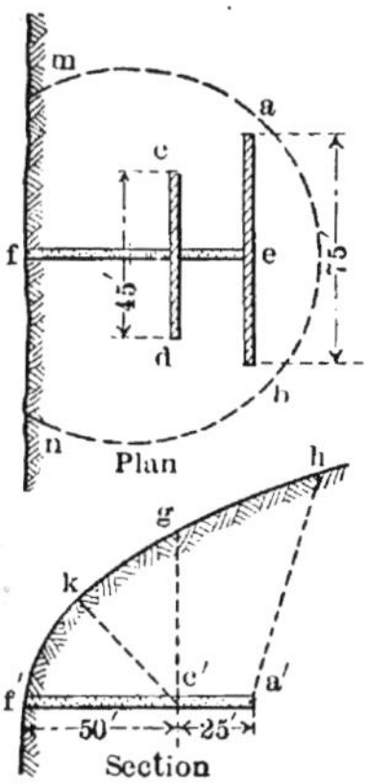

FIG. 11.—Blasting by tunnels.

Flat Hole; Surface.—In loosening huge banks of placer gravel in California before hydraulicking, small adits have been used with cross-cuts at their ends to hold the explosive. From a breaking stand-point, these adits correspond to flat drill-holes with chambered ends. The same method has also been employed for breaking great masses of rock in quarries or excavations. Often a shaft has been sunk as an entrance to the explosive chamber instead of an adit. Sometimes two cross-cuts from the adit may be made for explosive chambers, as shown in Fig. 11. There only the crosscuts *cd* and *ab* would be packed with gunpowder or low-power dynamite, while the adit itself would be blocked with timber or masonry bulkheads wherever it met the crosscuts. Elsewhere it would be packed with sand. For firing, electric fuzes or caps would be placed in the explosive at intervals of about 10 ft. Finally they would all be connected by wiring in order that they might be fired simultaneously by electricity c^1k being the line of least resistance. The chamber *cd* would break out the cone gc^1f^1, and the chamber *ab* would break out the prism ha^1c^1g, the plan of the line of fracture being *mabn.*

The same breaking equation, $pa = TS$, applies as in the case of drill holes, the factor a being the area of the cross section of the explosive taken along the axis of the crosscut.

Uppers.—Uppers are seldom used on the surface but are common in underground work not only in tunnel headings and raises, but also in overhand stoping. In excavating overhand stopes with square-set timbering, it is sometimes more efficient to drill the back with uppers as at *B*, Fig. 7, instead of the flats at *C* used in Butte practice. In the great stopes of the Portland mine, at Cripple Creek, Colo., where the pay shoot was in places 120 ft. wide and 400 ft. long, the ore hard and the back strong enough to stay up across the vein for several sets ahead of the timbermen, it was found that the fastest breaking was accomplished by drill ng uppers from piston drills set on tripods, one drill being used in every set across the stope.

CHAPTER III

COMPRESSED AIR FOR MINING

In drilling with piston rock drills a high pressure gives a stronger withdrawing force on the bit which tends to prevent sticking in fissured ground and thus greatly increases the speed of boring. In hard, tough ground, like specular hematite or certain intrusives, a high air pressure is necessary, if it is desired to strike a blow, severe enough to cut the rock, with a light portable machine. In a certain mine, using 40 drills in hard and fissured ground the rock broken per machine was increased about 20 per cent. by the simple expedient of advancing the air pressure from 75 to 100 pounds. A low pressure system requires larger pipes to deliver the same power and heavier pumps and hoists in the mine to accomplish a given amount of work than an equivalent equipment working under high pressure.

The economical limit of pressure depends in a given case on commercial considerations, costs of fuel, labor and supplies, which in turn are governed in considerable degree by the mechanical efficiency of the plant. The high pressure limit, except for haulage purposes is about 120 pounds.

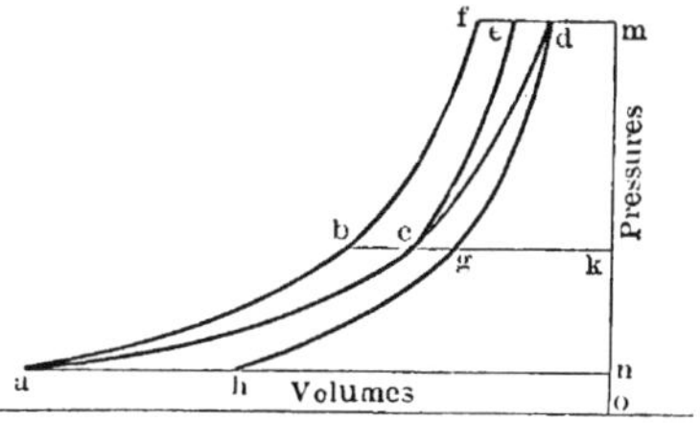

FIG. 12—Relations of volume and pressure in air compression.

It is wasteful to heat the air during compression to a higher temperature than that of the mine, as radiation in the pipe-line will cool any warmer air before it reaches the motor. A proof of this statement follows: Let V and P be respectively the volume and pressure of free air at the beginning of compression, and in the theoretical indicator card, Fig. 12, in which O is the origin of coördinates, let the abscissa of point a be V and the ordinate be P. Let V and P be the volume and pressure of air at any point of the stroke, during its compression by a reciprocating piston. Then if the temperature due to the heat of internal friction is retained in the air, we have adiabatic compression and get the curve $a\ b\ f$, the equation of which is

$$p=PV^y\left(\frac{1}{v^y}\right),$$

the value of y being 1.406 for dry air and somewhat less for the ordinary atmosphere, and p being the resultant pressure and v the resultant volume.

If the temperature is kept constant during compression, by removing the internal heat as fast as generated, we have isothermal compression and get the cruve $a\ cd$, the equation for which is $p=PV\left(\frac{1}{v}\right)$. Finally, the work lost by cooling the air, from the final adiabatic temperature to that of the free air, is measured by the area $a\ c\ d\ f\ b$, the total work of compression for one stroke of the piston being area $a\ f\ m\ n$.

Theory of the Intercooler

Although isothermal compression is the ideal, practical difficulties prevent its attainment. The air can be cooled in the compression cylinder by a water spray, but this method requires too slow a machine to compete with dry compression and external cooling. It can be easily shown, mathematically or by an indicator card, that water-jacketing the compression cylinder has practically no effect in cooling the air, although it is useful in keeping the bearing surfaces cool enough for lubrication.

In Fig. 12, the adiabatic and isothermal curves get farther apart as the pressure increases, so that the work lost by adiabatic compression increases at a faster ratio than the pressure. To avoid this increase for high pressures, a compression in two stages, with a surface intercooler between the high- and low-pressure cylinders, is frequently used. Unfortunately, few of the standard machines have a large enough intercooler to insure that the compressed air, entering the high-pressure cylinder, is as cool as the free air entering the low-pressure cylinder when the machine is running full speed. It will aid the intercooler, if the free air is sucked into the low-pressure cylinder from the coolest available place.

In the diagram, Fig. 12, K is the pressure at which the air leaves the low-pressure cylinder to pass through the intercooler and enter the high-pressure cylinder. The following cycle then takes place with a perfect intercooler. In the low-pressure cylinder the air is compressed adiabatically from a to b, reduced in the intercooler to the yolume at point c and then compressed adiabatically in the high-pressure cylinder from c to e, the total work of compression being the area $a\ b\ c\ e\ m\ n$. Thus the saving of work by the use of the intercooler is represented by the area $c\ e\ f\ b$, from which must be deducted any work expended in circulating the cooling water. In the design of the machine, the ratio of the diameters of the low-pressure cylinder and the high-pressure cylinder are taken so that the area *a b k n* is equal to area *c e m k* for average conditions.

There need be little difference in the efficiency of the steam ends between high- and low-pressure compression. With a cross-compound air end, the steam end can also be compound and for a single-stage air end the machine can be tandem-compound. The air-pressure governor has now been perfected and for the usual variable loads of mine work, is indispensable for any pressure, though it requires a duplex machine to avoid a stoppage on dead center with no load.

Preheaters

In the case of the air motor, the compression process is reversed. The air on entering the motor in the mine has the pressure and volume of point *d* (Fig. 12) and in a simple, unheated motor cylinder will expand adiabatically along the line *d g h*. Should the air be preheated to the volume of point *f* it will then expand along the adiabatic line *f b a* with a gain of work, over the unheated case, equal to area *a h d f*.

With two-stage expansion, the air may be preheated before entering the low-pressure cylinder to *e*, then expand adiabatically to *c*, next pass through an interheater so as to reach *b* on entering the high-pressure cylinder and finally expand adiabatically to *a*. Heating during expansion, like cooling during compression, gains in its relative effect on the efficiency, the higher the pressure. Aside from its gain in work, heating is often necessary to prevent freezing of the exhaust when the air is damp and cold on entering the motor.

Owing to the small size and portability of rock drills preheaters are for this service out of place, but for large hoists and pumps, with high-pressure air, they are always to be recommended. In the operation of the preheater the compressed air passes through a vessel containing heated tubes of sufficient radiating surface for the purpose. These tubes may be heated by a coal, coke or oil fire, but, since smoke contaminates the atmosphere of the mine, steam-heating is often both convenient and economical. In an air heater it is possible to utilize steam more efficiently than in the best condensing engine, for both the latent and visible heat of the steam are absorbed by the air and turned into work without frictional losses greater than the motor would suffer with unheated air. With steam heating the only important loss is that due to radiation in the supply pipe from the boilers, and by proper covering this can be made small. In the 500-gal. Dickson pumps, installed in the Anaconda mines at Butte in 1899, the air was successfully heated by steam in both the preheaters and the interheaters for the compound cylinders.

High-pressure pipe-lines, though smaller in diameter, require more care to keep them tight than lines for low pressure, and the velocity of exit of air from a leak varies directly as the square of the pressure.

The loss of power from the common practice of blowing out powder smoke with the air hose is the greater the higher the pressure, for the ventilating efficiency depends only on the quality of free air discharged. With pipes properly proportioned for the quantity of air to be delivered the frictional line losses will be moderate with either pressure, if care be taken to avoid unnecessary bends and to use gate valves instead of globe valves.

The compressor should discharge its air into a receiver the cooling action of which will not only at cone reduce the volume to that which it will have in the mine, but will also precipitate any extra moisture and keep it from entering the pipe-lines. A good device for the surface receiver is a condemned boiler, set in a wooden tank in which is water circulating through the boiler tubes, while the compressed air fills the shell. Underground the receivers need only be plain steel shells for storage, but they must be numerous and large enough to preserve the pressure constant under the variable power requirements. Preheaters in use serve as receivers.

When air is used for haulage it needs a special piping system to hold the requisite pressure of 1000 lbs. upward. This piping also serves as a receiver and accumulator of air between locomotive chargings so that the compressors can be run under a constant load. It is evident that the piping system will need a lesser proportionate capacity as receiver the greater the number of locomotives supplied, for each charging will involve a less relative displacement of air. Under the usual traffic and air pressure a pipe line of 6 to 12-in. dia. is amply large, both for distribution and storage of air, without placing tank receivers at the stations.

The air ends of compressors for haulage systems should be at least 4-stage, of moderate speed and with ample intercooling surfaces; for Fig. 12 shows how fast the power loss due to inefficient cooling increases with the pressure. Until recently the locomotives were single-stage and had consequently a low efficiency and capacity; but the new compound, Porter locomotive obviates these troubles and gives air-haulage a chance for extension beyond its present special field of gaseous or dusty coal mines.

CHAPTER IV

PRINCIPLES FOR CONTROLLING EXCAVATIONS

The art of timbering is not synonymous with that of the control of ground as some suppose; a good carpenter can frame timber better than any miner, but unless he places it underground as directed by the latter, his accurately jointed sets are liable to prove worthless for the purpose intended. The subject of ground control naturally divides itself under two topics: I. The control of the roof of an excavation; II. The control of the sides and floor of an excavation and of the whole overlying formation. Both topics will be considered separately before their inter-relation will be discussed. In practice, we have not only to consider the freshly broken surfaces of an excavation, but their future conditions after exposure to the weathering action of the mine atmosphere.

Control of the Roof

(*a*) *Roof over a Horizontal Room.*—This case is the simplest and occurs in mining horizontal seams or beds. Let *abb'a'*, Fig. 13, represent the cross-section of a rectangular room excavated in a team of the thick-

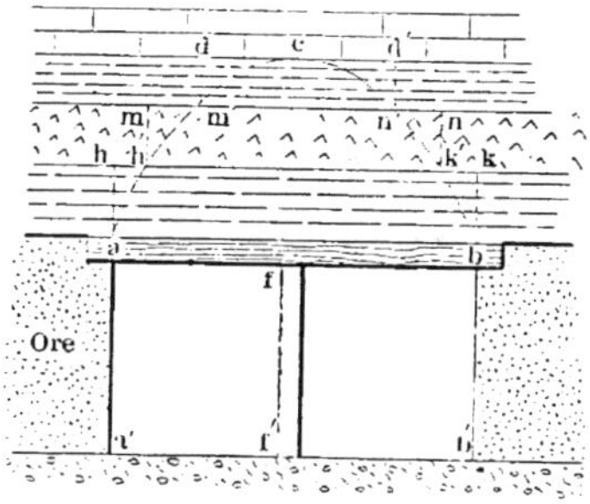

Fig. 13.—Homogeneous or horizontally-bedded roof.

ness *aa'*. Then the support of the roof over the opening *ab* depends upon the immediately overlying formation. The structure of the last falls usually under one of the five following cases: (1) homogeneous, (2) horizontally bedded, (3) weakly-consolidated, (4) non-conformable, (5) broken.

With case (1) or a *homogeneous* roof stratum, either massive or in a sufficiently thick bed to act as such, the lines of vertical pressure far above *ab* tend to combine themselves into resultants which follow a surface *acb* and throw the downward pressure onto the walls at *a* and *b*. The resultant surface takes the form of an arch, over a tunnel, or of an arch with domed ends, in a room of limited length. This means that the sub-arch block *acb* is all the weight that has to be supported to maintain the roof intact, and that its stability depends first, on its strength as a beam of continuous width to bear its own weight across the span *ab* and second, on its being held in place by the tensile strength of the rock area along the arch, or potential surface of fracture, *acb*. In case (1) the natural arching is usually sufficiently convex so that the sub-arch-block has sufficient depth *cf* to make is self-sustaining as a beam across *ab* except in soft rocks like certain shales which may not only need the support of a cap like *ab* but also must be lagged.

In old mine workings where the sub-arch block *acb* has fallen out so that the shape of the natural surface of equilibrium *acb* can be discerned, it appears as an arch whose proportions vary with the width of the room and the nature of the roof. Fayal gives as working rules for limited areas like rooms:

If w = width of room (as *ab* in Fig. 13);
h = height of arch (as *cf* in Fig. 13);
If w is less than 6 ft., h may be as much as $2w$ (Fayol's first rule);
If w is more than 6 ft., h may be as much as $4w$ (Fayol's second rule).

In railroad or mine tunnels, a homogeneous roof can be made self-sustaining by excavating it, at the start, along the natural arch form. In the rooms of coal seams, however, or in iron-ore beds, the sub-arch block must be sustained intact until the mineral beneath is removed and the room abandoned. The tensile strength of the arched surface is seldom sufficient to accomplish this unaided, except in narrow rooms. In wider rooms, a cross-beam *ab*, or one or more props like *ff'* must be put in whose strength, however, need only equal the difference between the weight of the sub-arch block *acb* and the tensile strength of the arched surface *acb*, provided that *ff'* is inserted before the surface *acb* has begun to fracture. Should the latter accident have taken place, the weight of the whole sub-arch block may have to be sustained by props and thus a heavy unnecessary expense be incurred.

With case (2) or where the roof is in beds thinner than the sub-arch block so that bedding planes like *hk* and *mn* (Fig. 13) intersect the surface *acb*, a different condition arises from case (1). It is evident that now the sub-arch block instead of being a single stone beam *acb* is divided into three stone beams *ahkb*, *h'mnk'* and *m'dd'n'*, so that for a self-sustaining roof, the lowest beam *ahkb* must be strong enough to sustain the we ght of the two beams above it, the central beam *h'mnk'* must sustain the top beam *m'dd'n'* and the tensile strength of the sub-arch surface *acb* must be, as before, sufficiently strong to hold up the whole sub-arch block. It is, therefore, likely that a room would need stronger props in case (2) than in case (1) because the lowest sustaining beam of case (2) has a depth at the middle of *ha*, which is only a fraction of the corresponding depth *cf* for case (1), and the cross breaking strength of a beam increases directly as the square of the depth. Also we now do not have a uniform tensile strength for one surface of fracture *acb*, but a different strength for each of the three beds which *acb* intersects. Hence for case (2) we have to acsertain both the cross breaking and the tensile strengths of all beds in the sub-arch block before we can ascertain how much propping is required to sustain the roof across a room of a given width. A roof of an elastic nature like slate may at first simply bow downward from an excess of pressure instead of fracturing as a beam. This may cause it to fail by shear at the abutments. For the maximum strength of a roof it is important to exclude water from the bedding planes in order to prevent the slipping and weakness caused by its presence.

Case (3) often occurs in coal mines where the roof stratum is "clod" or a kind of soft shale containing concretions of considerable size. A common device is to leave the upper layer of the coal seam under it which then acts as the lowest beam *ahkb* of case (2) to partially sustain the clod-stratum. Where all the seam must be removed beneath the clod, the roof can only be kept intact by excavating the mineral with little or no blasting and keeping the supports close to the working face. Props, cross-pieces and lagging may all have to be used. If the clod stratum is thicker than the room's natural arch, masses may fall out from above the surface *acb*, after the sub-arch block has been taken down, so that roofs must be arched higher than the clod in order to stand permanently unsupported.

Where the roof is a weakly consolidated stratum of more uniformly sized stones, l ke a conglomerate, the problem of support is similar. Practically the whole weight of the sub-arch block must be held in place by artificial supports, and in addition the beam *acb* itself must be reinforced by cross-beams of props or by both. Where the roof stratum is so weakly consolidated as to be incoherent it requires close lagging, and where quite loose an advance can only be made by driving fore-poles ahead of the timber sets right up to the working face. Loose sand, if dry or only moist, can be sustained, like loose gravel, by close fore-poling, but if it is wet enough to flow freely like quicksand the case is hopeless except by the use of some such system as that of the pneumatic shield recently employed in the Hudson river tunnels at New York.

It is evident, however, that while the quicksand roof of a railroad tunnel might be penetrated and sustained by the expensive pneumatic shield and its follower, a cast-iron tube lining, such a device would be commercially unpractical for ordinary ore deposits. For the latter the only hope for overcoming quicksand is sufficient drainage so that the sand loses its fluidity and takes the compact condition of its merely moist state. If drainage of the quicksand covering is not feasible and the ore body cannot be mined by some subaqueous method, it is worthless, as was recently proved for a huge hematite deposit under a swamp on the Mesabi range, Minn., which was abandoned after wasting a large sum in attempting to open a mine in it.

Even if the bed or pocket of quicksand does not rest directly on the ore body, but is separated from it by a rock stratum, great care has to be taken against it. The only safe plan is to open the mine excavations of small size and with sufficient support to keep the rock roof intact, for otherwise vertical cracks may develop reaching to the quicksand. When the quicksand once begins to flow into the mine, the results may

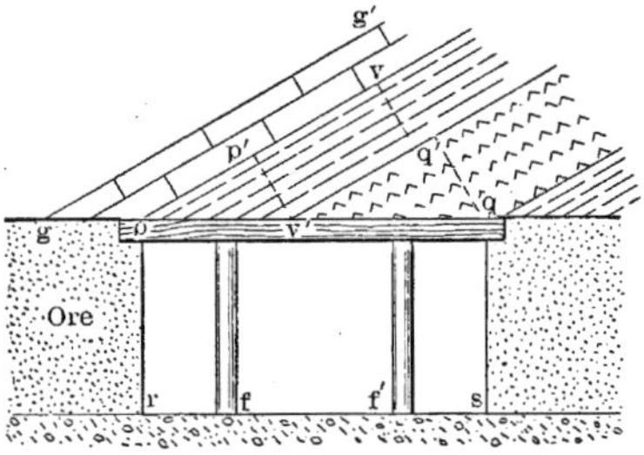

FIG. 14.—Non-conformable roof.

be far-reaching for its escape from its matrix may mean the collapse of the latter and consequent disastrous movements of the whole overlying cover.

In case (4) the *non-conformable* roof strata may dip in any direction with reference to the underlying mineral seam. If the roof strata strike along the long axis of the room as in the cross section of Fig. 14, then it is evident that conditions will not produce an arch of fracture as in Fig. 13. The upper stratum $g'gpv$ is entirely above the room opening and bridges it slantingly from one side to the other, while the two lower strata, $vpv'q'$ and $q'v'q$, have their lower ends unsupported and projecting like cantilever beams. Then the natural surfaces of fracture will be normal to the bedding planes and will be qq' for the lowest and $q'v$ for the middle stratum. The tensile strength of surface qq' must be enough to hold the weight of projection $q'v'q$ and any unbalanced pressure from above, while surface vq' must hold its end $vpv'q'$ and any weight above. A line of props at t strong enough to sustain the excess of strain over the resisting strength of surface qq', will hold up the roof without a second line at t provided that surface $v'p'$ is strong enough to sustain the weight on it from the projection $p'pv'$.

As the dip of the roof beds increases, the strain on the surface of fracture $qq'v$ becomes more tensile than cross-breaking until with vertical beds the strain is all tensile. In the last case, the weight to be sustained by each stratum is its block below a natural arch of fracture across the room, which is differently proportioned for vertical beds than is *acb* of Fig. 13 for horizontal beds.

Where the strike of the inclined beds of the roof is *across* instead of *along* the room beneath, we have a mixture of the cases illustrated by Figs. 13 and 14. Each bed can first be considered separately as forming a single sub-arch beam whose side elevation is *acb* in Fig. 13. Each bed must then be calculated separately both for the self-sustaining power of its sub-arch beam, across the span of the room, and for that of its tensile surface *acb*. A bed may then be artificially supported if necessary, by prop ff' or cap *ab*. If Fig. 14 be assumed, for this case only, to be the longitudinal section of the room whose cross-section is Fig. 13, we see that a cross bed like $vpv'q'$may have the same breaking-off action on a lower bed $q'v'q$ as has just been discussed in the last paragraph, and supports must be modified accordingly.

The *broken* roof of case (5) may arise from planes of faulting, fracturing, jointing, etc. If the breaking planes are parallel or in one general

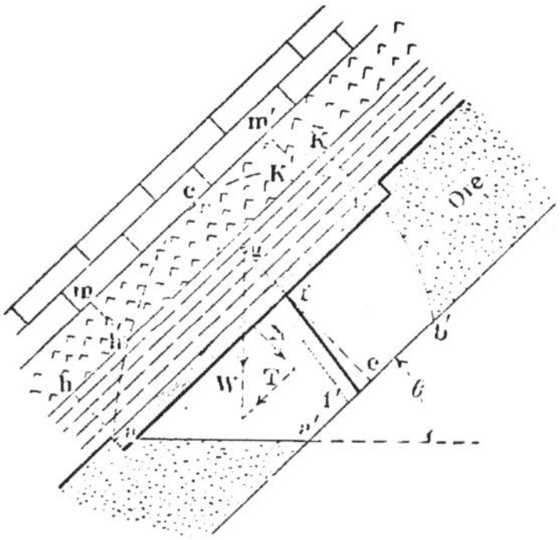

FIG. 15.—Roof-over inclined room.

direction, we can handle the roof as suggested for case (4). If the planes are in several directions so as to cut the roof into monoliths, the support of each block will have to be studied separately. Where a roof monolith is of indefinite height, we may illustrate it by Fig. 13 with *ab* its length and *acb* the section of its natural surface of fracture, which will be of dome shape, so that only the support of the sub-arch portion *acb* has then to be considered. When, however, the roof monoliths are broken also by a plane in a horizontal direction, like *mn* in Fig. 13, so as to become free blocks like *amnb*, they can only be kept in the roof by sustaining their entire weight artificially, and fore-poling will have to be used for excavating beneath a roof surface containing them.

(*b*) *Roof Over an Inclined Room.*—This case occurs in mining seams on a dip which may vary up to 90 deg. from the horizontal. Let *abb'a'*, Fig. 15, represent the cross-section of a room in a seam of the thickness *bb'*, which has the usual horizontal floor *aa'* for tramming. It is evident that the principles of roof support similar to the previous case of horizontal rooms apply here, but the action of the superincumbent weight in the roof is affected by the angle of dip. Thus in the diagram of Fig. 15, if W = superincumbent weight; θ = angle of dip; N = normal pressure on roof; T = tangential pressure on roof; then

$N = W \cos \theta$ . . . . (1)
and $T = W \sin \theta$ . . . . (2)

For homogeneous strata the weight of the overlying formation would be thrown onto the pillars at *a* and *b* and the potential surface of fracture would be the arch *acb*. Thus the span *ab* has only to sustain the normal pressure of the sub-arch block *acb* acting both in tension on the surface *acb* and in cross-breaking strain on the beam *acb* as described for case (*a*) in Fig. 13. The back of the ore should also fall on the arch line *bdb'* instead of a straight line from *b* to *b'*. A prop to hold up the roof will be subjected to the least pressure and be of shortest length if it is placed in a line *gfe* drawn normal to the hanging wall from the center of gravity of the sub-arch block at *g*. Because of possible shrinkage of prop or movement of ground, however, which would cause a normal prop to fall out, the usual practice is to incline it about 10 deg. downward from the normal line as *ff'*. The sub-arch block *acb* can also be sustained by a cap *ab* from the back to the floor, or by both prop and cap. With the roof-strata bedded parallel to the seam, the surface of fracture assumes the stepped-arch form *ahh'mm'k'kb*. In comparing the strains and the support of the bedded roof, as well as of the weakly-consolidated, of the non-conformable and the broken roofs with those of the homogeneous roof, the same differences arise as already explained for case (*a*).

Control of the Overlying Formations

It is evident that when part of a bed is removed, the balance left as pillars must sustain the whole overlying formation. There are three factors that enter into pillar calculations, the roof, the pillars or sides and the floor. The stability of the room does not depend alone on the strength of the pillars as columns for an excess of pressure may force a sound pillar into a roof or floor of insufficient compressive strength and cause a settling. This happens with materials like clay which are hard when dry and become soft when moist, so unless they can be kept dry during mining, the pillar calculations must guard against their moist state. An excess of pressure on a plastic floor will cause it to spread laterally and rise from under the lower periphery of the pillar, thus exerting a horizontal rending force on the latter which tends to disrupt its edges.

Any downward bowing of an elastic roof over the rooms must be compensated for by an upward bowing over the interior of the pillars. This causes an oblique pressure at the upper edges of the latter which tends to shear them off as the roof bends more and more. The obliquity of this roof pressure on the pillar edges is also often increased by a rolling floor.

Thus a mine floor and roof act not only vertically on the areas of contact with the pillars, but also laterally, while the bowing of the roof produces strains parallel to the strata that tend to separate them along their bedding planes, and thus weaken the cross breaking strength of the roof. For these reasons, a mine pillar will stand best and can be made of the minimum volume when its base and capital meet floor and roof in broadly spreading tangential curves, which are concave in profile.

Sometimes the roof and floor beds, in direct contact with the seam, are themselves quite hard, but so thin that they bend and transmit the pressure on the pillars to an adjoining soft stratum and force it out through any fissures that may be in the roof or floor. When the pillars themselves are too weak for the pressure, what is called a "squeeze" (failure of pillars) begins, by a shelling off of the outer surface, and later a collapse occurs, which may be a gradual sinking, with elastic strata like some coals and shales or a sudden fracturing in masses with hard blocky rock-like limestone or quartz. Some substances, like coal, pyrite and easily weathered rocks, loose strength on exposure to mine air and this fact must be considered if durable pillars are to be made of them. The minimum fraction of a bed necessary to leave for pillars may be thus calculated:

Let x = fraction of area to be left in pillars;
h = depth of cover in feet;
w = specific weight of cover in pounds per cu. ft.;
s = ultimate compressive strength in pounds per square foot of least resistant stratum adjoining pillar;
m = factor of safety,

Then, weight held by 1 sq. ft. of seam $= hw$, and compressive strength of corresponding pillar $= xms$, hence $hw = xms$.

$$\text{or } x = \frac{hw}{ms} \qquad (3)$$

If excavation is inclined at an angle as in Fig. 15, then the pressure is

$hw\cos\theta$, so that $hw \cos \theta = xms$ or $x = \frac{hw \cos \theta}{ms}$ .... (4)

It is difficult to get the real strength of the floor, pillar and roof beds because the beds themselves are seldom free from planes of weakness which would not be appreciable in the small blocks that must be used in the testing machines for compression, tension or shear. For this reason the factor of safety m of equations (3) and (4) is taken at from 2 to 10, varying with the nature of the strata and of the mine layout.

It is only by close watching on the changing conditions that movements of the formation over wide excavations can be prevented even in well laid out mines. An incipient "squeeze" of pillars may sometimes be checked by building up stone-filled wooden cribs along their edges, but this remedy may merely shift the pressure and transfer the "squeeze" elsewhere. Often it is better to localize rather than to attempt to support a squeeze and this can be affected by allowing the roof to cave over the disturbed section, assisting the fall where necessary by blasting the roof and pillars. The volume of roof thus made to fall will be that under the dome of fracture as *acb* of Fig. 13, the span *ab* in this case not being the width of a single room, but of the whole disturbed section. If the seam is thin in proportion to the height of the falling dome, the broken rock, as it occupies more space than when solid, will fill up the space under the surface of fracture and form a sufficient support to prevent any further strain on the overlying formation.

The caving of the roof over the disturbed area is also a remedy for "creep" (oozing of roof or floor into excavations), but if the ground surface is to remain intact a safer plan is to fill the excavation solid with rock. Where a supply of fine material like mill tailing or sand can be obtained cheaply, the filling is best done by mixing it with water and running it into the workings through pipes by the flushing system of the Pennsylvania anthracite regions as described in Examples 58 and 59.

The caving of the roof, locally, by blasting can be easiest affected by reversing the methods already explained for roof support. If pulling or blasting out all artificial supports does not bring down the roof, any rock pillars in the area should be drilled and blasted by simultaneous firing. The next procedure is to drill holes into the roof so as to cut a groove around the springing line of the dome *acb* in Fig. 13. The work of the drill men around the edges of the excavation will be safe and the circumferential groove can thus easily be widened and carried higher until the central bell of the sub-arch dome has so much of its sustaining surface *acb* cut away that it drops out.

Effect of Caving on Overlying Objects

In working superimposed beds simultaneously, it is necessary to determine the proper relative position of pillars in the various beds.

Pillars must also be located, in caving mines, where it is desired to protect valuable surface structures. In modern coal mining, both the longwall and usually the room and pillar method involve the caving of the excavations.

How far up an underground subsidence will reach depends on a number of conditions, such as area, height and manner of making of excavation, nature of overlying formation, presence of faults and dikes, etc. By Fayol's second rule, the height affected by subsidence would not exceed four times the width of the excavation, but this only holds good for a limited area whose sub-arch roof block can scale off at leisure. When large areas are excavated, complex stresses arise which are apt to cause sudden irresistible strains on the roof which cause it to develop long cracks and fractures analogous to faults. If the overlying strata contain many strong rock beds, these may act as beams which rest on the broken caved formation beneath them and prevent any effect above. Thus at Sunderland, England, where half of the strata are hard rock, coal seams have

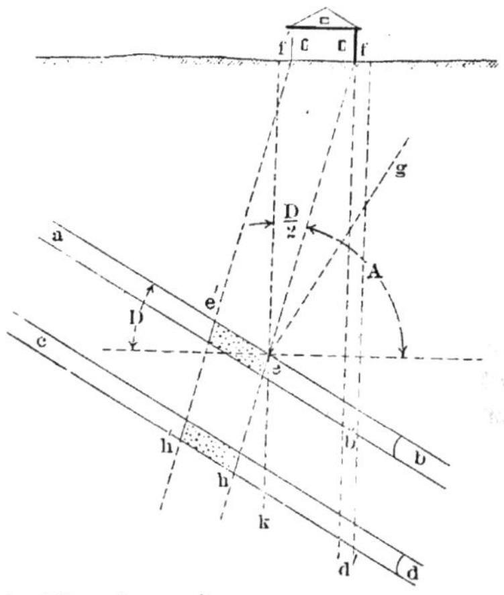

FIG. 16.—Effect of excavation on overlying bed and on surface.

been mined and caved at the depths of 1600 ft. without affecting the surface. In the Transvaal gold beds, dipping at around 40 deg., caves may occur over areas of several acres at depths over 1000 ft. without surface movement. With a formation of soft friable strata, like shale or glacial drift, however, there is nothing to arrest a subsidence beneath, and under such roofs the effect of caving coal mines, 2000 ft. deep, has depressed surface structures.

Fayal's third rule applies to excavations of large area and is "where the area is infinite and the beds are chiefly sandstone with a dip less than

40 deg., the height of the zone of subsidence is less than 200 times the height of the excavation." This means that the caving of an excavation, 6 ft. high, would not affect the surface if over 1200 ft. be'ow it. The third rule is based on the height of excavation rather than on its width, like the other rules, and depends on the principle already mentioned that the strong strata tend to rest solidly, ultimately, on the caved ground below.

Subsidence does not break strata perpendicular to their bedding planes. For defining the disturbed area over excavations under unbroken stratified formations two rules are used, the first for slightly and the second for steeply dipping roofs. Thus in Fig. 16,

if D = dip of roof strata in degrees
A = dip of angle of fracture,

for roofs under 30 deg. dip Richardson gives,

$$A = 90 \text{ deg.} - 1/2\, D \qquad (5)$$

which signifies that the plane of fracture $e\,f$ (Fig. 16) of bed ab lies half way between the vertical and the plane eg (normal to the dip line of the roof).

For roofs over 30 deg. dip Hausse gives,

$$\tan A = \text{cotan}\, 2D + 3 \text{ cosec}\, 2D \qquad (6)$$

Formula (6) gives for a 30-deg. roof only a slightly larger angle of fracture than formula (5), but as the dip gets steeper the difference between the two formulas steadily increases while a maximum A is reached with formula (6) when D is between 50 deg. and 60 deg. as shown in the following table:

	Angle A, degs.	
Dip D	Formula (5)	Formula (6)
0 deg.	90	90.0
10 deg.	85	85.2
20 deg.	80	80.5
30 deg.	75	76.2
40 deg.	70	73.0
50 deg.	65	70.8
60 deg.	60	71.0
70 deg.	55	74.0
80 deg.	50	80.8
90 deg.	45	90.0

These formulæ can only be considered as general guides to the probable location of the plane of fracture and they must be modified in practice by a consideration of the surface topography, of the structure of the formation and of natural breaks like joints and faults. Where thick dikes cut across the roof strata, the plane of fracture is more apt to follow along the surface of the dike than to break it.

Protection of Surface.—The practical use of these formulæ is shown in Fig. 16 where it is desired to protect the building at *fb'* when mining the veins *ab* and *cd*. Here *h'f'* and *hf* are planes drawn parallel to the plane of fracture *ef* and their intersection with the beds defines the inside boundaries of the pillars *c'e* and *h'h*. The margin of safety to be left around these inside limits of the pillars for "draw" varies with the importance of the building and how closely the strata have been observed to follow the fracturing formulæ.

Protection of Overlying Beds.—Where the veins *cd* and *ab* are being mined simultaneously, it is evident that the pillars to be left in *cd* to protect pillar *c'e* must not be the ground vertically beneath, as *hk*, but that enclosed between the same planes of roof fracture as *h'h* with a due allowance added for "draw." In excavating, also, the direction of the roof fracture *ef* must be taken instead of the vertical plane as the guide to relative operations in the upper and the lower beds. Thus for safety the bed *ab* would be stopped "ahead" (measuring from the plane *ef*) of bed *cd;* except in the case where *cd* was being filled, when the slight subsidence of the floor of *ab*, caused by the settling of *cd* (when "ahead") on its filling, would render the breaking of *ab* easier.

In mining the superincumbent parallel anthracite seams of the Lehigh Valley Coal Co., by the room and pillar system, the pillars must overlay each other when the parting is thin. A neglect of this precaution, with the usual parting, is liable to result in the squeezing of the overlying pillars down into the rooms of the seam below. When the parting is over 40 ft. thick, however, it is only necessary to have the panel pillars (at ten-room intervals) of adjoining seams superincumbent, and to lay out the entries and room axes of both seams approximately parallel to each other; in this way the work in different seams can be pursued more independently and just as safely.

Shaft Pillars.—The same principles and formulæ can be applied to the design of pillars for protection of shafts. In Fig. 16 the vertical shaft *fd* will need a pillar in each seam extending to the intersection with the plane of fracture passing through the shaft collar at *f*. Thus the minimum upper limit of these pillars must be at *e* and *h*, which for considerable depth, would mean many hundred feet away from the shaft. But this involves only a moderate loss of ore because the pillar may be narrow and need extend only a short distance down the dip to *b* and *d*. The distances *b'b*, *d'd* and the width of the shaft pillar along the strike of the seam may be estimated by formula (4).

For inclined shafts following the mineral seam, the protecting pillars should be continuous strips on each side with break-throughs only for the loading stations. The width of these strips, if estimated by formula (4), should increase gradually from the surface downward. Although this last requirement is seldom fulfilled in practice, it gives the minimum loss of ore for the maintenance of a stable roof.

Support of Excavations

Curved Sections.—A tunnel section may be supported by the circular lining 1 (Fig. 17) against external pressure from any direction since the portion of the ring taking the ground pressure will be an arch to transmit its load to the balance of the ring acting like arch abutments. If we consider only the keeping open of a given area with the least material, the circular lining may be replaced with advantage by the elliptical, when the pressure is greater in one direction than in another, by placing the long axis of the ellipse parallel to the direction of greatest pressure. Thus if the greatest pressure comes from roof or floor, the ellipse should be vertical as 2 in Fig. 17, and if from the sides, the ellipse should be horizontal as

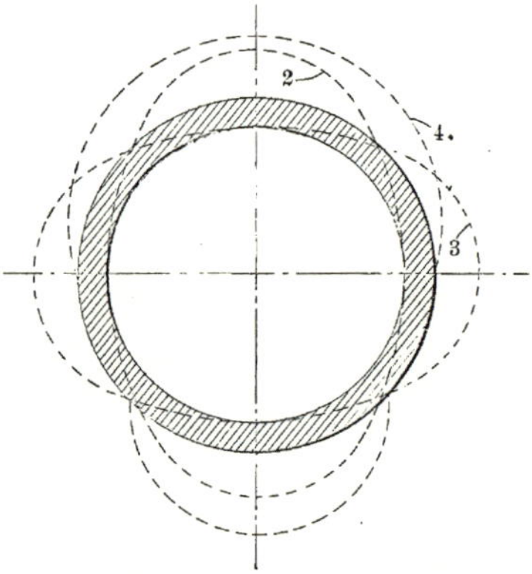

Fig. 17.—Tunnel sections.

in 3. The circular lining is most economical when the external pressure is equally distributed, or where it comes in an oblique direction, for an oblique ellipse would be generally unsuitable for use. The oblique pressure is apt to occur when driving along the strike of inclined beds. Other considerations, besides economy of lining, usually prevail in practice so that circular sections are less used than elliptical ones, which fit cars more closely in transportation tunnels, or egg-shaped ones, like 4, which have a lesser hydraulic gradient for water conduits or drains. The material most used for curved linings is cast iron or some kind of masonry, though steel shapes are also formed to fit, and timber polygons to approximate curved sections. To merely support the ground, it is clear that only that part of the tunnel section need be lined which has weak walls so that we see in practice linings on the roof alone, on the roof and one side, or on three sides, the ends of the lining resting in each case against abutments of solid rock.

Rectangular Sections.—The greatest available area in transportation tunnels for the minimum volume of excavation is obtained by using the rectangular instead of the curved section. Ordinary brick or stone masonry, having little tensile strength, is unsuitable for lining any part of the rectangular section subject to cross-breaking strains. Therefore it is not used except for side walls. Timber or steel beams and re-enforced concrete are the common linings for rectangular sections. With a weak roof and strong sides only the piece *ab* (Fig. 18), which is called a cap or a "quarter-set" is put in; with both roof and one side weak the cap *ab* and the post *bc*, called a "half set," are needed; with roof and both

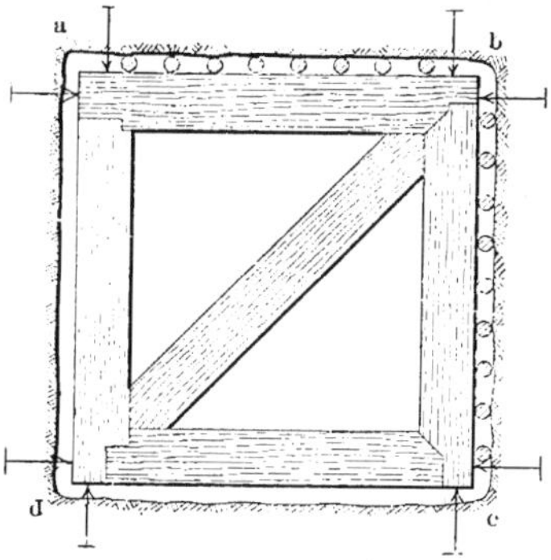

FIG. 18.—Tunnel lining with timber

sides weak a cap and two side posts, or a "three-quarter set," is used; while with four weak walls a cap, two posts and a floor sill, or a "full set," is required.

The attempt will not be made here to discuss methods of framing except as they are affected by ground pressure. With a predominating vertical pressure a good joint for square timber is at *a* (Fig. 18), and for a round timber at *b*. Where the main pressure is horizontal a joint for square timber is shown at *d* and one for round timber at *c*. A rectangular frame can resist pressure acting parallel to its sides, but tends to collapse under oblique pressure. It is to make them more stable under oblique pressure that tunnel sets have outward-battered instead of vertical posts.

Stope Sections.—In keeping open the large stopes of some metal mines with the framed cubical cells of the square-set system, the same precautions of properly designed joints and of uniformly spaced points of contact with the surrounding rock must be observed. Miners have found to their sorrow that it is useless to attempt to keep open square-setted stopes under oblique pressure unless diagonal braces (like *bd*) are inserted, parallel to the direction of pressure, for transforming the unstable squares of the frame into stable triangles.

Zones of Fracture and Flowage

Wooden or steel frames will only keep the peripheral surface of excavations intact against the pressure of loose pieces or sub-arch blocks like *acb* in Fig. 13. For the support of the overlying formation, even masonry is only of limited commercial utility, therefore rock pillars or filling with waste must be relied upon. Beyond a certain depth, or below the "zone of fracture" of geologists, we have the "zone of flowage," where no opening can be maintained permanently owing to the inability of any fraction of the rock, left as pillars, to sustain the superincumbent pressure.

Transposing formula (3) we have for h' (the depth of the zone of fracture):

$$h' = \frac{smx}{w},$$

but for the zone of flowage both m and x are $= 1$ and substituting these values we have

$$h' = \frac{s}{w} \qquad (7)$$

From equation (7) it is evident that the depth h' depends solely on the compressive strength of the basal rock and the specific gravity of the overlying formation. Assuming the specific weight w of the earth's crust to be 150 lb. per cu. ft. and the compressive strength of the basal rock to be 3,000,000 lb. per sq. ft., we have by substitution in (7)

$$h' = \frac{3{,}000{,}000}{150} = 20{,}000 \text{ ft., or about 4 miles.}$$

In those localities where the surface rock is so porous as to contain a considerable proportion of water, w might be less than 150 lb. and the depth of the zone of fracture correspondingly increased.

CHAPTER V

PRINCIPLES OF MINE DRAINAGE

Those miners who talk much of pumping but little of drainags resemble those old-fashioned doctors who spend all their time on remediee and neglect diagnosis. Instead of studying water conditions beforehand as a basis for drainage equipment, a too common way is to try to fit the pumping plant to the in-flow after it has appeared. This policy may mean a drowned mine, and weeks of delay for the installation of larger pumps and the clearing from water; it may mean a set of makeshift pumps of the wrong size and of low efficiency which may really be wholly unnecessary owing to the feasibility of natural drainage.

Problems of drainage involve chiefly the four sciences of meteorology, geology, hydraulics and mechanics. From the first we may determine the quantity of rain likely to fall on our mine watershed; from the second the conditions affecting the behavior of underground water in the rocks; from the third the laws governing the pressure and flow of water; and from the fourth the mechanical methods of unwatering.

Estimate of Water to be Drained

There are multitudinous mineral deposits, each with a special problem of drainage of which only some general features will be discussed here under three cases: I. Deposits in unconsolidated rock; II. Deposits in stratified rock; III. Deposits in massive rock. For any type the water encountered in mining operations will depend on four factors: (1) the area of contributory watershed; (2) the moisture falling on watershed; (3) the moisture percolating the surface of watershed; (4) the facilities for underground water to enter the mine.

Case I. Deposits in Unconsolidated Rocks.—In Fig. 19 is shown a cross-section of a gentle syndinal rock trough *ab* filled with alluvium up to the surface *cd*. It is proposed to lower the "water table" or ground water level *wt* down to sump *s* in order to mine a placer deposit extending from *a* to *b*. The conditions which determine the quantity of water to be handled depend on two items; namely, the quantity of ground water, and its velocity of entrance into the workings. For the first item we have to calculate the area, rainfall and percolation of the contributory watershed, while for the second item the fact that it will be affected by the method of drainage will have to be taken into consideration. The area and rainfall are also the basis for the calculations of the water-supply engineer, but the latter reckons rather with the run-off than with the percolation which concerns the miner.

With underground conditions as represented in Fig. 19, the area of the contributory watershed evidently extends in width from *e* to *f*, and in length from the sump *s* to the head of the valley, if *cd* is a river trough, or to the bounding contour of the watershed if *cd* lies in a lake basin. In general the contributory watershed is all the ground area that drains toward the surface lying over the sump, wherever the surface is connected with the sump by pervious strata as in Fig. 19. The depth

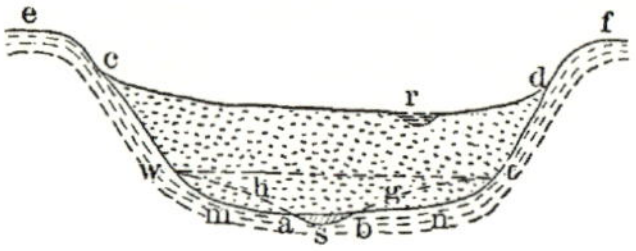

FIG. 19.—Drainage in unconsolidated rock.

of current rainfall is recorded for most localities in civilized countries at the government meteorological stations; and in solving drainage problems, these records should be scrutinized for the maximum, mean and minimum rainfalls both by months and years. From this data, we have the rainfall in the wettest year or season in contrast with that of drouths, but it is important also to note what part of the moisture falls as snow and the melting seasons of the latter.

The whole rainfall, however, does not concern the miner. He is concerned only with that fraction of it which sinks into or percolates the ground after evaporation and run-off have taken their tolls. Then if

area of a watershed	$=A$ sq. ft.
depth of moisture falling on a watershed	$=D$ ft.
volume of moisture falling on a watershed	$=Q$ cu. ft.
volume of moisture running off from a watershed	$=R$ cu. ft.
volume of moisture evaporating from a watershed	$=E$ cu. ft.
volume of moisture percolating a watershed	$=P$ cu. ft.
fraction of moisture Q evaporating, or evaporation factor	$=e$
fraction of moisture Q running off, or run-off factor	$=r$

we have,

$$Q=E+R+P \qquad (1)$$

but $E=eQ$ and $R=rQ$

so substitute in (1) and

$$Q=eQ+rQ+P \text{ or } P=Q\ (I\text{-}e\text{-}r) \qquad (2)$$

but $$Q=AD \qquad (3)$$

hence $$P=AD\ (I\text{-}e\text{-}r) \qquad (4)$$

Evaporation is dependent on the state of the atmosphere and the covering and texture of the soil. The atmosphere affects evaporation by its changes in humidity and in movement. Both dryness and high winds hasten evaporation which is usually compared for different atmospheres by observing water surfaces. Thus, in the United States the mean annual evaporation varies from 40 in. in the Middle Atlantic states to 50 in. on the Gulf of Mexico, and 95 in. at Yuma, Arizona.

In the same locality the rate of evaporation which is approximately equal for all bare soils, is greatly increased by a cover of vegetation. Thus a 5-year trial at Geneva, New York, with an average rainfall of 23.7 in., gave its evaporative factor (*e* in Equation (4)) as 0.64 for bare cultivated soil, as 0.71 for bare undisturbed soil, and as 0.85 for sod. Not only the heat of summer but its vegetation increases evaporation, while the ground surface in winter acts much like bare soil unless covered by snow or ice, the daily evaporation rate of which in New England is .02 in. and .06 in. respectively. A less proportion of severe rains is evaporated than of drizzling rains, for as a given area has only a limited rate of evaporation any excess moisture must either run off or percolate.

The common method of estimating the run-off is from measurements of the quantity of water flowing in the streams of the watershed. When the bed of a stream is once mapped in section, a record of its surface-height readings renders possible a calculation of its sectional area which, combined with corresponding readings of a current meter, gives the data for computation of flow. The percentage of rainfall found in streams, evaporation being neglected, depends both on the slope of the surface and on its covering. For gently rolling land as in Iowa, the run-off factor (r in Equation (4)) is 0.33, for the rougher surface of the Middle Atlantic States it is 0.40 to 0.50, while in the mountain states of Colorado and Montana it is 0.60 to 0.70. The surface covering most favorable to a heavy flow is frozen snow over which over 90 per cent. of the rainfall may run into the streams, while the melting of the winter's snow by warm rains causes the freshets and floods of spring. Where the surface is irregular so that the rainfall collects in ponds and swamps instead of reaching streams, the run-off is lessened, and the evaporation and percolation is correspondingly increased.

The beds of surface streams must be relatively impervious, for if they were freely percolated by water, there would soon be no visible flow. Where a stream's bed is is partially porous, much of the water sinks to the first impervious stratum and there forms an invisible stream called the underflow which often contains more water than its parent overhead. Where a stream has not naturally a channel of impervious rock or clay, the tendency is for it to stop the pores of a sandy or other pervious bed with sediment; especially is this so in alluvial valleys like that of Fig. 19, where there might be no visible stream at all had the river at r not a clay-coated bottom.

For our drainage problem of Fig. 19, we have now discussed how to ascertain the area of watershed A, the depth of rainfall D, the evaporative and run-off factors e and r, and by substituting these values in equation (4) we have the percolation P. The result from solving Equation (4) can be compared with the following table which gives for the percentage of total rainfall percolating various surfaces:

sand	=60 to 70
chalk or gravelly loam	=35
sandstone	=25
limestone	=15
clay or granite	=15 and less

We do not have to provide at s for the drainage of volume P, but only for that portion of it which is not drained off elsewhere, does not reascend to evaporate at the surface or is not held in the pores of the subsoil. The drainage elsewhere would be nil in a lake basin with impervious bottom, but in the usual self-draining basin it would constantly tend to lower the water table.

The lake-basin condition is well exemplified at both Bisbee and Tombstone, Arizona. These camps lying in the Mule Mountains, where the annual rainfall is under 12 in. and the evaporative factor large, would be casually reckoned as having dry mines, but the very opposite is the case. The ore bodies in each camp are found in limestone and shale beds which are so folded as to form, with the adjoining intrusive rocks, an impervious basin which catches all the rain percolating the surface over a large watershed. The present water in the basins represents the accumulation of years, so to lower it has taken more pumping than would be necessary in a very wet valley whose ground water was dependent solely on current rainfall. At Bisbee in 1906 the water level had been permanently lowered, for three years' pumping by several companies had reduced the inflow at the Calumet and Pittsburg shaft from 3000 to 1500 gal. per min.; but at Tombstone in 1911, where nine years of pumping of the Contention shaft had little affected the original flow of 3000 gal. per min., it was deemed unprofitable to struggle further and the pumps were pulled. The excess of water in the Tombstone basin probably comes from an adjoining watershed through underground channels.

The loss of ground water by evaporation increases with a damp soil and a high water table. Consequently, in self-draining ground the evaporation is greater if the rainfall is evenly rather than sporadically distributed. Evaporation, however, usually affects the water table only slightly as compared with the capacity of the formation for the storage, surrender and passage of water.

In Fig. 19, the section of the water-storage area between *wt* and bed-rock is not the whole area *wmnt*, but this area multiplied by the factor for "voids" or the proportion of intergranular spaces in the formation. The void factor depends less on the size of rock grains than on their uniformity, and varies from 0.2 to 0.5. Yet another item must be included, in estimating the quantity of water that must be drained to lower *wt*, and that is the factor for surrender or yield which depends on the capillarity or fineness of the grain of the formation. The yield factor is almost nil for clay, 0.5 to 0.6 for porous soil, 0.6 to 0.7 for sand, and nearly 1.0 for clean gravel or boulders. A low yield factor means not only the retention of rainfall in a porous formation until it is saturated, but a long delay before a heavy shower begins to be noticed underground. From the above, if

W = volume in cu. ft. of water-bearing formation tributary to sump s

W = volume in cu. ft. of water the water-bearing formation yields tributary to sump s

x = factor for voids in formation

y = factor for water-yield of formation

then $W = xyW$. (5)

To free our placer *ab* from water, it will not be necessary to lower the water table to the profile *wmnt*, but only to the profile *whabgt* where *wha* and *bgt* are the profiles of the hydraulic gradient toward the sump *s*. The hydraulic gradient increases with the fineness of grain, though very small in gravel, it is 30 to 50 ft. per mile in sand and in a large basin, it would thus considerably decrease the volume of water tributary to sump *s*.

The hydraulic gradient for a given formation can be directly measured by digging two wells in the same line of water drainage, at some distance apart, and then recording their water levels. The hydraulic gradient will then be the difference of water level divided by the distance between the wells.

Hazen gives as a formula for the velocity of passage of ground water

$V = KD^2S$ (6)

where V = velocity in ft. per sec. of flow through ground pores

K = 0.29 (a constant)

D = diameter in mm. of sand grain "effective" (*i.e.*, 90 per cent. of grains must be larger than D). Formula (6) is inapplicable when d is less than 3

S = sine of slope of the hydraulic gradient

Then if B = area in sq. ft. of a vertical surface enclosing mine openings extending from water surface in pump to water table. Height of surface B should be small for use of formula (6)

f = volume of water in cu. ft. entering placer per sq. ft. of area B

F = total volume in cu. ft. entering placer over total area B

k' = fractional factor for voids in walls of area B

It is evident that

$$f = k'V \text{ and } F = k'BV$$

Substitute for V from Equation (6) and

$$F = kK'BD^2S \qquad (7)$$

In practice the possibility of keeping the placer dry enough to permit miners to work would of course depend not only on the means of drainage available to keep sump s clear, but also on f or the rate of inflow at the mining face. As f increased beyond a certain figure, the miners would find themselves working in a heavy spray and standing in a gurgling pond. In such a case, unless the inflow could be controlled by a cofferdam, subaqueous mining would have to be resorted to.

Case II. Deposits in Stratified Rock.—An example of this case is shown in Fig. 20, a cross-section of a coal seam A in a synclinal basin. Beneath the coal is a thin layer of clay B resting on a sandstone C, and

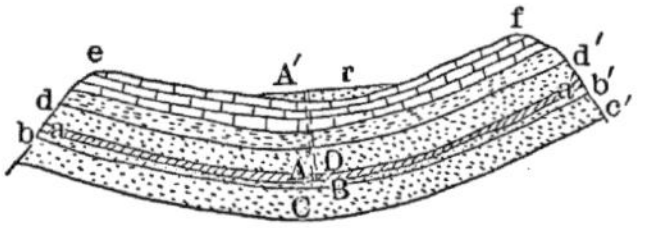

Fig. 20.—Drainage in stratified rock.

above it are strata of sandstone D, shale G, and limestone H. Along the surface runs a river r over a valley-filling of alluvial soil. Then the percolation into a coal mine at A will depend not only on the coal itself whose bedding and joint planes may be somewhat permeable, but on the nature of the adjoining rocks.

Clay and shale are not only relatively impermeable but plastic, and tend to close tm any openings made inyaving aorogenic movements. Sandstones vary in their structure, some h duebh texture as porous as free sand, while the grains of others are closely cemented and almost impermeable. Limestones, especially if dolomitic, abound in irregular channels and pot-holes, often large enough to contain underground rivers or ponds. Should the rocks of Fig. 20 be subjected to metamorphism, their permeability would be much diminished, or perhaps entirely destroyed, as pores and bedding planes were obscured, until we approached as the limit the massive formation of Case III. Clay and shale, when metamorphosed, become dense and strong slate or schist, sandstone solidifies into impermeable quartzite, and limestone changes into crystalline marble.

From these considerations, it can be seen that the stratum of shale at G, provided it has not been pierced by orogenic movements or human hand, acts as a screen to keep out any water which may percolate into the limestone from the watershed *ef*. As the strata outcrop, however, beyond the summits *e* and *f* of the synclinal basin, the coal seam will be exposed to percolation from watersheds *ec* and *fc'*.

As long as the impermeable clay floor *B* of the coal is uncracked, the watersheds contributory to the coal seam will extend only from *e* to *b* and from *f* to b^1 and not all of their percolation will reach the coal, because the shale stratum *G* will seal off any surface water that may enter the limestone layer *H* between the crests *e* and *f* and the roof of *G*. Should the floor *B* be cracked or feathered out in places, it may be serious from a drainage standpoint, for the coal seam will then be open to a flood from the hydrostatic water in sandstone *C*. Thus in the rock formation of Fig. 20, the ground water would not occur in a connected body as in Fig. 19, but each porous zone would contain its own pool separated from the others by an impermeable stratum. The equivalent of the water table *wt* of Fig. 19 would be found here in the limestone *H*, but it would circulate there in irregular open channels instead of in intergranular pores. The contributory watersheds having been thus measured, we have only then to gather the other meteorological and physical constants, as explained for Case I, in order to solve Equations (1) to (7) for the drainage of Fig. 20.

Case III. Deposits in Massive Rock.—In Fig. 21, let *cd* be the cross-section through a fissure vein in massive rock, which is either of igneous

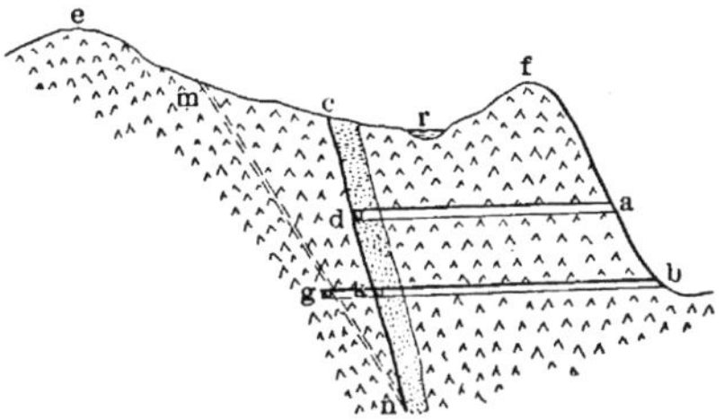

FIG. 21.—Drainage in massive rock.

origin or so metamorphosed that its sedimentary pores and bedding planes are practically obliterated. This formation, then, instead of being quite porous like that of Case I or irregularly porous like that of Case II, is in its original condition, more or less impermeable, but in mining regions it has usually been so cracked by earth movements as to abound in openings which grade from wide fissures, both long and deep, to such minute fracture planes as those of the Bingham copper porphyry which scarcely pass seepage water. When rainfall can only percolate the surface of Fig. 21, through irregularly spaced crevices or joints instead of through a porous zone, there can be nothing like a general ground water level except within areas whose crevices are all connected. Thus each crevice system has a height of water table varying according to the size and nature of its contributory watershed. The mineral veins themselves have often trunk channels along their walls which receive water from numerous branch cracks and fissures.

The watershed tributary to vein *cd* of Fig. 21 will not extend laterally from *e* to *f* as in Case I unless all the intermediate fissure systems lead to the vein, but it may cover a much wider area owing to the possible juncture of subterranean streams with *cd*, which streams in mountainous regions may be under a high hydrostatic head. In fact, only the mapping of the region's underground water channels, and this could seldom be done except in an extensively developed district, would enable an engineer to satisfactorily solve Equations (1) to (7) as in the two previous cases. In mining *cd*, care would have to be taken on the hanging side, for by the tapping of natural blind crevices or by allowing the hangwall to move and crack, the river *r* might be precipitated into the workings.

It is probable that some of the hot water found in mining such igneous formations as the Comstock lode comes, not from rainfall, but directly from the occluded moisture of cooling magmas. According to the nebular hypothesis, all surface water had originally an igneous origin. The miner who operates in a region of magmatic water cannot estimate its quantity beforehand, as in the case of meteoric inflows, but must simply handle it as it appears.

As mines get deeper and rock pressures become greater, fissures and other open spaces tend to close up and long before the bottom of the zone of crust fracture is reached, at a depth of something around 4 miles, there is little free water in the rocks. At the Calumet and Hecla Copper mine, Mich., in a conglomerate lode bedded between amygdaloids, the maximum water flow is at 1800 ft. along the 38 deg. dip, while at 3000 ft. the water flow is insufficient even to supply the drills.

Control of Water

This topic naturally divides itself into surface and underground control.

Surface Diversion.—It is much better to keep water out of a mine than to use the most approved method of drainage after its unnecessary entrance. Surface run-off is kept out of a mine ditching around shafts and vein outcrops as *C* in Fig. 21. Often it is best to refrain from stoping a vein out quite up to the surface in order to keep rain out of the workings. A stream above the mine, which seeps badly into its bed or whose bottom may be cracked by caving operations, can often be diverted to another channel or carried in a flume over the dangerous stretch.

Underground Diversion.—In Fig. 20 the penetration of the impermeable shale layer *G* by the shaft *AA'* will eventually drain the wet limestone layer *H* into the mine at *A* unless some precaution is taken. Two remedies suggest themselves, the first, a concrete shaft-lining from the surface down to a sealed footing in the shale; the second, the usual pervious shaft-lining provided with a water ditch or "ring" around the shaft, in the roof of the shale, which catches the water from above and leads it to a sump, which has means for drainage, on the same level.

In ore deposits in hilly regions, an impervious floor sometimes has below it a sandstone or other porous stratum which dips toward an outcrop, on a hillside at a lower level, and is thus self-draining. In such a case, diamond-drill holes or a winze through the floor of the mine sump into the porous stratum will effectively drain the workings.

Natural Dams.—Rock barriers are highly useful in the control of water in mines. As already explained, where tight strata cut off the mine from wet formations, such natural seals should be left undisturbed if possible. Pillars of mineral are often left between adjoining mines to keep their water systems distinct, and in many states a barrier about 50 ft. wide must be left unmined around the boundary of coal properties.

The Lehigh Valley Coal Co. is now mining, near Hazleton, Pa., a synclinal trough containing parallel anthracite seams which extends for several miles and dips for about 3000 ft. vertically in that distance. The trough has been divided into three drainage basins by leaving a transversal barrier pillar of coal, 100 ft. wide, below each. The barriers are at altitudes of 1084, 1250, and 4000 ft. respectively and each has its own unwatering system. Each barrier is pierced by boreholes lined with pipes whose valves can be opened to drain the basin above into the one below in case of an emergency.

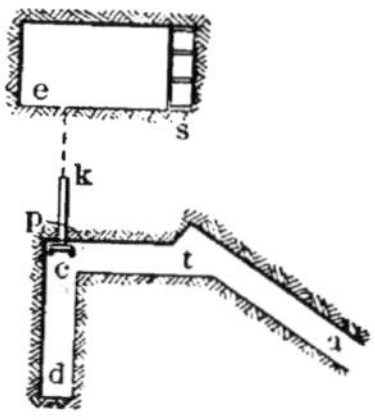

FIG. 22.—Tapping sump at Iron Mt. Mine, Montana.

It is often necessary to penetrate water barriers in order to drain old mines, and where the dammed-up water is under a high head it is best tapped by drilling. Boring long holes for tapping can be done in any direction by a diamond drill. A customary safe-guard against heavy pressure is to bore the first few feet of the hole large enough for a pipe lining *ck*, Fig. 22, whose exterior is made to fit the rock tightly by a packing of lamp-wick, wound spirally, or of cement. When it is necessary to regulate the flow from the hole, a valve *v* is put on the lining pipe whose end must then be anchored to posts *p*. Such a valve on a drill-hole flow, small anyhow, allows the cautious emptying of old workings where a sudden release of water might damage the shaft or other important pillars.

For short distances, tapping can be well performed with a percussive drill and a typical recent case can be cited at the Iron Mt. Mine, Montana, where the new drainage adit was connected with the old shaft-workings which contained some 100,000,000 gal. of water under a head of 900 ft.

When the face of the tunnel *ta*, Fig. 22, arrived near the old shaft *s*, a 6x7-ft. cross-cut *tc* was run for 30 ft., parallel to the shaft station, and from *c* a taildrift was carried back for 30 ft. to *d*. Next a 3-in. percussive drill was set up at *c* and a hole drilled in for 10 ft. to admit a 4-in. pipe lining *ck* which was then well cemented and anchored in. Drilling was proceeded beyond the lining with a 1 3/4-in. bit when, at *e*, 23 ft. from the collar, the point holed through. On loosening the chuck, the bit was shot back by the water pressure against the end *c*, and was followed by a swift stream of water, but as a low dam had been erected across the crosscut at *t*, the men climbed over it and safely reached the adit mouth, a mile and a half distant.

Artificial Dams.—Many dams are built underground: for making sumps out of old headings or stopes; for regulating the flow to the pumps; for isolating the water of abandoned workings; and for confining water to

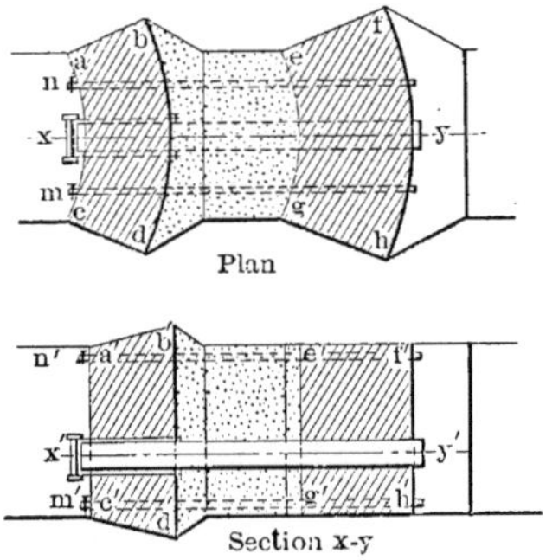

FIG. 23.—Artificial dams.

certain localities, as in the case of flooding mine fires or of filling seams by the flushing system. Mine dams differ from those on the surface in the fact that they often stop openings of small height relative to the pressure of water to be sustained. In such cases, mine dams must have a solid footing all around their periphery instead of just at the base and sides like a river dam. Favorite materials for dams are wood, brick, stone and concrete.

A diverting dam whose crest is higher than the water surface it sustains can be built light and like a surface structure; but precautions must be taken to successfully sustain a high water-head (which causes a pressure of 0.434 lb. per sq. in. for each foot of height), and the arch is a favorite form for this purpose. Fig. 23 shows a composite plan and section of a dam, to back up water at *y* across a heading or shaft, which is made of two arches, *ab* and *cd*, with a filling between of puddled clay or concrete. It will be noticed that the heading walls are cut out to give indented skew-backs for the arches except at *e′f′* and *g′ h′* where a plastic roof and floor might make indentation unnecessary if the swelling wood construction, to be described later, were used.

Both drain-pipe at *m* and air-escape at *n* are provided with valves and sealed tightly in the structure. The manhole pipe *xy* is anchored to the dam and is as essential during construction as afterward. With moderate pressure one arch, like *ch*, is enough; and to build it of wood, each piece should be the length of *ch* and tapered wedge-shape like an arch stone. A tight joint can be made between wood and walls by tarred felt and small wedges, and the pipes can be sealed in with wedges. When of masonry, the arches are laid over wooden centers, the under one of which is left in permanently if the dam is across a shaft. Masonry dams are kept tight by a concrete or clay backing, and as the latter needs to be confined under heavy pressure, the double arches of Fig. 23, with clay between, are then especially suitable. Flat wooden dams are often used and usually they are held by posts with ends hitched into the walls. The wooden lining is made of several layers of planks and, with walls too soft for its support by posts at intervals in hitches, the lining itself may be extended into brick-lined hitches cut in the walls, and its central portion be backed by a timber set whose battered posts are set in the direction of pressure and rest in hitches cut in the heading's walls.

At the Chapin iron mine, Michigan, dams have been helpful in the control of a big inflow of water. The Chapin ore-body is a hematite lense appearing in cross-section about like *cd* in Fig. 21. It is enclosed by slate walls but has an extensive dolomite formation about 100 yards' distant on the hanging side. The author found on his visit in 1908 that the flow at the 1000 ft. level had not been appreciably lessened in spite of pumping 2000 to 3000 gal. per min. for the previous seven years. The water proceeds from channels, in the dolomite hangwall, which are believed to connect with two small lakes several miles to the northeast. If the originally impermeable slate hangwall, that cut off the ore from the water-bearing dolomite, had not been cracked for over 300 ft. from the surface (by the caving operations), the drainage problem could easily have been solved by keeping shafts and cross-cuts entirely in foot-wall.

The No. 2 Hamilton, vertical shaft, then used for pumping, had been sunk in the hanging dolomite and great difficulty had been encountered in driving the 1000-ft. and lower cross-cuts because of the water crevices encountered. In starting the 1000-ft. cross-cut a compound station pump was first installed; but, nevertheless, the first water crevice struck had to be dammed with masonry, and the pressure gauge showed a static head there from a water level within 300 ft. of the surface. Next, a branch drift, some distance back from dam No. 1, was begun; but this also struck a crevice and had to be dammed.

A second branch drift was then started and dammed (after only a short advance) and a second compound pump installed at the station. This last dam was fitted with a water-tight iron door opening outward, so when drifting was continued beyond it (to make a chamber for diamond drilling) the excavated earth could be passed back in boxes. With the diamond drill, the space yet to transverse to reach the vein was searched for a cross-cut opening free from crevices; but as none was found the cross-cut had finally to be finished anyhow by the aid of strenuous pumping.

Damming by Deposition

E. B. Kirby has devised a method (U. S. Pat. No. 900,683) of sealing the rock crevices of mine workings by the deposition therein of sediment. Finely divided clay is preferable but other materials may be used such as sand, mill tailing or slime, cement, saw-dust, horse manure, chopped hay or fiber. The injection of the water bearing this material in suspension may be made by force-pumps, or by stand-pipes extending far enough toward the surface to furnish the necessary pressure.

The suspended particles, when put in a cavity containing water in motion toward exits in the mine, are seized and carried toward such exits, settling, accumulating in, and choking at various points the contributory passages. The moving currents automatically select those passages which are discharging water into the mine and require sealing; they disregard other passages because the water is not in motion in them. The choking which occurs in the outflowing passages is gradual and at those most favorable points where the passages are smallest and the flow most diffused. In fact large passages cannot be thus choked but must by dammed.

When the flowing passages are choked the process ceases even though other passages are still open. If by the choking of one or more passages the current is deflected to others, the deposition is there automatically resumed. At any choked locality the water pressure holds the choking particles firmly in place and produces a perfect seal by shutting off the threads of water in every contributing passage.

Adits.—These are tunnels run in from a low surface point to drain underground workings. In Fig. 21, it is evident that the adit *ad* would drain all of vein *cd* above level *d* and that adit *bk* would drain everything above level *k*. The water-bearing fissure *mn* cuts the vein at *n*, and to drive an adit at the level *n* would obviously be impractical with the given topography; but nothing would hinder the extension of adit *bk* to the fissure, and the running of a drift *g* along its footwall, as far as necessary, to intercept all the water drainage into the mine at *n*. This scheme was employed to supplement the lower adit of the Horn Silver mine in Utah.

The following remarks apply to all drainways, whether adits proper, debouching at the surface, or merely interior tunnels emptying into a sump. The minimum grade for long modern adits with unlined ditches is 1/4 of 1 per cent. The carrying power of water channels can be thus estimated:

If

Sine of slope of hydraulic gradient of water flowing in channel	$=S$
Area of cross-section of water flowing in channel	$=a$ sq. ft.
Velocity per sec. of water flowing in channel	$=v$ ft.
Wetted perimeter of the containing surface of channel	$=P$ ft.
Constant, increasing with smoothness of containing surface of channel	$=c$
Quantity of water flowing per sec. in channel	$=q$ cu. ft.

then from Merriman's "Hydraulics"

$$v=c\sqrt{\frac{aS}{p}} \tag{8}$$

but $q=av$

$$\text{hence } q=ac\sqrt{\frac{aS}{p}} \tag{9}$$

In those cases where adits are only to be used for drainage, a circular section is often preferable; because it carries, when running half full or more, the most water for a given volume of excavation; is stable against external pressure; and is readily adaptable to masonry lining, which, being smoother than wooden sets, gives a larger value for c in formula (9), and consequently passes more water. Where the water deposits sediment, the egg-shaped section of Fig. 17 (4), used for sewers, best enables a uniform carrying power to be maintained as the height of water fluctuates.

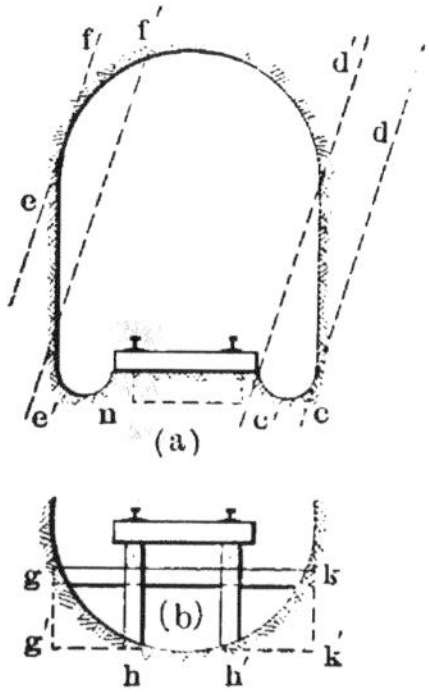

FIG. 24.—Adit sections.

When adits serve for haulage as well as drainage, the economical shape is usually oval or rectangular. The oval shape is best for weaker walls with external pressure mainly vertical, and it can easily be lined, where necessary, by masonry. The rectangular shape is common where the adit follows some flat stratum like a coal seam with a strong roof, that will stand without arching; or where most of the length has to be supported by timber or metal sets which are ill adapted to curved sections.

A compromise section for an adit is shown in Fig. 24 (a) with a self-sustaining arched roof and a flat bottom to give a cheap footing for the track ties. With a moderate amount of water it can be carried in a side ditch which is easier to watch and clean than one under the track. Where the adit of Fig. 24 (a) is in a narrow vein of width from f' to d', the ditch is best placed along that side whose cutting-out, to give space for the adit, will admit the least water from the walls. The tightness of the ditch's bottom rock against seepage should also be considered, if there are to be workings underneath it, and sometimes a wooden or concrete lining may be necessary as commonly it is for sumps. Where the adit can be placed between vein walls as *ef* and *ed* without cutting them, it is usually best to have the main ditch along the footwall at *c;* and connect it, if necessary, by cross ditches to an auxiliary ditch along the hangwall at *e'*.

Both ditches and swamps should be covered in hot mines like those of the Comstock lode in order to prevent any unnecessary humidifying of the air.

The new Roosevelt adit at Cripple Creek, Colo., will be over 3 miles long and used only for drainage. This gold mining district lies in an igneous formation, and as it occupies an area of about 8 sq. miles, it is estimated that each foot in height of its ground water means 60,000,000 gal. of water. The adit was started with a section like Fig. 24 (a), 10 ft. high and 6 ft. wide, but it has been changed to one 6 ft. high by 10 ft. wide to give space for a curved ditch, 6 ft. wide by 3 ft. deep, and a narrow track along one wall.

Where side ditches are inconvenient or inadequate, they can be replaced or supplemented by a central ditch *nc'* (shown dotted in Fig. 24 (a)) cut under the rails. For heavy flows, however, the whole bottom of the adit may be utilized. In that case, it should be cut round, as *ghh'k* in Fig. 24 (b), in order to obtain the cheapest rock breaking and the maximum carrying power for a given sectional area; unless a flat-bedded formation makes the excavation of the larger square area *gj'k'k* just as economical. The track ties may be spiked to stringers which are set on posts or brick piers, *h* and *h'*, of sufficient height to keep the ties above the high-water flow. In double track adits, three rows of stringers on piers are sufficient if long ties are used. Where the track is far above the rock bottom and the adit is narrow, cross beams like *gk*, hitched into the walls, may be the cheapest supports for the stringers.

Adits are especially advantageous in mountainous regions of steep slopes, where a great height can be drained with a short adit. The only drainage expense with adits is for interest and maintenance, and if well constructed, they are not subject to the breakdowns of mechanical apparatus at critical moments. When the adit mouth is some distance higher than the stream into which it drains, the escaping water can be effectively utilized for power. In a wet district of large producing mines whose drainways can easily be connected, it is often advisable to drive a very long adit for general drainage.

Notable among such modern American adits are, in Colorado, the Roosevelt and the 5-mile Newhouse at Idaho Springs; and in the anthracite region of Pennsylvania, the 5-mile Jeddo-Basin in Luzerne Co., the 1-mile Oneida in Schuylkill Co., and the 1 1/2-mile Lausanne near Mauch Chunk. The last named drains 13 miles of underground tunnels and 14 different collieries.

Siphons.—In mining flat coal and other seams, convex rolls often occur in the floor of the gangways which dam up the drainage. It is feasible to pass a low roll by deepening the water ditch; but a high roll, unless it is advantageous to also cut the whole gangway through it to obtain a uniform haulage grade, is often better surmounted by a siphon. A siphon consists of a vertically-curved pipe with both ends set in sumps, of which the outlet sump must have the lowest water-level.

Mine siphons are usually made from welded iron pipe and water can be carried horizontally in them for considerable distances provided they are tight. The limit of vertical lift, from surface of intake to highest point on the pipe of any siphon, is the height of the water barometer minus the total loss of water head, due to internal friction, etc., in the siphon itself. This limit is usually below 26 ft. Several rolls can be passed by one siphon if escape valves for air are put on the pipe at the high point of each vertical bend. It is also possible to drain several sumps or "swamps" along a gangway with one siphon, by running a branch pipe, with a valve on its end, down from the main siphon into each swamp. A siphon is best rigged with a valve at inlet and outlet; and with its highest point joined by a small pipe, with valve, to a water-barrel from which it can easily be filled before a run.

Mechanical Unwatering

Apparatus for mechanical drainage can be grouped into two classes. First, those moving water in buckets, and second, those moving water through pipes. In the first class, water-cars are moved horizontally by the same tractors as ore-cars, while tanks or kibbles are hoisted in shafts or slopes by similar engines to those used for hoisting ore in skips. The second class includes all types of pumps. The first class is often preferable for intermittent unwatering, even if it has a higher operating cost, for where the existing ore-hauling and hoisting equipment can be utilized to move the water-buckets, the heavy expense of installing pumps is obviated.

CHAPTER VI

SURFACE SHOVELING IN OPEN CUTS

EXAMPLE 1.—MOA AND MAYARI IRON MINES, CUBA

Drag-line Excavators on Shallow Flat Placers Without Mantle.—No soil surface exists over these ores; indeed, the ore itself is the soil, upon which grow either pine forests or a characteristic tropical jungle. The deposites at Moa constitute a surface-mantle varying in thickness from a mere film to 121 ft. and occupying an area of 60 sq. miles. The area of more than 8000 hectares of ore drilled showed an average of 18.83 ft.; the Mayari deposit is a trifle thicker and shows an area sufficient to contain more than 600,000,000 tons of commercial ore. The thickness of the ore-mantle is affected by local causes, assisting or delaying the breaking-down of the serpentine bed-rock (which experts agree to be the mother of this ore), erosion by streams, and other causes. The ore lies directly upon the serpentine, and mining will be somewhat unfavorably affected by the fact that the gradation from ore to rock is not at all regular, but very rough so that in cleaning the bottom of an ore-body with any sort of automatic machine, chunks of serpentine are liable to be broken off and lifted with the ore, unless care is constantly exercised. This ore is a clayey limonite containing nodules of magnetite and hematite near the surface. It shows, as chief constituents,

Iron up to 46 per cent.
Alumina up to 15 per cent.
Silica up to 7 per cent.
Free water up to 30 per cent.
Combined water up to 15 per cent.

Prospecting.—By reason of the character and condition of these ores exploration can be carried on by a process that is simple, accurate, rapid, and cheap. Ordinary 2-in. auger-bits are forged on one end of 4-ft. sectional rods, and other being fitted to receive a sleeve-nut, 5 or 6 in. long, into which another 4-ft. section may be screwed. As a hole is driven down by the auger-bit, additional threaded sections are screwed on the rod, making it any desired length. On each end of each rod, except where the bit is shaped, to a backing-nut screwed down hard, in order to prevent the rods from working too tightly into the sleeve-nut when turned into the resisting ground, which would render it difficult to release quickly. In most cases ore can be bored by this simple tool with comparative ease, and when hard blocks and boulders are enconntered, they are sometimes cut by the substitution of a cutting chisel-bit for the auger-point: in other cases the men will move a few feet away and drive another hole, experience having shown that a very short distance will usually be sufficient to avoid a boulder. The hole is started through the drier top soft ore or nodules on the surface, a little water is poured in, the bit lifted and driven down by the combined strength of two men, and then turned in the ore. The work is a combination of churning and boring. Every few feet the tool is lifted, the ore adhering to the bit is cleansed off by pressing a stick into the point of the bit and then revolving

FIG. 25.—Drag-line excavator at Mayari, Cuba.

the tool, and saved for analysis, and all sludge that has collected above the bit is scraped off. Were it not for the peculiarity of this clayey ore of standing without caving, this system of drilling would be impossible, and it would be difficult for the engineer to follow and check the depth of holes by dropping down a measuring-rod, or by inserting a bit with which to test the material at the bottom. It is not uncommon to check grades, of properties previously drilled, by inserting bits in the old holes and reaming out a sample from the sides of the hole.

The price paid the borers begins with from 1.5 to 2 cents per foot for the first 10 ft. of depth, and increases by the addition of a like sum per foot for each succeeding 10 ft. of progress following. In ordinary ground, each borer will earn from $2.50 to $3.00 per day; in other words, a pair of borers will complete from 10 to 13 holes, averaging 20 ft. deep, per day. One 81-ft. hole was drilled in two long days by two men.

In no other way it is possible to explore such an area except at great expense and in a long time. No system of tunnels, pits, or other openings is so well suited to this work. It is well enough to sink pits occasionally, to check by actual observation certain facts that seem patent from the drilling, or to answer any questions that may arise.

To those accustomed to vein-mines or to the great replacement-deposits of the Mesabi iron range, borings, varying from 100 to 300 m. apart may seem utterly inadequate to prove grades and tonnages. In early examinations of the Mayari field original borings were spaced every 100 ft., but as the work proceeded the ore was found to be so regular in analysis, texture, and thickness that holes were gradually spaced at intervals up to and even exceeding 1000 ft.

Excavating.—With no over-burden to be removed, the deposit situated close to the sea, with stream-valleys cutting through the ore-beds and running directly to deep water, and with an average thickness suitable for about one shovel-cut, these ores may be mined at low cost by ordinary steam-shovel. The drag-line excavator is being tried at Mayari (see Fig. 25) and has advantages there, as the deposit of ore is comparatively thin and the floor quite rough. Also, its radius of action is far greater than that of a steam-shovel, which must be moved frequently. The is no question of the relative efficiency of the two machines if the shovel can get one or two full cuts in clean ore, but such opportunities are comparatively rare.

Example 2—Mesabi Iron Range, Minn

(See also Examples 7 *and* 46.)

Steam Shovels on Rolling Lenses with Mantle of Glacial Drift.—Hidden as the great deposits of the Mesabi are by a thick mantle of drift-it is no wonder that new bodies are even to-day being discovered after 16 years of active exploration, when it is remembered that the produc, tive range is 100 miles long by one-fourth to two miles wide.

The long, flat basins which hold the ore are the outcome of gentle folds, transverse to the range and cut up into basins by cross-anticlines.

The ores are mostly soft, hydrated hematites with subordinate, soft limonite. In the great, open pits the occurrence of the ore in continuous beds of different colors and grades is noticeable.

The beds differ much in texture; quite common are layers of broken joint-blocks of hard hematite from 1/4 in. to 2 in. thick, which generally occur alternated with continuous layers of hematite sand or dust. The pore space is considerable, which is shown not only by the speed with which surface water sinks to the drainage shaft in an open pit, but also from 12 cu. ft. to 14 cu. ft. per ton allowed for this ore in place, as compared with 8 cu. ft. to 9 cu. ft. for specular hematite.

Of the coarser joint-blocks it is possible to use 60 per cent. to 75 per cent. in the blast furnace mixture, while of the dust 33 per cent. has been the practical maximum. By selection and judicious mixing about 75 per cent. of the high-grade ore mined can be loaded as Bessemer, and much of the remaining high grade is utilized for Bessemer pig, after mingling with lean, low phosphorous ores from other ranges. Much of the lean non-Bessemer ore is necessarily removed in open-cut work, and this has been stored in stock piles for some future use.

The Mesabi ore deposits are enormous, and single bodies are known to contain from 20,000,000 to 40,000,000 tons. The deposits cover great areas, and, owing to the drift mantle of 10 ft. to 150 ft., many

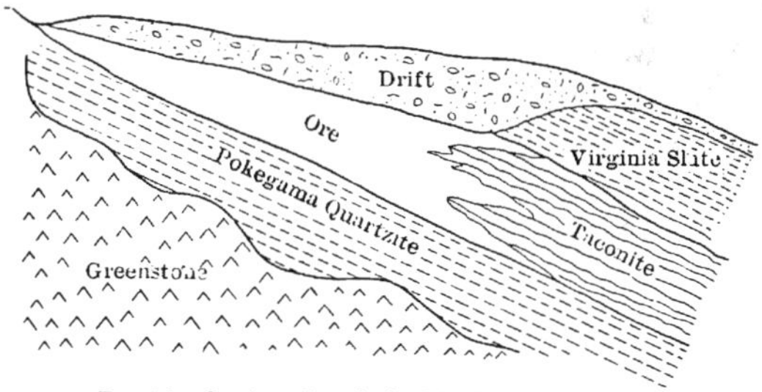

FIG 26.—Section of ore body, Mesabi range.

different adjoining mines might be on one continuous ore body without its being known. As a rule the important ore bodies are several acres in area and have a thickness increasing from the periphery to a maximum of 200 ft. at the center.

The bottom of an ore body, where resting directly on Pokegama quartzite, may be smooth, but when on taconite it is generally irregular and often stepped up on the trough side. (See Fig. 26.) At the bottom of some large deposits are beds of "paint rock" and limonite, forming an impervious basement. In other cases the good ore rests directly on the porous taconite and the basement of the water circulation must be looked for farther down, as, for instance, some dense layer of quartzite.

Of three prevalent methods of ore extraction on the Mesabi, *i.e.*, open-cut, underground mining and surface milling, steam shoveling is easily first, underground mining second and milling last. In 1902 the second and last systems produced 46 per cent. and 7 per cent. of the total output, respectively, but they have since dwindled in relative importance, for in 1909 they only produced 15 per cent. of the yield of 29 1/4 million tons.

In a new mine the surface topography and the prospect drilling (which has previously prospected the deposit with holes at 200-ft. intervals) will enable an intelligent choice of systems to be made. With suitable conditions the open-cut method is cheapest, the milling second and the mining dearest. The cost of mining is about $1 per ton, or twice as much as average open-cutting.

For open-cutting a deep mantle may be stripped if the ore beneath is proportionately thick, and the common rule is, roughly, that a foot of drift can be removed for each foot of ore. With a desirable maximum track grade of 3 per cent., and a possible one of 5 per cent., the proper

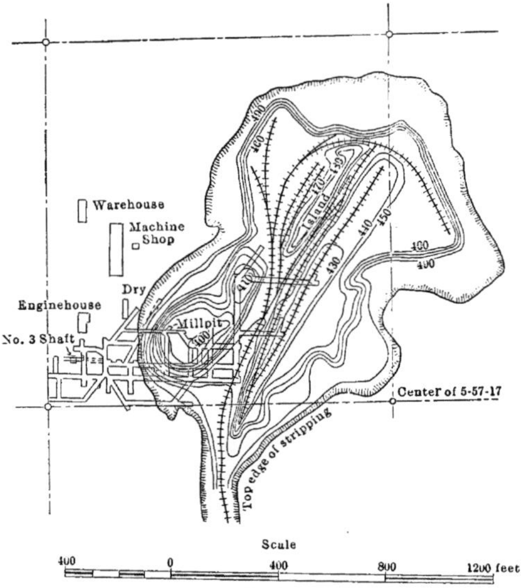

FIG. 27.—Fayal open pit.

layout of trackage to secure all the ore is the first consideration. A side-hill body means an easy approach; if, also, it has an area in proper proportions as to total depth and width so as to allow for suitable benches we have an ideal condition.

There must be an available and adequate dump ground, and the annual ore output must be sufficient to cover the extra interest on the increased investment of this over other systems. Lastly, the lean layers in the ore must be harmlessly situated. If one is on the bottom it can be left, or if on the top it can be removed with the strippings; but if intermediate, so that it cannot be separated in digging, it makes the system less practical.

Drainage.—In beginning an open-cut mine a drainage shaft is generally sunk to the lowest point of the ore having two compartments for a cage-way and a pump-way. At the bottom is placed a station pump of size proportional to the water flow, and on the surface a boiler plant of sufficient capacity to furnish steam to run a small hoist, a dynamo engine for electric lights and the surface pumps which are ocasionally needed to drain ponds formed on some impervious layer in the pit.

Stripping.—The general layout of the digging operations is planned entirely from the results of prospect drilling. There are two possible trackage schemes, one a cut longitudinally along the long axis of the deposit to be worked outward to both sides, and the other a cut in an elliptical ring, from which work proceeds both inward and outward.

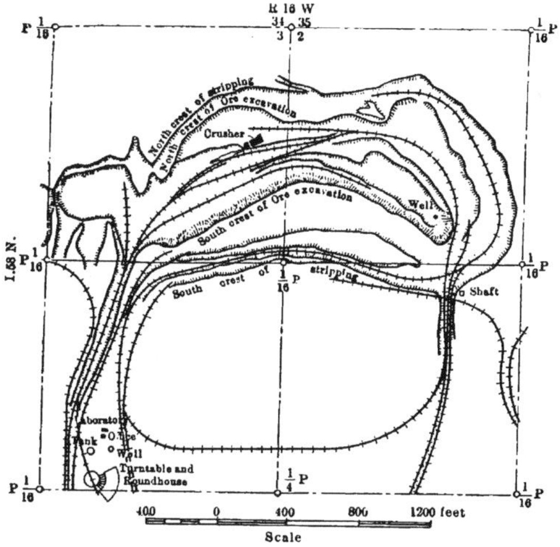

FIG. 28.—Biwabik open pit

The topography determines the choice of methods, as in early plans of the Fayal and the Biwabik open-cuts which incorporate ideas from both systems. (See Figs. 27 and 28). Most stripping has been done by contract, but recently some of the operating companies have started doing it direct in order to better utilize ore-digging equipment. The only different apparatus needed for stripping is the wooden dirt car, which has a shallow body set on a central longitudinal hinge on its truck so it can be dumped to either side. It holds 6 cu. yds., has automatic couplers but no brakes, and is made by the Russel Foundry Co. of Detroit. Though contractors often used a 3-ft. track gauge, it has been found feasible to use the standard gauge for the dirt cars so as to be uniform with that of the ore cars and shovels. The contract price for the great Eveleth bodies has been 30 cts. a cubic yard, which for a depth of 10 yds. would mean nearly $50,000 an acre.

Locomotives.—Steam locomotives are used which are commonly of a 60-ton size, with six drivers and 19-in. by 36-in. clyinders, made by the American Locomotive Works. They handle 11 of the 6-yd. dirt cars, four 25-ton or two 50-ton steel, bottom-dump ore cars. Some larger locomotives are in use, and it is planned at the Fayal mine to restrict the 60-ton size to service around the shovels. Longer trains will then be made up for haulage to the terminal yards by 100-ton locomotives.

Shovels.—For digging, the favorite shovel is the 90-ton size, with a dipper handling 2 1/2 yds. of dirt, or 4 1/2 tons of ore, either of the

Fig. 29.—Openpit work, Mesabi range

Marion, Bucyrus or Vulcan make. This is mounted on two 4-wheeled trucks on a standard-gauge track set 20 ft. from the car track, which allows a cut of 30 ft. wide at the bottom (see Fig. 29). The shovel works best under a bench 20 ft. high, as higher ones are apt to cause trouble by caving. This makes it necessary to keep the shovel at considerable distance from the face, thereby sacrificing the digging efficiency. With work well arranged this shovel can dig and load 150 6-yd. dirt cars or 50 50-ton ore cars in 10 hours. Records much higher than this can be made, as one shovel timed by the author loaded 50 tons of ore in 4 2/3 minutes.

After removing the bulk of the stripping the surface of the ore body must be cleaned before ore can be dug. Formerly this was done by hand shovels and barrows, but most of this hand cleaning has now been superseded by the following method: When the last stripping cut is being made a scoop scraper is chained to the shovel dipper, which drags the dirt on the ore surface over against the next stripping bench. In 1909 nearly 19 million cu. yds. was stripped on the Mesabi, of which under 30 per cent. was handled by contractors.

Loosening.—Though much of both overburden and ore is soft enough to be dug by the shovel direct, it makes quicker work to have the benches first loosened by explosives. The holes are bored in gravel or ore with hand churn drills to the same depth as the height of the shovel bench. They are placed 15 ft. to 20 ft. apart along a bench and staggered, two abreast. The drills are of 1-in. to 3 1/4-in. octagonal steel, with 1 1/2-in. chisel points, and are operated by two or four men, according to depth reached. In ore, 24 ft. per man is drilled in 10 hours, but in the drift, with its boulders, only 10 ft.

The drills are lifted by a movable, steel cross-piece, one for each pair of men, held to the shank by wedges. A small intercepting boulder may be dislodged by a squib, but the larger ones are drilled. The finished hole is squibbed several times with 60 per cent. dynamite to make a chamber for three to eight kegs (25 lbs.) of 1/4-in. black powder, which is not loaded until the hole has been dried by pouring in sifted sand.

Spoil-banks.—A part of many of the open-cut mines was originally, or is still, worked by caving, and the resultant surface depressions can often be utilized as convenient dump holes for stripping. The large area of wild, rolling land around the ore bodies makes it easy to find a dump for the balance without too long a haul. The common method is to lay the switch on a side hill and dump the dirt down hill until the dump becomes high enough to move the track sideways to the edge, where the dumping process is repeated. To keep the dirt from clogging the track it is customary to level the dump by a plow-like scraper with a V-shaped nose, mounted on a heavy truck and adjusted for different heights.

At Coleraine the Oliver Co. has adopted a new scheme for disposing of dirt from the Canisteo mine. The initial dump was raised by degrees to a height of 50 ft. along the shore of Trout lake, and a trestle resting on piles was then built to support the track on the edge of the dump facing the lake. From this trestle the dirt cars are dumped sideways into the lake, and the dirt is kept from accumulating by the use of water jets played upon it from above. The scheme saves removal of the terminal track, the only extra expense being for wash water, which is raised by a special pump located below the dump.

Labor.—In open-cutting the following force is employed: For shovel 9 (runner, craneman, fireman, four pitmen and two track cleaners); for trains, 3 (engineer, fireman and brakeman); then there are the blasting and dump bosses and their helpers, also the engineer and pumpman at the drainage shaft. The general force comprises the superintendent, day foreman and night foreman, the sampler, surveyors, assayers, accountants and clerks.

Shops.—For keeping the heavy machinery up to its work extensive repair shops are maintained by the different operators. For instance, the Oliver Company's mines at Hibbing have a blacksmith shop with six forges and a steam hammer, a machine shop with planer, boring mill, drill and wheel presses, lathe, shaper, etc., large enough to handle any heavy repairs and renewals for shovels or locomotives. There is also a foundry for medium-sized castings; the large castings are shipped in the rough and finished to dimensions at the mine machine shop.

Prospecting.—The early mines of the Lake Superior iron ranges were started on ore out-crops showing through the glacial drift, which had been followed up by test-pitting. In the last few years, churn drilling has been in vogue in drift and soft ore, while in hard ore or rock, the diamond drill is used.

On the Mesabi range, prospecting has been very systematic, and it is estimated that over 30,000 holes have been drilled from the surface. The unit of area is the 40-acre lot, and (where not adjoining ore-bearing ground) if nothing is found by drilling one hole near the center, and one at each of the corners, the lot is considered as barren. The surface boundaries of the iron formation have now been so well mapped by the government geologist that there is no further excuse for wild-catting on impossible areas. If ore is penetrated, a hole is put down at each 200-ft. interval, and records are kept to determine not only the outlines of the ore body, but also the depth and assay of each of the rock and ore strata penetrated.

The Mesabi prospecting is started with a churn drill which is of the portable type run by a gasoline or steam engine. A boring sample is taken for every 5 ft. of depth and assayed for SiO_2, Fe, Mn, P, and Al. (The Al assay has lately been added to determine if the sample is from aluminous paint rock or from the ore proper.) The bit is only wide enough to go inside a 3-in. welded pipe casing, which is kept within 5 ft. of the bottom of the hole. When 5 ft. below the bottom of casing has been drilled, and the sample taken, the casing is forced down to the bottom. The sample it obtained by running the sludge through two settling boxes, from which the slime is saved and sent to the assay laboratory, where it is dried and cut down and, after taking enough for analysis, about 1 lb. is kept in a tubular box for reference.

On reaching the hard schist, called Taconite (see Fig. 26), below an ore body, the hole's casing is seated there and the churn replaced by a 1 1/4-in. diamond drill, taking a 7/8-in. core. If ore is again penetrated, the hole is enlarged sufficiently to let the casing descend. This is done by lifting the casing off the rock and then blasting the top of the 1 1/4-in. hole with a stick or two of 60 per cent. dynamite, fired electrically. The pieces are then cleaned out by the churn-bit and sand pump, and the blasting repeated until the hole is enlarged sufficiently to drop the casing down to the ore. The churn drill is then used until the bottom of the ore is reached. This alternating churn and diamond drilling is kept up until the basal quartzite is reached, beneath which no ore is ever found.

The whole diamond drill core is not saved, but only 1 ft. or so of representative rock for each 5 ft. of core, and this is stored in core boxes, packed in special cases along with the churn drill samples, in a fire-proof vault in the office building. The drilling is done for the mining companies by contractors at a cost for churn drilling of 50 cents to $1 a foot. The diamond drilling costs about twice as much per foot, but it is economical in taconite, owing to the slowness of the churn. To prevent the contractors using the faster and more profitable (to themselves) diamond drill in soft ore, they must show rock cores for all such work.

Sampling and Assaying.—One of the most complete systems is in vogue at the Oliver Co.'s mines on the Mesabi. For underground work, a grab sample is taken from each skip hoisted, and a day's hoist is assembled and cut down. Both underground and open-pit faces are sampled by the bosses (by grooving) frequently enough to insure intelligent stoping.

For open-pit shipments, a separate sample it taken from each 10 cars of 30 to 50 tons each. This is done by grabbing 10 to 12 handfuls off the surface of each car, from spots spaced equally along diagonals or two longitudinal parallel lines. The assembled samples from 10 cars are then reduced in the pit by the sampler to 10 lbs. (by coning and quartering), and sent in a sack to the assay laboratory.

In the laboratory the 10-lb. sample it first bucked by hand to pass a 2-mesh sieve, and then cut down on oilcloth by coning to 1 kg., and the balance used for determining moisture. The 1 kg. is crushed in a laboratory jaw crusher and quartered to 100 grams, which is finally dried, bucked by hand to 100-mesh, and stored in a sample bottle.

All shipments are analyzed for SiO_2, Fe, Mn, and P, and the drillings for Al in addition.

The sampling and analysis of a Mesabi shipment must be completed before the train reaches Two Harbors, which often allows only four hours. The result can then be telephoned to the dock master and he can assign the shipment to its appropriate pocket on the dock. In this way, cars of different grades to make a desired compound can be dumped into the same pocket, and by the time the furnace is reached, the several transhipments will have produced a uniform mixture.

Surveying.—For the Mesabi open pits, the Oliver Co. pursues the following survey system: Before the contractors begin stripping, the whole area is blacked out in 100-ft. squares and stakes set across the deposit on the 100-ft. lines at 20-ft. intervals. The level of each station is then taken, and, by holding a tape between neighboring stakes to make the longitudinal 20-ft. marks, the remaining levels can be observed at the corners of each 20-ft. square. When the drift has been removed down to the ore, the new level of each 20-ft. point is observed, and then the total volume of stripping can be accurately calculated. Recently the company has started doing some stripping on company account, and here levels are taken only at 40-ft. intervals.

Subsequently the ore removed during any given period can be easily calculated by securing new elevations of the same points used in the stripping survey. The ore reserves were formerly computed from the drill-hole records by treating each body as a whole. Owing to the fact that the Mesabi ore runs in thick layers of large area, but differing analyses, it has been found possible, with the same drill records, to estimate the volume of each ore layer separately, and this has made the results of the reserve calculations much more instructive and useful.

Example 3.—Utah Copper Mine, Bingham, Utah

(*See also Examples* 37, 41 *and* 43.)

Steam Shovels on Steep Sidehill Lenses with Rock Capping.—The Bingham orebody seems to owe its existence to the impregnation of the shattered zones in a monzonite-porphyry intrusion that was forced up vertically through the surrounding quartzites. The mineralization is not confined entirely to the porphyry, for in the Ohio Company's ground and in the Starless group shattered areas in the quartzite itself have been strongly mineralized. In the course of time this porphyry ore succumbed to erosion and oxidation more rapidly than the less shattered quartzites, and so it now forms the bottom of a gulch and extends up to the top of the divide between Main Bingham and Carr Fork canyons.

In the upper part of the sulphide ore at Bingham the copper in the ore has been leached from the walls of the seams, while between the seams in the unreached center of the porphyry the ore is unaffected and assays fairly well in copper. This fact would indicate that the ore approached the present richness before secondary enrichment. Somewhat deeper the ore becomes richer and then in turn drops in grade until it falls below workable value. The principal copper mineral in the sulphide ore is chalcocite, disseminated in extremely fine particles.

Leaching has progressed concomitant with erosion, and so the orebody is covered with a capping of oxidized porphyry from which some of the copper has been carried away in solution. Still much of this carries well above 1 per cent. and in places as much as 2 and even 2.5 per cent. This copper of course occurs in oxidized form, and therefore is not so amenable to concentration as the sulphide ore. At present all the capping is looked upon as waste no matter how much copper it may contain. This line of demarkation between capping and ore is not distinct, but it roughly parallels the surface, causing difficulty in steam-shoveling the ore on account of its slope, which is too flat to permit the loosened capping to run down to the shoveling terraces and too steep to allow easy access without many terraces. The capping averages 70 ft. thick.

Owing to the cost of underground mining, the Utah Copper company now obtains most of its ore by means of steam shovels. But, owing to the fact that the property is cut in two by a gulch whose sides approximate a slope of 30 deg., the orebody is not especially adapted to steam-shovel mining. Indeed, because of the rough topography, already 16 miles of track have been necessary—even when a switch-back system for entering the terraces and grades varying from 2 per cent. to 6 per cent. are used. Besides, on account of the slope of the gulch and the line of merger between capping and ore, there is much mixing of overburden and ore. In addition, the disposal of this capping has necessitated the purchase of dump ground involving a maximum haul of 1 1/2 miles.

The difference in elevation from the creek level to the highest ore is 900 ft., so that nine terraces have been necessary and a tenth is now being surveyed. The elevation of these are: *I*, 6825 ft.; *H*, 6750; *G*, 6687; *F*, 6600; *E*, 6536; *C*, 6415; and *A*, 6340 ft. Stripping tracks are placed at whatever elevation the lay of the ground dictates. Stripping is done on all of these levels, but all the ore above the *F* line is to be shoveled down to the *A* pit and loaded there, it being cheaper to handle the ore twice than to load it into cars on the upper terraces. Indeed, in *A* pit much of the ore is being shoveled from a bank 230 ft. high (see Fig. 30), a condition quite dangerous for the shovelmen.

Both the ore and the capping require blasting in order to loosen them for the steam shovel. This is done in three different ways, according to the varying conditions. In case there is sufficient room a "gopher" is used, consisting of a drift, 2 1/2 ft. square, driven into the bank for 30 to 45 ft. and with a cross drift about 15 ft. long, driven to each side at its end. This is loaded with black powder, some dynamite being used to explode the charge. In case that there is not quite so much room for blasting, churn-drill holes are resorted to. In drilling these holes No. 3 Keystone churn drills are used. There are two of these, one working on capping and the other on ore. The holes have a diameter of 6 in. and are cased generally only down to solid rock. A crew consists of a driller at $5 per 12-hr. shift, a tool sharpener, as the helper is called, at $4, and a man to get coal, water, etc., at $1.75, and drilling and loosening costs about 3/4 cents per ton.

The depth of these holes varies considerably, as well as their spacing from the bank, but when a hole is to be drilled 50 ft. deep it is generally placed so as to have a burden of 35 ft. at its bottom. In some cases holes 85 ft. deep have been used; these are loaded generally in three places, the bottom only being sprung. In case the bank is low, horizontal holes, drilled with a 3 1/2 in. piston machine to a depth of 25 to 26 ft., are used. These are sprung 2 or 3 times so that they will hold three boxes of 30 per cent. dynamite. Almost all blasting is done with a battery, and three or four caps are used. Still misfires occur even with these precautions, but they are rare. Most of the capping and ore breaks quite fine, owing to the fracturing that has occurred throughout the mass

FIG. 30.—Shoveling 230-ft. bank, Utah Copper mine.

of porphyry, but occasional bulldozing of boulders is necessary. This is cheaper than block-holing the boulders.

After each blast the bankmen, by means of ropes, climb down the sides and dress down the bank. These men also do all blasting and gophering. Their work is especially dangerous, and they are paid $3.75 for 10-hr. shifts. The difficulty of their work can be seen in Fig. 31. The men work with a rope near at hand to grab in emergencies.

The banks in capping are carried at an angle of about 40 deg., as this is the best slope for working it; the ore bank 230 ft. high is carried at a somewhat steeper angle. The capping is shoveled into 4 1/2-cu. yd.

dump cars and run in trains to the waste dumps, while the ore is shoveled directly into 50-ton railroad cars ready for hauling to the mill at Garfield.

The routine of handling the shovels is quite similar to that in Example 2, and so will not be described. A 65-lb. rail is used and a standard broad-gauge track; in advancing the shovel 6-ft. lengths of track joined by fish plates are used. A shovel crew consists of a shovelman at $175 per month, a craneman at $125 per month, a fireman at $2.50 a day, and six pitmen at $1.75. Four of the pitmen work on the track and odd jobs, while two tend the jack screws and help advance the shovel.

FIG. 31.—Trimming bank, Utah Copper mine.

The shovels work from 60 to 65 per cent. of the time. The rest of the time is taken up mainly in waiting for cars, blasting and other similar delays. As yet, few shovels have been buried by caves, and there are not many break-downs. The company has 10 shovels; eight weigh 95 tons and have 3 1/2-cu. yd. dippers, while two weigh 65 tons and have 3-cu. yd. dippers. Seven are Marion shovels, two Bucyrus and one Vulcan, the latter being one of the first bought. The shovels work two 10-hr. shifts. The large shovels consume about 2 1/2 tons of coal per shift and the smaller ones, 2 tons. Seven of the shovels are working on capping all the time; two shovels on ore, and one partly on ore and partly on capping. About 6500 tons of ore (2 1/4 to 7 per cent. moisture) and 8000 to 10,000 tons of capping were mined in a day in June, 1909, with about 650 men at this work which cost 19 cents per ton of ore and 34 cents per cu. yd. of capping removed. In 1911 with 19 shovels, ore and capping are handled at the rate of 50,000 tons daily.

Example 4. Nevada Consolidated Mines, Ely, Nevada

(*See also Example* 44.)

Steam Shovels on Rolling Lenses With Rock Capping.—The ore occurrence at Ely is not nearly so similar to that at Bingham as many seem to think. At Bingham the intrusion of monzonite porphyry was laccolithic in character, while at Ely the intrusion was more in the nature of a dike. It cuts persistently across the bedding planes of the different horizons of the limestone country.

The whole area has been subject to much mineralizing action, and the monzonite has been much kaolinized in places so that it is much weaker than at Bingham. In fact most drifts in the Ely monzonite have to be timbered, while at Bingham the drifts stand without any need of support.

The orebodies at Ely are due essentially to secondary enrichment, as is clearly indicated by the sudden change, within a vertical distance of 5 ft. from capping carrying only a trace of copper to ore carrying 1.5 per cent. copper. This secondary enrichment occurred at water level, and so the top of the orebodies is fairly flat, rarely undulating through a vertical height as great as 30 ft. in the orebodies near enough to surface to be worked by steam shovels, but being somewhat greater in the deposits covered by capping 30 ft. or more thick.

This flat character of the top of the orebody, the fact that it bears no relation to the undulations of the surface immediately above it, and the general rolling character of the surface at the places where the capping is thin enough to permit economic stripping with steam shovels, makes Ely an ideal camp for the use of shovels.

The depth of the capping varies throughout the district, being only 30 ft. thick in some places and over 700 ft. in others, depending partly upon the height of the local ground-water level when that particular orebody was formed, but mainly upon the amount of erosion that has occurred subsequently to the formation of the orebody.

The principal ore mineral is chalcocite, but some bornite and some melaconite is also found. The original mineral seems to have been chalcopyrite. Owing to the important influence of secondary enrichment, most of the copper minerals occur along the fracture planes and in the more porous rock.

The orebodies occur where there are shattered zones in the porphyry and also where kaolination has allowed surface water to filter easily through the porphyry. Because of the direct connection between facility for leaching and richness of underlying orebody, it would appear that there is much less likelihood of finding orebodies under the limestone capping than where the porphyry comes to the surface.

The Nevada Consolidated Copper company has three No. 5 Keystone churn drills busy in prospecting—one at the Boston-Montana-Liberty orebody and two at the Ruth, all in ore. The holes are placed 200 ft. apart and are sampled in 5-ft. sections. The entire pulp is dried without decanting any of the water, a sample from a 5-ft. depth generally requiring four tubs. After drying, all the pulp is mixed and then quartered down. The depth of capping at the Ruth is 300 ft., and at the Boston and Montana, from 80 to 100 ft. Few of the holes require casing. At the Ruth the holes vary from 500 to 700 ft. in depth, while on the Boston and Montana they are only 200 to 250 ft. deep. In holes deeper than 350 ft. about 25 ft. are drilled in 12 hours, but in the shallow holes more than 110 ft. have been drilled in 12 hours, the amount ordinarily varying between 40 and 60 ft. in holes 200 ft. deep. In deep holes the first casing used is a 7 5/8-in., next 6 1/4 in. and finally if required 4 1/4-in. casing, but in shallow holes or in ground the nature of which is known generally an 8-in. hole is started. The cost of drilling in the Ely district varies considerably, but at one mine churn drilling cost $1.87 a foot for 44 holes averaging 250 ft. in depth. This included the cost of changing set-ups, pulling casing, repairs, lost tools, etc. Complete records of the character of the ground penetrated, the amount of depth drilled each shift, and the cost are kept. The holes are plotted, and on sections the assay of each 5-ft. section is recorded. A double record is kept of each hole. From these sections it is possible to estimate the respective advantages of steam-shovel and of underground mining.

The Copper Flat orebody is admirably adapted to steam-shovel work as the capping is only from 35 to 160 ft. deep (the average being about 90 ft.), the orebody 210 ft. deep, and the topography gently rolling.

For the bench a vertical height of 50 ft. has been found most admirable for safety and for intensive shoveling, while with a horizontal width of 50 ft. the loading track is not so likely to be buried by the blast, thus avoiding serious and costly delays. Of course, higher banks can be carried, a height of 200 ft. or more being possible, but this would necessitate tunnel-blasting, requiring a heavy tonnage of explosives, a much wider bench, and more care to avoid exposing the shovel and crew to danger from a treacherous bank.

A study of an actual section (Fig. 32), through a bank of the steam-shovel ore-pit, shows that the slope between the benches from the upper edge to the toe of the talus below is a little greater than 1 to 1, varying between 1.04 and 1.43, or an average of 1.18 to 1. The talus, or broken

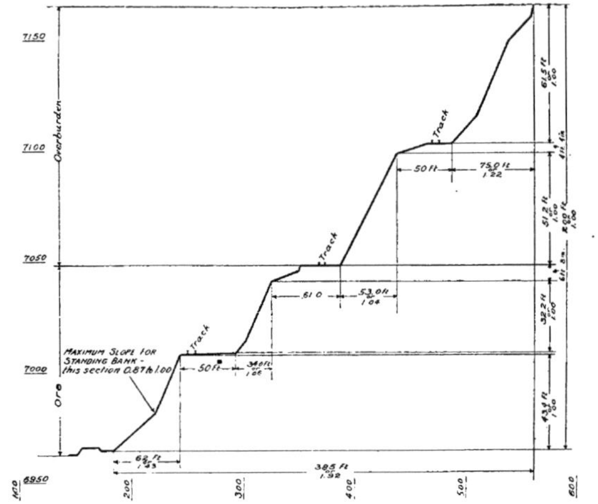

FIG. 32.—Bench-diagram, easy slope, Nevada Consolidated open pit.

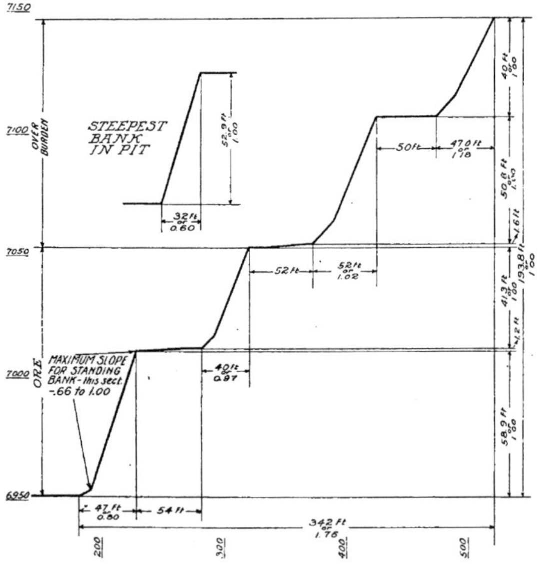

FIG. 33.—Bench diagram, steeper slope, Nevada Consolidated open pit.

material, that has become loose and has fallen to the bench below, as shown in the diagram, will always be found at the foot of a bank, the quantity varying in amount according to the condition of the standing ground. In the analysis of a steeper section, as shown in Fig. 33, the ratios of the corresponding banks are somewhat less, varying between 0.8 and 1.18, or an average for the four banks of 0.99 to 1. The steepest bank in the shovel-pit is one in the zone of sulphide ore, shown above in Fig. 33, 52.9 ft. high, and standing at a ratio of 0.6 to 1. This bank is freshly cut and will stand at this ratio for only a short time, when disintegration will cause it to crumble. It will be seen, then, that an average slope for this height of bank and material will average closely the ratio of 1 to 1, a little steeper in the zone of sulphides and a little flatter in the oxidized material above.

The general slope from the bottom to the upper edge of the excavation is, of course, much larger since the added width of the bench on which the shovel operates nearly doubles the horizontal distance. In Fig. 32 this ratio over all from the edge of the top to the toe of the bottom bench, will be seen to be 1.92 to 1, and in Fig. 34 1.76 to 1 for four benches. With the addition of more benches the ratio will not remain the same, but will grow larger by a decreasing amount. Using the ideal section (Fig. 29), with 1 to 1 slopes, 50-ft. benches and 50-ft. heights, the table and formula below are suggested by E. E. Barker.

Nunber of benches.	Horizontal distance.	Vertical distance.	Ratio of slope.	Ratio Difference.
2	150	100	1.50 to 1	....
3	250	150	1.66 to 1	0.16
4	350	200	1.75 to 1	0.09
5	450	250	1.80 to 1	0.05
6	550	300	1.83 to 1	0.03

From this table the formula below is deduced:

$$S=\frac{na+(n-1)b}{nc}\text{ or, substituting}$$

$$S=\frac{4\times 50+3\times 60}{4\times 50}=\frac{350}{200}=1.75\text{ to }1$$

Where S equals the slope ratio; a equals the base of the individual slope triangle; b equals the width of bench; c equals the vertical height of bank; n equals the number of benches.

Again, with $c=60$ ft.; $a=50$ ft.; $b=60$ ft.; $n=6$, we have

$$S=\frac{6\times 50+5\times 60}{6\times 60}=\frac{600}{360}=1.66\text{ to }1$$

In the selection of the width of bench, the deciding factor is the slope taken by the blasted bank. Ordinary broken material will repose at a slope of about 1 1/2 to 1, but the impetus given the broken rock in a blast usually causes the slope to form at approximately the ratio of 2 to 1. Since the drill-holes are placed about 10 ft. from the edge and the blast loosens the ground for about 10 ft. more, the bank, when blasted, lying at a slope of 2 to 1 assumes the position of the dotted line at the top of Fig. 34, still leaving 20 ft. clear on the bench below, which gives ample room to safely accommodate the loading track without danger of being covered. From the data at hand and with conditions as given

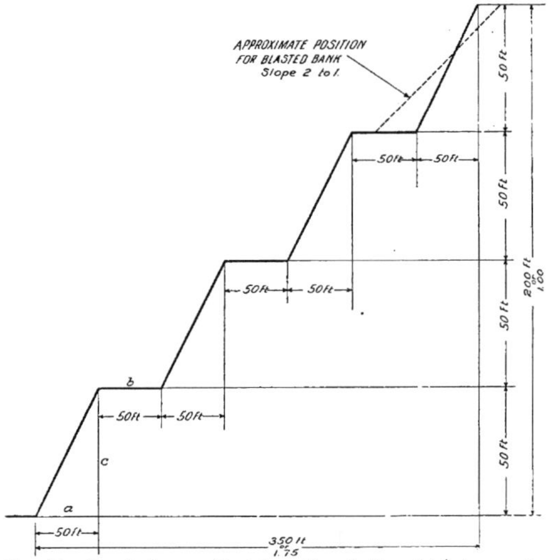

FIG. 34.—Bench-diagram, ideal slope, Nevada Consolidated open pit.

above, the section with the minimum bench and the maximum height and slope for economical operation, is the one shown in Fig. 34, with a general slope of 1.75 to 1, or a corresponding ratio for the number of terraces required. Already the steam-shovel operations cover an area of many acres. The pit is roughly oval in shape and the tracks are extended in ovals around the sides, so that only in starting a terrace is it necessary to load the cars singly.

LOOSENING ORE FOR SHOVELS

The ore and the capping require blasting to loosen it, but on the east side the ground is almost soft enough to shovel without blasting. Only churn drills are used in preparing the bank for blasting. After the ground along the terrace has been roughly evened, so as to permit the drills to move readily from one set-up to another, drilling begins. These holes are drilled about 10 ft. deeper than the steam-shovel terrace so that the bottom will surely be loosened. The holes are placed so as to have a burden of from 25 to 50 ft. of ground at their bottom, according to the nature of the ground, and the holes are spaced approximately the same distance apart as they have burden on them. At present there are seven distinct kinds of ground to be blasted. This variability also effects the loading of the hole. The hole is first sprung or chambered, 3 or 4 times with 40 to 100 lb. of 40 per cent. dynamite, then the hole is loaded with from 750 to 2000 lb. of explosive. Several different grades of explosives are used—in soft ground Dupont FF black powder, in harder ground Champion powder (a mixture between black powder and dynamite), and is still harder ground 40 per cent. dynamite (in winter 40 per cent. Trojan powder is used); but more 40 per cent. dynamite is used than other kinds of explosive. The blasting is done with electricity, and three XXXXX detonators are placed in each hole. The placing of these holes requires considerable experience and judgment; so the nature of the face is studied and examined for slips before being drilled. From 1500 to 3000 cu. yd. are moved at a blast, and about 1500 to 3000 tons of ore when blasting in ore. A churn drill will sink from 40 to 70 ft. of 6-in. hole in a 12-hour shift in preparing the bank for blasting, and at times as high as 120 ft. has been drilled in a shift. The boulders are bull-dozed. Two No. 5 Keystone churn drills work on capping, and only one on ore.

Standard-gauge equipment is used in steam-shoveling. The capping is loaded into dump cars of the Oliver type. Two sizes are used—a 6-cu. yd. and a 12-cu. yd. car according to conditions. The larger cars are equipped with standard air-brake apparatus, while the 6-cu. yd. cars are not. The larger cars are therefore better for long runs and require less "spotting" while being loaded; the small cars are better on curves and where much dumping on trestle is done. The large cars are handled in trains of five, and the small ones in trains of eight. The cars by actual measurement hold 5.4 and 10.9 cu. yd., respectively.

Bucyrus shovels are used entirely. There are three 95-ton shovels with 5-cu. yd. dippers; one 95-ton, with 3 1/2-cu. yd. dipper; and one 70-ton shovel with 3-cu. yd. dipper. Two of these shovels are working on ore, and the rest on capping. About 10 min. are required to load a 50-ton car when running regular, but numerous delays increase this to a much lower average. The ore is run directly to the mill in these same cars, but in time some way of screening the ore before the mill bins are reached will have to be arranged. As it is, the mining expense is increased considerably owing to the want of storage capacity at the mine and mill.

The tracks are laid at from 3 per cent. to 4 per cent. grade and in mining the ore 4 miles of track are used. Over this single-track system 220 to 230 trains a day are run. The ore-train yard is at the mouth of the pit, but the capping has to be hauled some distance. Formerly this was nearly 1 1/2 miles, but at present the dumps are much nearer and one is within a quarter of a mile of the pit.

In steam-shovel operations the company used six 16x24-in., saddle-tank American Locomotive Works locomotives and one 14x22-in., saddle-tank locomotive of the same make. These locomotives use about 3 tons of coal in nine hours. Locomotive engineers are paid $4.25 for 9 hours; firemen, $3 for 9 hours; switchmen, $3.25 for 9 hours. Trackmen and common labor is paid $2 for 9 hours. The powder boss gets $5 for 9 hours. Shovel and churn-drill crews are alike in number and cost to those of Example 3.

About 5000 tons of ore and about 3000 cu. yd. of capping are being moved a day, or 150,000 tons of ore and 90,000 cu. yd. of capping per month. On the pay roll there are 267 men, including men in the machine shop, repair men, in short every one connected with steam-shovel mining. At present about 12 acres of ground have been stripped, but this particular ore body has an area of 18 acres.

In 1910 the Company's average cost of shoveling ore was 15.4 cents per ton and of removing waste was 40.6 cents per cu. yd. Of the stripping cost 15 cents was apportioned to each ton of ore extracted so that the total cost of mining the ore was 30.4 cents per ton including repairs, maintenance, and general expenses.

Example 5.—Eastern Pennsylvania and Illinois

(*See also Example* 49, 51 *and* 59.)

Clam-shell Cranes and Wheeled Dipper-dredges on Coal Seams with Thin Mantles.—There are a number of places in the anthracite fields of Eastern Pennsylvania where flat coal beds outcrop so near the surface that they may be stripped for 100 feet before the cover attains an unprofitable thickness. The Hilldale strippings have a section from the surface down about as follows:

Soil, 4.5 ft.; coal *A*, 4.5 *ft;* rock 3 ft.; *coal B*, 1 *ft.;* rock 5 ft.; *coal C*, 2 *ft.;* rock 3 ft.; *coal D*, 12 *ft.*

Coal bed *A* has been exposed so long it is worthless, except where it has a rock cover. Bed *B* is a somewhat rusty good coal. Bed *C* is almost 2 ft. thick and could not be worked at a profit underground; in the stripping operations, however, it is, like bed *B*, a source of income. Bed *D* is the coal aimed for, and is as good as coal can be, although carrying a slate parting.

The method of stripping followed by Mr. Kinsley, the contractor, is as follows: First, the top soil and poor coal *A* is removed from the top rock by means of a clam-shell bucket and locomotive crane. The boom on this crane is 42 ft. long and can place the top material where it will be out of the way once for all. The dirt is first removed ahead of the crane in the direction it is moving; next from the side of the crane where the mining is to be carried on. The rock and coal are then broken down to the coal bed *D* by blasting. The coal from *B* and *C* is picked out by hand, while the rock is wasted and piled back by the bucket to form the track for the coal car. The coal bed *D* is next broken by powder and loaded into the bucket by hand. The bucket is then raised by the crane and the coal dumped in the car. Anthracite coal is so brittle it breaks in handling, and while it was at first intended to use the bucket to pick up the coal, it was found inadvisable to do so. This method of stripping is equivalent to making a side cut along the crop for the crane track, then excavating below this cut, and filling in. The fill will furnish the road for the next side cut when coming back from the boundary line of the property, as it does for the coal cars in going forward.

Near Danville, Ill., there was a 35-acre bituminous coal bed, 8 ft. thick, which was approximately horizontal. This coal bed outcropped on all sides of a flat-topped hill, and had an overburden from 38 to 40 ft. thick, composed of soil, clay, gravel, and about 20 ft. of shale. It was decided to strip to the coal with a single cut and to take a wide cut so as better to provide for the efficient mining of the coal. The plant decided on was a steam shovel, made by Bellefontaine (Ohio) Machine Co., having a dipper of 2 cu. yd. capacity, and mounted on a movable platform, provided with a Jeffrey belt conveyer for disposing of the material. The platform, which was 30 ft. wide, was mounted on four trucks that were moved as desired on two tracks. The machine, as it appeared in operation, is shown in the frontispiece. After excavating the material above the coal with the bucket, it was swung to a large steel hopper and discharged. From the bottom of the hopper the material was carried on a steel cross-feeder to the lower end of the belt conveyer. The latter was a 40-in. wide belt traveling on a steel arm 105 ft. long. The arm was supported by wire ropes from a tower built above the platform to a height of 48 ft. By this arrangement the waste clearance at the outer end of the arm was about 60 ft. above the tracks.

The machine is said to have had no difficulty in excavating heavy pieces, stumps, and logs, and depositing them in the space where the coal had been mined. After the overburden had been removed the coal was quarried and loaded into cars on a track laid in the place from which coal had been taken previously. As the shovel made a cut, the débris was deposited in the cavity made by removing the coal. The tracks for the machine were laid on top of the coal, which was mined so as to leave a bench for the shovel to come back on.

CHAPTER VII

SURFACE MINING

Example 6—Puertocitos Mine, Cananea, Sonora, Mexico

(*See also Examples* 18, 34 and 45.)

Quarrying Sidehill Lenses with Rock Capping.—The ore deposit is on a sidehill (see Fig. 35) and consists of streaks of malachite through a bedded limestone. On the surface the metallic contents have been leached so that a worthless limestone capping must first be removed before attacking the ore of which two-thirds can be rejected during sorting. The capping and waste are dumped from small cars down

Fig. 35.—Openpit mining at Puertocitos.

chutes into pockets above the railroad whence it can be hauled to nearby dumps or used as filling. The ore is dumped into similar pockets and is then hauled 10 miles downhill to the smelter for about 12 cents per ton.

In removing the capping, drill holes are put down only to its bottom, but in the ore regular benches are laid off, 22 ft. high. The drilling is by hand with three men to a crew. Down to a depth of 9 ft. the hole is drilled by hammers, there being two hammermen. This takes about half a day. As that is the limit of economic work with hammers, from that point the hole is churned down by the three men to a depth of 22 ft. This takes a day and a half more, or two days to complete a hole 22 ft. deep. These holes are sprung with dynamite, and then loaded with from four to six kegs, 50 lb. each, of black powder, since the holes are generally drilled with a toe of ground at the bottom about equal to the depth. The holes are also spaced about equally, although the manner of breaking of the holes previously blasted determines the position of the next holes. Each blast breaks, on an average, about 600 tons of rock. Electric blasting is used in firing these holes, and several are blasted at a time.

The tramming distances are short, but owing to the large amount of sorting necessary, the block-holing and the sledging required in breaking up the pieces for sorting, the cost of placing ore on the cars is about $1.70 a ton, including stripping and all other charges. Still, if much powder were used so as to avoid the sledging, the rock would be broken so fine that much ore would be lost that might otherwise be sorted out, and the fines might become too low grade to be sent to the smelter.

The mine is sending to the smelter an average of 140 tons of ore per day, and employs about 175 men. These are all Chinamen, as it is hard to get Mexicans to work outside during the winter months, for at this altitude the climate is rigorous and there is often snow.

Example 7.—Mesabi Iron Range, Minn

(See also Examples 2 and 46.)

Milling or "Glory-hole" System for Flat Lenses under Glacial Drift (Vertical Chutes).—The milling system, having little machinery, costs much less to start than open-cutting by steam shovels, but the cost of ore per ton in the extraction is greater on account of underground development, tramming, hoisting and lighting. It is adapted to deposits where the overlying mantle is not so deep as to necessitate underground mining and where facilities of approach and the general shape of the ore body are unsuitable for open-cutting. Like the latter, milling cannot be worked so cheaply in winter and is, therefore, best adapted to deposits which need only make shipments during the open season on the lakes. Milling can often be used as an auxiliary to open-cutting to remove such parts of the stripped ore body which are so deep or so obstructed as not to allow of easy railroad grades from the pit's approach. Many mines first opened on a small scale by milling have since been extended for open-cutting by steam-shovels.

In starting a mine on this system it is only necessary to strip sufficient area so that the top diameter of the final funnel-like ore pit will be adequate to allow the base of the ore to be reached by a series of concentric, descending benches of economical height and width. While stripping, a working shaft *s* (Fig. 36), with two skip compartments and one cage compartment, is sunk in the wall rock alongside the ore and a crosscut *c* run from the foot wall to a point under the proposed mill, where a vertical raise r is made to the surface. By extending a drift *d* from the foot of this raise along the ore body and raising from it at appropriate intervals (50 ft. to 100 ft.) as many mill holes can be started as the area stripped will permit.

The stoping begins at the top of of a mill by circling it with an underhand bench *b*. The height of a bench is governed by the depth of the hole that can conveniently be bored with the piston air drill used; usually between 10 ft. and 20 ft. The width of a bench depends upon the

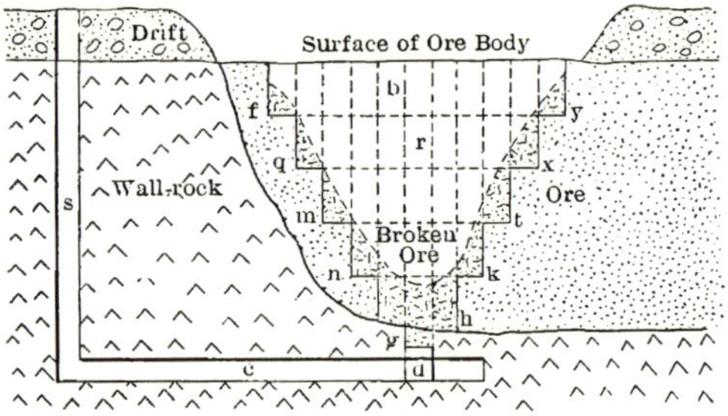

FIG. 36.—Milling or glory-hole system.

economical burden that one line of holes will carry when chambered by squibbing, and also upon the width that can clear itself after blasting, mostly by gravity.

When the benches have reached the bottom of the ore body and the funnel is completed along line *f-q-m-n-g-h-k-t-x-y* (if there is only one mill) the broken ore, if the benches are cut down much farther, will no longer slide. Then the benches have to be cleared, and for this purpose a steam shovel can sometimes be efficiently used. In Fig. 27 the great "mill pit" was formed by cutting down and uniting a number of mill funnels. At the bottom of each mill is a chute gate, from which the ore is drawn into cars to be trammed to the shaft, whence it is hoisted in a skip to the surface. As the Mesabi ore is friable, there is seldom trouble from the clogging of chutes by boulders.

In other districts where steam shovels for stripping are not so easily available as on the Mesabi or where the topography or repair facilities are unsuitable for their operation, the surface waste over the ore body can be removed by several other methods.

The first to suggest itself would be to carry the raise of each mill-hole up to the surface and drop all the waste through it for disposal on a dump, near the exit of shaft or adit, before attacking the ore beneath.

This would involve handling all the waste in adit or shaft and often a cheaper way of stripping would be by such rapid methods as horse scrapers, drag-line excavators, hoist-cableways or hydraulicing.

Sometimes the broken ore is hoisted from the bottom of the open pit by a clam-shell or other self-filling bucket suspended from a hoist-cableway stretched between derricks on the surface, and as this system saves the cost of tunnel and shaft it is often cheaper for small deposits.

Example 8.—Traders' Mine, Menominee Range, Mich

(*See also Examples* 38 *and* 46.)

Milling or "Glory-hole" System for Vertical Wide Vein Without Mantle (Vertical Chutes).—This ore body is 100 ft. to 200 ft. thick and has little or no overburden. It extends conformably to the enclosing beds of the Traders' formation (named from the mine) for over a mile, and dips 60 deg. south, the foot wall being of greenish and the hangwall of reddish slate. Though a few high-grade pockets have been extracted, the bulk of the mines' output of 1,500,000 tons has been low-grade ore of 41 per cent. iron and 0.015 per cent. to 0.018 per cent. phosphorus. The iron mineral is high grade, but it occurs only in bands in a hard jaspilite matrix. The annual output is but 125,000 tons, as work goes on only during the season of open lake navigation. The Antoine Ore Co. is the operator. For a daily output of 900 tons to 1,000 tons there are 125 men, of whom 64 are machine men (16 drills on each shift), and most of the balance are trammers. The low mining cost of 30 cents per ton is less than that for steam shoveling on the Mesabi where the stripping is heavy.

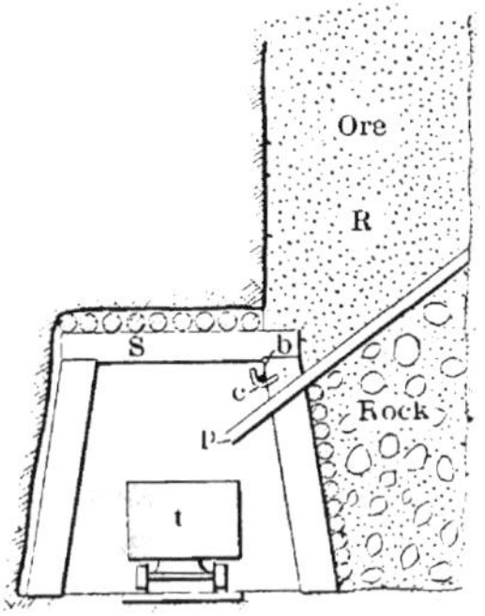

Fig. 37.—Chute for milling system.

Development.—The open pit was opened by a 4-compartment shaft in the foot wall, whose sump is 150 ft. from the collar and 225 ft. below the apex of the vein's outcrop. It is 7 ft. by 23 ft. inside of timbers and has a 5-ft. by 7-ft. cageway, two 6-ft. by 7-ft skipways and a 3-ft. by 7-ft. pipeway, all dividers and wall plates being 12 in. by 12 in. From this shaft a crosscut runs on the 80-ft. level to the main drift, which extends for 1,450 ft. along the center of the vein. From the main drift are turned off crosscuts at 100-ft. intervals, and where they strike the foot wall a vertical raise 6 ft. square is put up to the outcrop (as in Fig. 36) for the beginning of a mill hole.

No timber is used in drifts, crosscuts or raises except for the chutes. The bottom of the raise *R* (Fig. 37) is rock-filled and floored with poles *p*, which are set to as to leave a loading orifice 2 ft. by 5 ft., controlled by a steel bar *b* resting in cramps *c*. Two drift sets, S, are inserted opposite the raise to complete the chute with 8-ft. caps and posts.

For each mill hole there are two 3 1/4 in. Rand drills on tripods, which bore down holes 8 ft. to 14 ft. long and are supplied with air by small pipes from the top of the pit. As the ore is broken in big chunks these have to be reduced to prevent choking the chutes, and this is done by bull-dozing rather than block-holing, as the extra powder is found less expensive than the extra labor for boring block holes. Either the chute-raise must be kept full of ore or the entrance of each chute at the base of the underhand benches must be covered over by a log-grizzly to arrest the descent of the boulders into the chute until they can thus be broken. The tram cars *t* are handled by two men and hold two tons, or one skipload.

Example 9.—Alaska Treadwell Gold Mine, Douglas Island, Alaska

(*See also Example* 16.)

Milling or "Glory-hole" System for Sub-vertical Wide Veins Without Mantle (*Hour-glass Chutes*).—The orebodies her occupy a huge syenite dike that has intruded the slate country rock for several miles. The dike is irregular in width, varying from 420 ft. at the Treadwell mine to 150 ft. at the Mexican and 300 ft. at the Ready Bullion, a half mile to the southeast, while in the interval between these three miles the dike is a mere stringer. The mineralization of the syenite was due to a subsequent intrusion of barren gabbro which now forms the hangwall of the syenite orebodies and in places is badly schattered. There is also a third intrusion of barren basalt which is found in the orebodies as a single dike above, but as several smaller ones at depth.

The chief ore is of two varieties: first, stringers of quartz and calcite occupying fracture planes in the syenite; second, crushed and broken syenite which has been saturated by mineral-bearing solutions. The largest orebody is the Treadwell shown in section in Fig. 38, which dips 70° and has a gabbro (greenstone) hangwall and a black slate footwall. The climate is wet but mild enough to permit continuous outdoor mining, so that the main openpit finally reached a depth of 220 ft. below the adit level and 450 ft. from the surface with a maximum width of 420 ft. and a length of 1700 ft. The large slides of waste rock from the footwall and the need of a thick pillar of rock to protect the underground workings from surface water caused the stoppage of the openpit at the 220-ft. level.

To develop the Treadwell mine below the Adit Level, a four-compartment vertical shaft (see Fig. 38) was sunk in the hangwall and stations as wide as the shaft and 40 to 60 ft. long were cut at each level. A main crosscut, as *C*, is run on each level to the footwall, 20 ft. wide for the first 100 ft. and 12 ft. in width for the balance of the distance. Beneath the floor of the station an ore bin, *B*, is cut out in the rock with a capacity of 500 to 1500 tons, to afford ample storage. On the main crosscut, on its hangwall end, is cut a station for the winding engines for the tail rope system of haulage. Directly opposite the sinking compartment, on alternate levels, a station is cut for the sinking hoist.

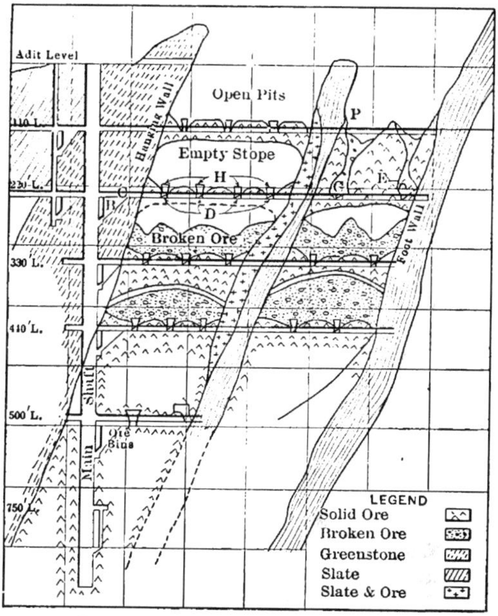

FIG. 38.—Alaska Treadwell mine, cross section.

When the main crosscut has reached the footwall, parallel drifts, *D*, are turned off, at right angles and about 60 ft. apart, to follow the strike of the vein. At intervals of 25 ft., raises are now put up on alternate sides of both crosscut *C* and drifts *D*. These raises are 15 ft. high, have a slope of 60 deg., so that the ore will run freely, and are fitted with special finger-chutes so the large quantities of ore can be easily run into the mine cars. At the same time as these drifts and raises, there are being run intermediate drifts (as *E*) directly above each drift *D* but separated from its back by a rock pillar, 10 ft. thick.

When a pit P is to be opened, a raise is put up from the nearest level and connected with the surface. This raise is started from an intermediate drift *E*, in general directly over a chute-raise. The chutes, 25 ft.

distant on each side, then serve as man-ways for the raise in course of erection, and the broken rock is drawn off through the middle chute-raise into cars. When the raise has been connected, the machine-drills are put to work cutting out a small stope of the bottom. Thus the raise when finished has the shape of an hour-glass, the top being formed by the open pit P and the bottom by a drawing-off stope *G*, covering three chutes and from 20 to 30 ft. high, the two being joined by the raise. The object of cutting out the pit-raises in this manner is, first, to obtain chute-capacity in case of their being hung up by large pieces of rock or by blasting; and, second, to afford an opportunity to break up any large piece of rock that may have been overlooked in the pit, which would stop up the chute unless it were broken to pieces small enough to pass through it.

Machine-drilling is seen at its best in these pits. The 3 1/4-in. diameter Ingersoll-Sergeant drills, set on tripods, are used in all the pits at present. The average number of feet drilled per machine in 10 hours is 36.35. The holes are drilled to an average depth of 12 ft., and each machine will break 69.69 tons of ore per shift of 10 hours. When the pits were smaller and the difficulty of setting up was not so great as at present, the average number of feet drilled was much higher, and the breaking capacity of a machine-drill was from 150 to 200 tons of ore shift of 10 hours. The pits are worked by drilling and blasting the ore from a series of benches or terraces around the chute-raise as a center, and when the ore is blasted the broken rock rolls down to the bottom. The small pieces are then broken by sledges, and the larger ones by placing sticks of powder on the surface of the rock, tamping with a little fine dirt, and blasting. For blasting holes, No. 2, or 40 per cent., dynamite is used, while for "bulldozing" No. 1, or 70 per cent., is best.

When the rock has been broken to the required size, it is drawn off, through the raises and chutes described above, into cars. These cars are hauled to the station ore-bins by horses, or by endless-rope haulage, where they are dumped. The ore is then loaded into skips and hoisted to the surface.

CHAPTER VIII

UNDERHAND STOPING

Example 10.—Disseminated Lead Field of Southeast Missouri

Underground Quarrying with Down Holes in Flat Lenses in Limestone.—Topographically the country is hilly, even rugged in places, and is traversed by a network of small streams. The elevation varies from 500 to 1000 feet above the sea level. The surface is well wooded with forests of oak, hickory, ash, and yellow pine, the first predominating.

The surface formation is of Cambrian sediments which abut against hills of Archean granite. The Cambrian consists here of the St. Francois limestone, which in places is 700 ft. thick, resting conformably on the Mine LaMotte Sandstone. The ore bodies lie entirely within this limestone, and its base represents the lowest points to which shafts have been sunk; the shaft depths varying from 90 ft. at Doe Run and Mine La Motte, to 600 ft. at Flat River. The greatest proved run of pay ore is that at Bonne Terre, whose length is over half a mile, width up to 200 ft., and height 25 to 100 feet.

Prospecting.—The great development since 1890 has been entirely based upon results obtained by systematic prospecting with the diamond drill, which was introduced at Bonne Terre by the St. Joe Lead Co., in 1869. The lead areas, underlying the country horizontally under shallow depths of homogeneous limestone, make conditions unusually favorable for this form of prospecting. The gently undulating topography offers no difficulty to the movement of the drills, while the warm climate permits out-of-door work for most of the year. A portable drilling outfit is used, with the drill and pump attached to a "agricultural" boiler on wheels.

The distance drilled in a shift is very large compared with vertical vein practice. Sixty feet of uncored hole in 10 hours is a common average for holes 500 ft. deep, and as much as 100 ft. is often run. The coring is not begun until the lead-bearing zone is approached, and then 20 to 30 ft. per shift is commonly done. The bit used is set to cut a hole 2 1/8 in. diameter, with usually four diamonds outside and four inside of the bit. The filler, with inside diameter of 1/2 in., to make the bit solid, is set with four diamonds more of two to three carats apiece, rounded carbons being prefered.

As much of the sludge never reaches the surface, being filtered off through cracks, sampling it would be of no value; so the core is entirely relied on, both for locating the deposits vertically and for giving the assay yield. The sludge, however, gives the drill man warning of the approach of the lead horizon. The cost of drilling 500 ft., with diamonds at $40 a carat, is not far from $0.50 per foot at Flat River.

In locating the first hole in a virgin tract an endeavor is made to locate the extension of the axis of an adjoining lead run. If there are no nearby ore runs, surface crevices and ancient diggings are looked for as a guide to the occurrence of a possible ore body underneath. If neither adjoining runs or surface indications are present, the only recourse is to lay the area off in 5, 10, or 20 acre squares, according to the probability of finding ore, and bore a hole in the center of each square. If one hole strikes an ore body, a circle of 209 ft. radius is struck off, and holes bored on this circumference 209 ft. apart to determine the axis of the lead run.

For traversing the surface broken ground which is sometimes 100 ft. deep, the St. Louis Prospecting Co. used a portable steam churn-drill outfit in the following way: To penetrate the surface soil, the bit's cutting edge is made 7 inches. This allows a casing of 5 5/8 in. inside diameter to be used that is inserted 2 ft. into the bed rock. The drilling is then pushed downward until a rock suitable for diamond drilling is encountered, with a bit of 5 1/2 in. diameter. This diameter gives room for the placing of an inserted joint casing of 5 in. inside and 5 3/8 in. outside diameter at the upper end, in case mud channels or opinings are struck so large as to make progress without the casing impossible; a speed of 10 ft. in 11 hours was the average made in this kind of work.

Development.—The more thoroughly the ore has been drilled the easier it is to lay out the shaft and drifts advantageously. The problem is similar to that of a coal seam development, but unlike the usual bituminous seam, the lead run is irregular both in thickness and in the level of its floor. The shaft is located from three considerations: 1, the lowest point of the ore body; 2, the center of the pay ore body; 3, the propinquity to a good mill site.

The customary way of horizontal advance is breast stoping, but drifting is used to reach pockets separated from the main run of ore, or to traverse barren places. To reach the top of the ore body to start a stope, either vertical or inclined raises are used. Fig. 39 shows the method of vertical raising. As soon as enough is excavated above the floor the two stulls are put in, a floor of poles laid on them, and the raise then continued in two compartments to the top, the chute being separated from the ladder-way by a pole partition. The broken rock is left in the chute for the men to stand on, the drill bar is wedged horizontally against the walls and the surplus broken rock is thrown down the ladder-way. When the raise is finished, the floor is blasted out and the rock loaded. A gate is only put in the chute at the bottom, if the chute is to be used for the lowering of rock after the completion of the raise, which often occurs in exhausting ore bodies above the main level in operation.

The inclined raises are run the same as drifts and as steep (45 deg.) as is possible without using timbers to hold up the broken rock that the men stand upon. Whether the vertical or the inclined raise is used to open up a stope depends on the shape of the ore body.

Stoping.—There is no timbering in the drifts or stopes, except sprags for supporting piping, and no filling system. The roof is held up by pillars which, however, can be laid out on no regular plan as in coal mining. As far as possible, pillars consist of lean ore. The width of the stope between pillar depends on the strength of the roof and varies

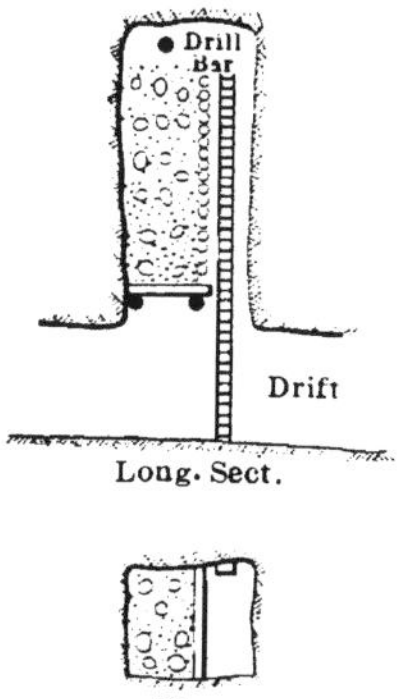

FIG. 39.—Raise, S. E. Mo.

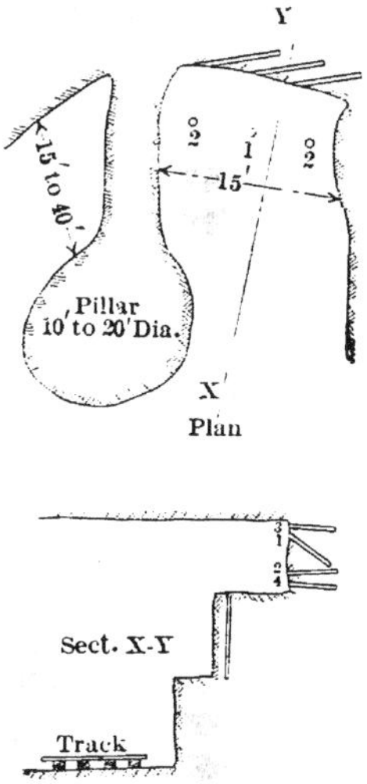

FIG. 40.—Stope, S. E. Mo.

from 40 ft. at the Desloge and Bonne Terre mines, to 15 or 20 ft. at the Theodora, and other eastern Flat River mines.

The underhand stoping system is used. A breast 6 ft. high is run from the top of raise along the roof of the ore body, which is tested for ore above by driving an upper air-drill bole into it occasionally, and as many benches quarried out beneath it as are necessary to excavate the ore to the level below.

There are two systems of breast stoping, one for narrow and the other for the wide stopes. In Fig. 40 a breast 15 ft. wide is shown, the holes are 6 to 8 ft. deep and are nine or ten in number in three or four vertical and horizontal rows. Each row is fired in order 1, 2, 3 and 4, and causes an advance of but 2 ft. in the breast, each side-cut being kept on the same side of the breast to turn off a round pillar. Two men with a machine can drill the blast a round of nine holes in a 10 hour shift, the drill bar being set up once for each vertical row of holes.

For the wider stopes a similar method is used; but an advance of 3 to 5 ft. instead of 2 ft. is scored for each round, the holes for this being placed nearer across the breast, the rows *A*, *B*, *C*, etc., Fig. 40, breaking from 2 to 4 ft. each. The machine men drill complete as many vertical rows as possible and blast them before going off shift.

When the face of the breast stope is 10 to 12 ft. from the side of the raise, a bench stope is started. The drills for this are set on a tripod and the holes are 6 to 10 ft deep. The holes take off 2 to 4 ft. of burden; and they are placed from 4 to 8 ft. apart along the bench, the center one being fired first and then the sides. The bench holes are farther apart in the wide stopes.

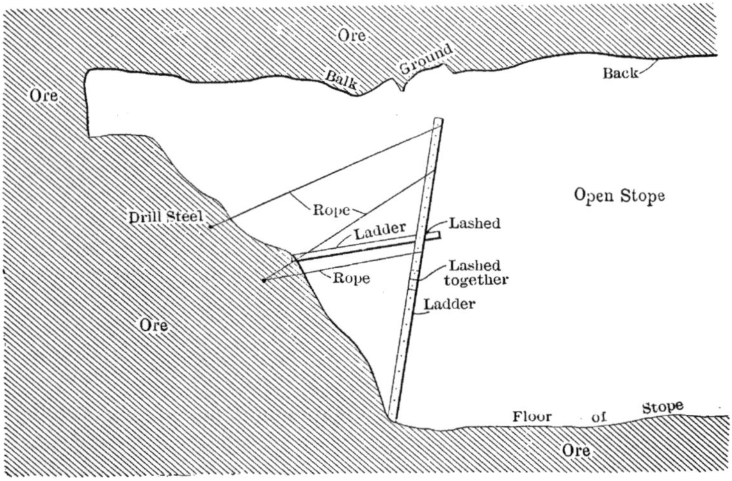

Fig. 41.—Ladder scaffold, Ducktown, Tenn.

In firing a stope the lower bench is fired and the bench holes the last of all; the holes being ignited simultaneously at the end of the shift. The fuses are cut of graded lengths. One drill can bores six to eight bench holes in a shift, and in the 15 ft. stope of Fig. 39, two bench drills can keep pace with three breast machines—which ratio does not alter appreciably for wider stopes, for breast as well as bench stoping gains in speed in them. A pound of powder will break three times as much rock on a bench as in a breast stope. One air drill in a Flat River producing mine with stopes 12 to 20 ft. high will drill enough to break 20 to 30 tons of rock in two shifts, while at Bonne Terre in stopes 20 to 60 ft. high, one drill will break over 40 tons in the same time. The explosive for stopes is 35 per cent. dynamite and 1 lb. will break from 1 ton of rock, in a narrow breast, to 4 tons on a wide and deep bench.

Though the roof is self sustaining, pieces are liable to shell off so that a roof man is necessary to bar down the loose pieces with gad and pick; the sides of the shaft have also to be occasionally inspected and barred down. The waste limestone is broken down, loaded, and put on the surface dump as it seldom convenient to store it underground.

Ladder Scaffold, Ducktown, Tennessee.—In the similar system of underhand stoping in the Tennessee Copper Company's mines, where the miners practically never see the back, which in an open stope is frequently from 70 to 90 ft. above them, it is evidently necessary to keep the roof well trimmed of all heavy, or "balk ground." To insure this, a crew of men is continually kept at work, looking after the condition of the roof. This work is extremely dangerous and ready resource is required to enable the men to gain access to the back. Fig. 41 shows the method of rigging ladders to reach the roof over the benches of an underhand stope, open to its full height and for a width of from 50 to 150 ft. The ladders are securely lashed together, and, as shown, stayed by ropes secured to the drill steels set into the rock face. A small stoping drill is frequently slung from the ladder and used to put holes in the roof where much balk ground must be slabbed down. Shooting the roof, is, however, a dangerous practice, as shattered rock is apt to be left to fall later, when the face of the stope has advanced and the back is inaccessible.

EXAMPLE 11.—DAVEY MINES, AMERICAN ZINC, LEAD, AND SMELTING COMPANY, JOPLIN, SOUTHWEST MO

Underground Quarrying with Horizontal Lenses in Limestone.—The zinc and lead ores of this typical sheet-ground formation vary from 16 ft. to 20 ft. in thickness, and underlie continuously a large area. The depth of the deposit is about 240 ft. The upper 6 ft. is considerably the richest portion of the sheet or bed and carries most of the "jack" or sphalerite. The lower portion contains most of the lead, which occurs in pockets or in irregular sheets.

The shaft of Davey No. 3 mine is 256 ft. deep and is 9x18 ft. all the way down. This size is larger than necessary, the shafts later sunk being 7x12 ft. To sink a 7x12 ft. shaft in limestone costs about $20 per foot. No. 3 shaft has two 9x5 1/3-ft. hoisting compartments, and a pipe compartment at each end. The timbering is very simple, consisting of 4x9-in. stulls set centered every 5 ft. 4 in., over which is nailed a lining of 1x12-in. boards.

The chert rock of the ore-bearing sheet formation breaks quite readily because of its brittleness and compactness. The system used for breaking ground is as follows: A machine drill set-up is made next the roof of the sheet, as shown in Fig. 42, so as to advance a heading face about 7 to 8 ft. in height, leaving beneath an untouched bench which will vary from 10 to 14 ft. in thickness to the bottom of the sheet. When the heading is advanced about 18 ft. the bench beneath is drilled and blasted. The placing of the holes in the heading face is shown in Fig. 43 (*a*), the firing being in the order designated by numbers. All the holes are 8 ft. deep, numbers 1, 2, and 4 being drilled horizontally, and No. 3 bedded at the junction of the roof and the ore. No. 5 is given

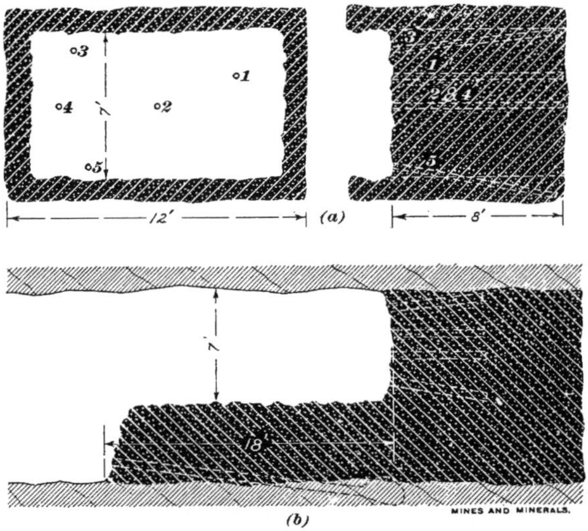

FIG. 42.—Blasting sheet ground, S. W. Mo.

a slightly downward pitch. Placing 3, 4, and 5 as shown will usually give a break which leaves room for a good set-up. The drilling is done with a 3 1/4-in., type E 24, Ingersoll-Rand drill set on a 7-ft. column. The long bench holes are put in with the drill mounted on a tripod; they will average from 16 to 18 ft. in length and are placed as shown in Fig. 42. The hole is then squibbed three or four times with 40 per cent. dynamite and finally the resulting chamber is filled with 50 to 100 sticks of the same explosive and fired by a fuse and cap. Only the most expert drillmen can put in these long holes but they have been found highly advantageous because they can follow one of the compact rock beds. On the contrary, vertical holes across the bedding planes would be difficult to drill on account of cracks and pockets. For the same reason they would be so porous as to render their exploding inefficient. It only takes three of these long horizontal holes to break off a bench 14 ft. high across a stope 40 ft. wide. Between two squibbings the hole is cleaned and cooled by blowing out with compressed air. Holes are squibbed only at night and those squibbed one night are blasted the next at the end of the shift. A drill on the bench can break as fast as four drills above on the heading, even though only one and one-half of the flat bench holes can be drilled in eight hours.

The illustration shows a hole after squibbing, when it is loaded with from 50 to 150 lb. of 40 per cent. dynamite and fuse-fired. An average of 35 tons are broken per 50 lb. of powder, giving a cost per ton of 12 cents.

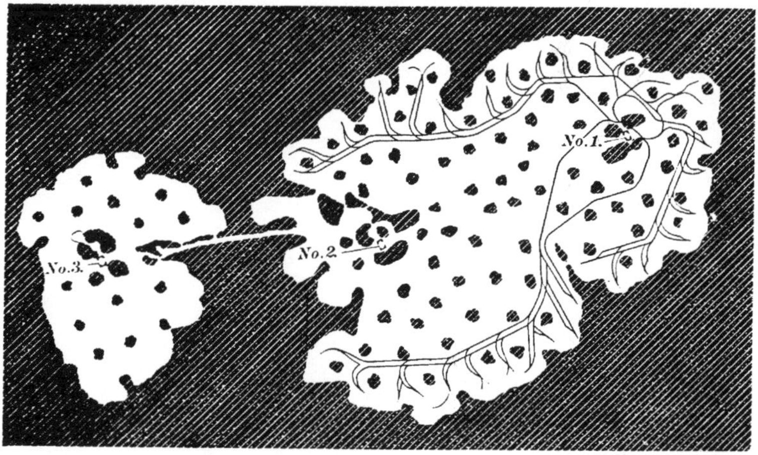

MINES AND MINERALS.

FIG. 43.—Stope plan, S. W. Mo.

The "dirt" is shoveled into tubs, pushed to the shaft, and hoisted in the usual manner, using 30x32-in. tubs holding about 950 lb. The hoisting is done by friction-geared hoisters placed on the floor of the derrick, one above each shaft compartment. These are 6x7-in. engines with 24-in. drums, of the English Samson pattern, manufactured by the English Iron Works, Kansas City. It takes 45 seconds for a car to make a round trip with a travel of about 280 ft. from the derrick floor to the shaft bottom. The average rope speed is about 1,300 ft. per minute, so that dumping a bucket requires about 20 seconds.

The total cost of mining for the past few months has averaged 67 cents per ton hoisted including the pumping-cost of 3 cents per ton.

There is no timbering as the roof is supported by pillars of ore 25 ft. in diameter, set at 40-ft. intervals as seen in the plan of the mine workings, Fig. 43; no definite pillar system is employed, as the pillars are set where the conditions of the roof demand. As a rule the roof consists of a solid flinty bedded rock averaging from 18 in. to 3 ft. in thickness. It seldom causes trouble except where pitted with large pockets known as "sand holes."

Plans are now afoot to make a very radical change in the method of handling the mine dirt. An automatic traction shovel made by the Thew Automatic Shovel Co., of Dayton, Ohio, will be installed with a dipper of 1/2 cu. yd. capacity, which will clean up anything within a circle of 20 ft. without moving the shovel. Power will be supplied by compressed air. Cars will be used instead of cans, and will be hauled to the shaft in trains by mules instead of hand tramming. Here at the shaft it is intended to put in an underground hopper which will discharge into 3-ton balanced skips. The hoisting engine will be moved to the ground and steel head-frames substituted for the present wooden ones. At present the average daily mine output (two shifts) is about 725 tons at No. 3, with a total from the four shafts of about 2,250 tons. Hoisting is now done both day and night, but it is hoped that the new hoisting system will get out enough dirt in one shift to run the mill for two shifts.

At the author's visit, the labor for the day's output of 1850 buckets (879 tons) was as follows:

Dayshift.	*Nightshift.*
32 machine drillers.	6 machine drillers.
20 muckers.	14 Muckers.
	2 shot-firers.

The high average output of 26 tons apiece for the muckers (which included an average tram of 150 ft. to the shaft bottom) was only attained by allowing them to earn high wages under individual contracts.

Pillar-Robbing.—It is generally the practice in this district to leave the poor ore as pillars and not attempt to recover it; but in one mine, with rich ore, where the roof was heavy and a large amount of timber had to be used to enable even half the ore to be extracted, the following method was used for robbing the pillars. Beneath the ore was a solid, compact limestone stratum. The shaft was sunk a few feet into this, and a sub-drift extended beneath the ore pillars with a 7- or 8-ft. roof. A raise was put up in the center of each pillar and the ore shot down into the drift below, and trammed to the shaft. This method gave a safe place in which to work and at the same time allowed nearly all the ore to be recovered from the pillars.

Gobbing and Saving Timbers

Theoretically the gob should be let in at either end of the section, at the center, making it possible to remove the stringers. The saving of stringers depends on how the gob is let into the stope, the weight of the ground on the section, and the condition of the stringer. Sometimes it would cost more to remove a timber than it is worth. It such a case no attempt would be made to save it.

In general, an average of perhaps 50 per cent. of the stringers can be saved in a sulphide stope, while in an oxide stope with light ore 75 to 90 per cent. of the stringers are saved. The square sets are not gobbed as they are used in the mining of the next section. Only the central portion is filled. When the next section toward the main drift *A* is mined, the pillar of ore to be sliced is 15x25 feet.

Requirements for Application of Method

From the method described, it will be seen that the requirements in order to work such a body of ore are: (1) There must be a solid back which can be easily supported. (2) The ore must contain little or no waste, as everything goes into the chutes, permitting of no selection. (3) Lateral and vertical pressure must be small in order to prevent the square sets from buckling before the stringers are put in; also to allow the mining of the whole section before the gobbing is commenced.

Cost of Mining Reduced

In regard to the reduction of the cost, it is best to compare this method with that which employs square sets alone. It is evident that less timber is used with this system. With the saving of 75 per cent. of the stringers, the working of several sections alongside of each other makes it necessary to run only one row of square sets for each section mined. There is a saving of perhaps 50 per cent. of the timber of that used in square-set system. The mining of the ore in the square sets *B* and *B′* would cost approximately the same as by the regular square-set system. The cost of mining a lead row of sets is higher than mining corner sets in a square-set stope, but this increased cost is offset by the fact that the ore from the lead row of sets falls directly into the chutes, making shoveling into a wheelbarrow and wheeling to a chute unnecessary.

In mining the core, the amount of powder used is reduced to about one-half. The cost of timber and timbering is also reduced to one-half, while the cost of breaking ore is reduced to one-third of square-setting.

Saving in Labor

There is a greater saving by this system in the handling of the ore than in the method of timbering. A large percentage of the ore is shot directly into the chutes and requires little or no handling except the breaking of boulders which are too large to pass through the grizzlies.

In square setting it is often difficult to place the chutes so that the miner can shovel directly into them. With the Mitchell slicing system the wheelbarrow is never used and shoveling is reduced to a minimum.

In working out the sill floor, the ore is handled by the ordinary method, as here the ore must be shoveled directly into the mine cars, unless worked from the level below, which is often done.

Increased Tonnage Obtained

The amount of ground broken per man per eight-hour shift, when using the regular square-set system, is from 5 to 6 tons.

In mining the pillars with the Mitchell system in sulphide ore, 12 to 15 tons are broken per man per shift, while in oxide ore in auger ground 25 tons per man per shift is not unusual. When once the mining of the core commences, the work is carried on rapidly, a core often being worked out in 8 or 10 days. It has been found convenient to mine these cores when there is any sudden demand for an increase in the output of a certain kind of ore, which is another valuable feature of the method.

Conclusion

The system can be worked on any section of ore provided that it contains no waste, is not too heavy, and is as large as 20x30 ft. There is flexibility in this method as it may readily be switched to square-set stoping in mining irregular portions of the orebody. It has not been found practical to mine a section more than 50 ft. thick. The system is new, and Mr. Mitchell is adding improvements which will make it a still more valuable method of mining.

The method has been a success, but owing to its rigid requirements, its field is quite small. It will suit only a few of the orebodies in Bisbee and therefore will not become an important factor in reducing the cost of mining there.

Example 12.—Calumet and Arizona Mine, Bisbee, Ariz.

(*See also Example* 23.)

Underground Quarrying of Panel-cores, or the Mitchell System for Flat Lenses in Limestone.—Because of the peculiar conditions under which most of the orebodies in Bisbee exist, square-set stoping in panels, as described in Example 23, has been the chief method of extraction.

In 1908 at the Calumet & Arizona mine, while working a heavy sulphide stope by the square-set method, a large mass of ore broke away from the back, and in order to mine it, long timbers were thrown across the top of the ore to support the back, after which the ore was taken out. From this slight incident a combination of the square-set and underhand stoping systems was worked out by M. W. Mitchell, the foreman of the Calumet & Arizona company. The system has given excellent results where the conditions have been favorable.

Recently some bedded ore deposits have been found in the Calumet & Arizona property Chalcopyrite, bornite and pyrite have replaced the limestone, the ore following the original bedding of the limestone and including little waste. These bedded deposits rarely exceed 50 or 60 ft. in thickness. The limestone hanging-wall is well defined, solid and easily supported. It is in these deposits that the Mitchell system has been employed. The greatest success, however, has been attained in the mining of the oxide ores when they contain little or no waste.

Method of Blocking Out the Ore

The orebody is first thoroughly prospected to ascertain its general direction, size, and limits, in order to determine whether this method is suitable. The theory of this system of stoping is to outline a block of

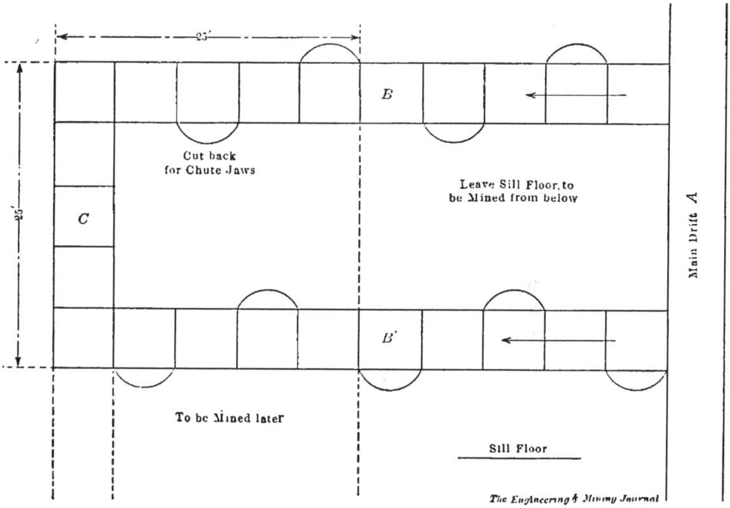

FIG. 44.—Blocking out ore by Mitchell slicing system

ore by means of regular square sets, allowing the included core to rest on its own base and then cut it out in slices from the top down after the roof or back has been properly supported. The method followed is illustrated in Fig. 44. Two lead rows *B* and *B'*, 15 ft. apart, of regular sill-floor-stope square sets, are run from the main drift *A* to the end of the section to be mined. These are connected by the square sets *C*. Regular 7-ft.

10-in. stope sets are carried up to the limits of the ore above the end sets *C* and above the sets *B* and *B'*. These sets now include on three sides a block of ore 15x45 ft. and as high as the ore extends.

Fig. 45 illustrates the method of framing used for the square-set timbers. The posts and caps are usually 10x10 in. with 8x10 girts. In the rows *B* and *B'* the ties or girts are put in across the drift with caps running parallel to *B* and *B'*.

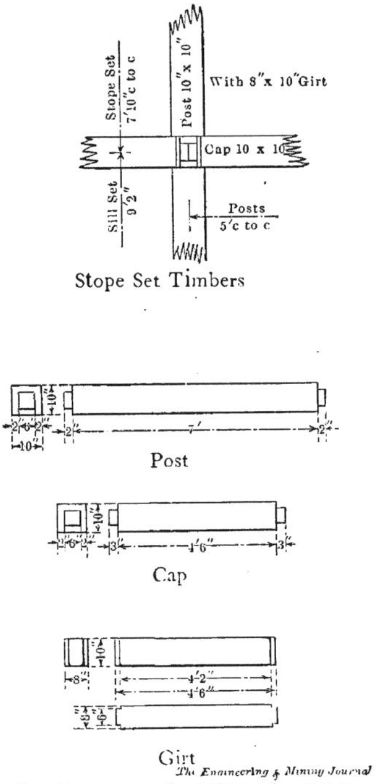

FIG. 45.—Framing square sets at Calumet and Arizona mine.

UNDERHAND STOPING

The stoping system proper now commences and is illustrated by Fig. 46, which shows a plan and two sections of the stope. The drills are mounted on columns or bars between the caps or posts of the square sets and holes drilled from the sides. When the ore is broken, stringers S_1 and S_2 are put in and Sagamore or so-called segment sets S_3, are put in between S_1 and S_2. In the second slice or bench and those following, stringers S_4 are put in without the segment sets.

In mining the second bench, and those below, the best practice is to mount the drill column between the stringers and drill vertical holes downward. The stringers on the top floor are 10x10 in. and framed like girts to fit the square sets. On the second floor 8x10-in. stringers are used, while 8x8-in. may be used on floors below provided that the ground is not too heavy. Segment sets are put in on the top floor only to support the back. On the remaining floors stringers alone are used with perhaps an occasional stull or spreader to reinforce them. The rows of square sets *B* and *B′* are used as chutes, grizzlies being put in to prevent large boulders from clogging the mouth of the chutes which are merely small openings cut back of every other set on the sill floor as shown in Fig. 44. These openings are cut just large enough for a chute, when the sill-floor lead sets are run. With a small amount of barring, the cars are easily loaded from these chutes.

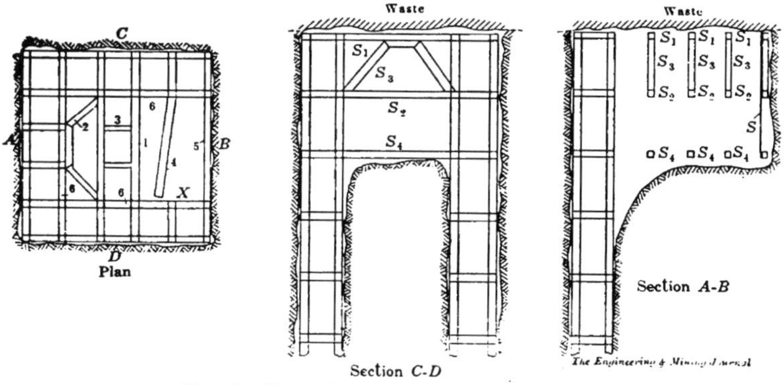

FIG. 46.—Plan and section of Mitchell slicing system.

PLACING OF TIMBERS

The plan in Fig. 46 illustrates some of the details of the method employed. No. 1 shows the stringer in place. No. 2 shows diagonal braces to hold the square sets in position. No. 3 shows temporary spreads which are sometimes used to reinforce the stringers when the ore is blasted. The method of putting in stringers is shown by No. 4. One end is put in against the posts and the caps of the square set in the same way an ordinary girt is put in. At *X* one cap of the square set is cut down 2 in. to permit the 2-in. tenon of the stringer to go into position. When in place a small piece of plank is spiked to the cap to hold the stringer. When the section is worked out and is ready for gob vertical planking is put on at the end of the section at No. 5 in Fig. 46, and the inside of the square sets is lagged, as shown by No. 6. When this has been done with the ore worked out to the level or to the bottom of the orebody the stope is ready for gob.

EXAMPLE 13.—SECTION—21 MINE, ISHPEMING, MARQUETTE RANGE, MICH

(*See also Example* 35.)

Underground Milling in Sub-vertical Vein with Back Caving.—The Section 21 mine (Oliver Co.) is 3 miles south of Ishpeming adjoining the Whitby open pit, which was exhausted some years ago by surface milling, and used an inclined skip for hoisting the ore. The west end of the present mine is a large trough of continuous non-Bessemer soft ore, which is worked by the room-caving system of Example 46. On the east

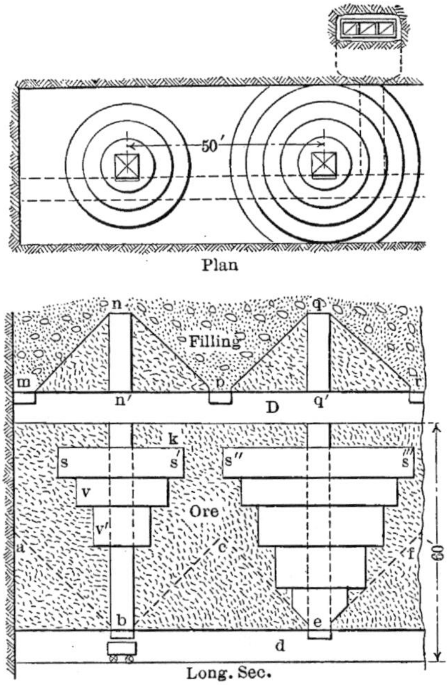

FIG. 47.—Stoping at Section—21 mine.

end the ore is mediumly hard and has a thickness of 15 ft. to 50 ft., lying in a highly inclined lense on a diorite footwall with a jaspilite hangwall and is worked by underground milling.

In Fig. 47 the drift *d* of the new level should be completed as soon as the stope above level *D* (60 ft. higher) has reached the contour *m n p q r*. Raises *b* and *e* (50 ft. apart) are then put up from *d* to *D*. After leaving a 6-ft. pillar *K* under level *D* breast stopes, as s-s^1 and s^2-s^3 are begun under the pillar *K* in each raise in order to start the ordinary annular underhand benches r, v^1, etc., of milling, these being cut down till the limiting contour for self-clearing (*a b* c *e f*) is reached.

The robbing of the pillars *m p* r still above level *D* can then begin by putting upraises n^1 *n* and q^1 *q* into the highest portions and milling down around these raises into the chutes *b* and *e*, which spout into a tram car *c* in *d*. As such ore pillars, when cut by a dike, are liable to slip along the plane of contact it is necessary to begin their extraction at the end of the lense at *m*, in order to minimize the danger to the miners of a collapse of the hangwall.

Finally, only enough of a pillar remains above level *D* to sustain the filling, and this pillar is drilled, as is also the pillar *K*. The next step is to blast both these drilled pillars (continuously by fuse-firing) until all the ore remaining above the contour *a b c e f* has sunk into chutes *b* and *e*. It is true that the filling also descends, but most of the ore can be drawn from the chutes before the filling appears.

In one case an ore lense extending 150 ft. above the level was worked as one mill by putting a raise to the top and starting the mill at that point. As the dip of the footwall had a pitch of 45 deg. it was easy to stand on it and inspect the hangwall to guard against accident to the miners who were cutting down the benches.

This system can only be safely worked where the ore and hangwall are strong enough to sustain themselves over the width of the vein and for the height and length required for economical milling. A weak back, however, could be sustained over the underhand stope by a V arch of heavy stulls or "saddle back" which was formerly used with success at the Fayal mine on the Mesabi to cover rooms 23 ft. wide and 60 ft. high, whose sides were untimbered. The footwall must also be steep enough to clear itself by gravity. Several levels can often be worked simultaneously by postponing the robbing of the pillars until later. The ventilation is good and little or no timbering or shoveling is required. The ore is not broken as ordinarily in caving systems, chiefly by hangwall pressure, but the next cheapest method, i. e., underhand stoping, is employed, which requires few expensive development openings.

In suitable ground the chief objection is in the loss of ore through contamination by the filling; but this does not preclude its use in the mining of many iron ores. It is, also, only adapted to orebodies that are homogeneous, as no sorting can be conveniently done underground.

CHAPTER IX

OVERHAND STOPING WITH SHRINKAGE. NO FILLING

Example 14. Wolverine Copper Mine, Houghton County, Mich

(*See also Example* 19.)

Stoping Amygdaloid Beds with Strong Walls (*No Chutes*).—For the amygdaloids and conglomerates of this district, the mining problem is to excavate practically the whole contents of beds, from 3 to 30 ft. thick, of indefinite depth, and of a length along the strike, depending on the mineralization, but seldom less than several hundred yards. Usually the beds have a greater dip than the angle of repose between broken rock and footwall.

Fig. 48.—Shaft station in Wolverine mine.

In developing the Wolverine (Fig. 48) amygdaloid with its strong hangwall the drifts are at 100-ft. intervals, and are carried 20 ft. high across the vein, for the length of the payshoot, which, including barren spots, is 3000 ft. An overhand stope is then started above a drift (see Fig. 49) and extends up to the 10-ft. longitudinal rib, under the level above. In the excavation, the only hanging wall support is a pillar p (15 ft. diameter), for each 75-ft. room; which is formed by cutting around until only a 10-ft. neck of ore is left which can be pierced by a single round of drill holes. In long ore shoots it is best to also leave a 15-ft. panel pillar, along the dip from level to level, every three or four rooms.

A room is let on contract to four men (two on each shift with one air drill) with quarterly settlements, the monthly advance to each man being $65. To simplify the calculation of contract-excavation, the vein is assumed to have an average width of 2 fathoms, so that only the distance stoped along the drift and up the footwall need be measured. For drifting, a sectional area of 2 sq. fathoms is deducted from the stope, and this is paid for at the rate of $5 to $5.50 per lineal foot, while the stoping itself is let at $7 to $9 per cubic fathom (216 cu. ft). At these prices, everything is furnished by the company except explosives (the powder being charged $17 for a 50-lb. box). One-third the wages ($30 a month) of the nipper boy and the wear of steel is also paid by the contractors. For wear of steel, each man is charged $1 a month, and in the quarterly settlement, any ost drills are put in at the rate of 25 cents per lb.

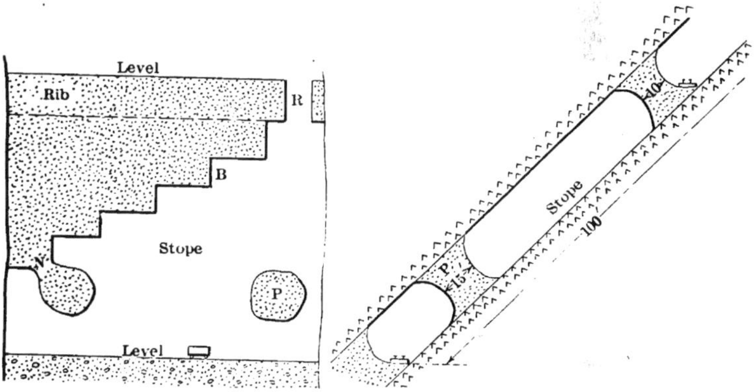

Fig. 49.—Stoping at Wolverine mine.

Shaft-sinking is also let on similar contracts, the price being around $16 per lineal foot for a section 8x17 ft. in the clear. To prevent sub-letting, all contractors are paid individually.

The muckers are paid $2.30 a shift, and work in pairs, each pair having a stunt of loading and tramming, from stope to shaft-station, 40 2-ton cars per shift. The footwall slope of 40 deg. is sufficient to cause the coarser broken ore to roll into the drift, whence it is shoveled into cars, as no chutes are put in. The waste from dead work is dumped into old stopes, though recently a little has been used for dry-walling to support some weak hanging wall. Only a small percentage of broken lode need be rejected in the rock-house above, as too poor for the mill.

The stopes are cut out in the usual horizontal benches *B* (Fig. 49) and water holes are drilled wherever possible. Six or seven of 8- to 10-ft. holes can be bored per shift with the No. 3 Rand drill in use. This permits an excavation of 35 to 50 cubic fathoms per month per machine, or about 1000 tons of broken ore. Twelve pounds of 40 per cent. dynamite will break a cubic fathom of ore. A raise *R* is put through the longitudinal rib at each room for ventilation.

At my visit, 37 drills were used on the day, 26 on the night shift. The drills are sharpened by a Ward Bros. machine, with a coke-heater and a forge blown by a special fan made by the Garden City Fan Co. of Chicago. The bits have + points up to 4-ft. length, and beyond that, chisel-points. The drills are generally run from tripods, set on a scaffold, or partly on a stull in the wider stopes.

For executives there are a captain and assistant (both on day shift) and a night shift boss, whose chief duties are to see that only good ore is broken down. Unprofitable portions of the vein are left as extra pilars, but the stope contractors are allowed something extra at the settlement for their consequent loss in volume, as is also the case if the stope has exceeded the assumed average thickness of 2 fathoms. For the muckers, there is a boss on each shift at every working shaft.

This mining system as described is suited to an ore-bed comparatively free from waste and having a hangwall strong enough to stand alone over wide areas and a footwall sufficiently steep to be self-cleaning.

Example 15.—Homestake Mine, Black Hills, South Dakota.

Sub-vertical Wide Vein With Strong Walls. No Chutes.—Many of the early miners at the Homestake were from Virginia City, Nev., and as in a great many other camps, the early mining is a record of Comstock methods—a desire to square-set everything. Even to-day there are timbered stopes still unfilled, and which will stand open, no doubt, long after the square-set timbers rot and fall apart. Later, when the system of "open-stope" for filled-stope mining was adopted, still clinging to the old idea of putting in the timber, the entire sill floor was square-set, supposedly for the purpose of keeping the haulage gangways from swinging. With this method no lagging was used over the timbers, except to protect the gangways, and it is said that the amount of timber that was crushed and broken in filling the sill floor was enormous. When the stope was drawn, this timber caused endless trouble.

The uselessness of timbering anything other than the haulage ways on the sill floor was, of course, soon evident, so the next change in method was to break out the sill floor, then shovel through the necessary gangways, timbering and lagging them, and packing rock around them as a protection; the stope was then carried up. This method is still used in many places, but the last stage in the development of the stoping practice has been to do away with the use of even this timber wherever possible.

The main ledge is so wide throughout most of its extent that the stopes must be carried across the orebody instead of along its strike. In other words, the hanging-and-foot walls are the ends of the stopes instead of their side walls. The width to which stopes are best worked parallel to the strike varies with the nature of the wall rocks, but it may be stated that as a general thing when the ledge attains a width of more than 80 ft., it has been considered best to lay out the stope across the orebody. In many places the orebody is over 400 ft. wide. The No. 1 North stope on the 700-ft. level was 60x520 ft. on the sill floor. This stope was worked with square sets. In working the level above the 900, stopes were carried 60 ft. wide from foot- to hanging-wall, and pillars of 60-ft. width were left between stopes. More recently, however, 60-ft. stopes and 42-ft. pillars have been adopted; the 1500 level from the Ellison shaft is being laid out on this plan. To a depth of 1100 ft., levels were carried at 100-ft. intervals; below that, they are 150 ft. apart.

Present Stoping Scheme

The usual method of approaching these cross stopes through timberless crosscuts is shown in Figs. 50 and 51. The ore is drawn into these crosscuts, shoveled into cars and trammed out in timbered drifts.

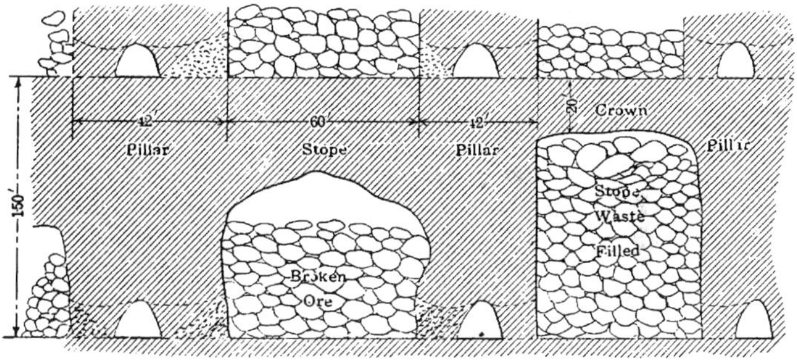

Fig. 50.—Cross-section of stope, Homestake mine.

The orebody is first developed by a drift, as shown. Laterals, or crosscuts, are then turned off at 102-ft. centers and run through to the walls. Simultaneously, the stope sills being cut out by driving across between the pillar crosscuts and breaking out to the full 60-ft. width. The crosscuts pass through the center of the pillars, and at 30- to 35-ft. centers connections to serve as draw holes are broken through to the stope. These crosscuts are connected with the footwall drifts, serving as main haulage ways. The stopes are then worked up, just enough ore being drawn so as to keep the drillers within reach of the back. One or more manways, depending upon the conditions, is carried up with each stope.

By maintaining the footwall drifts and tapping stopes through crosscuts in the pillars, timbering is practically eliminated in the first mining stage.

The stopes are usually carried up to within 20 ft. or so of the level above; the back, or crown, being removed after the stope has been completely emptied of ore and filled with waste. The crowns are taken up in small sections of 24 or 30 ft., using square-set timbers. On the 300 level, No. 1 Pierce stope, the crown was being taken out at the time of my visit. This stope is about 100 ft. wide and 200 ft. long, and the crown was probably 30 ft. thick. A hole was broken through on the footwall, where the ore is generally of better grade and the rock benched back

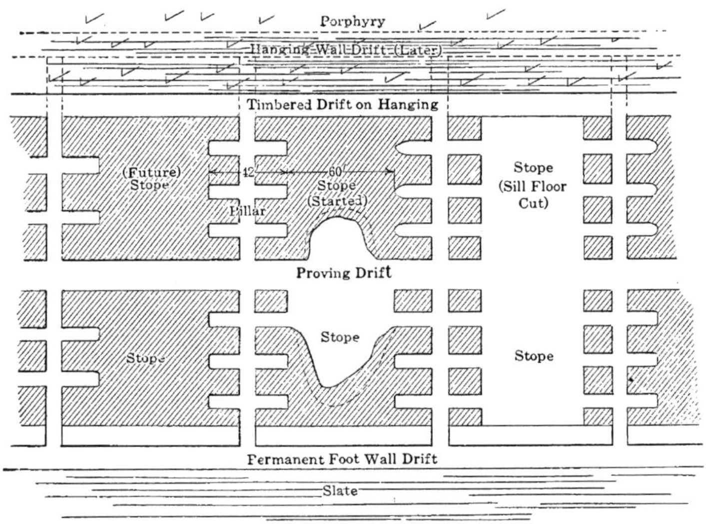

FIG. 51.—Plan of stope, Homestake mine.

toward the hanging. Finally about three sets of timber were put in next to the footwall, and under the remainder of the crown, which was then carefully worked out to the hangingwall. In breaking out the crown, the ore is left across both ends of the stope so as to form a supporting arch, until the timbers are under the hangingwall portion. This work is dangerous and requires careful watching.

It is planned to take out the pillars, even in the lower workings, by working them in small sections of square-set timbered stopes. This will be the second stage of mining at the Homestake. After the ore is drawn from the primary stopes the sides against the pillars are laced up with lagging set vertically, to which slabs laid horizontally are nailed. A section of lacing is put up, then waste run in from above until this section is filled; another section of lacing is placed, more waste run in, etc. Doubtless, by the time the pillars are removed the slabs will in many cases be rotted, but they serve to catch up the waste as square-sets are put in the pillar or secondary stopes. In all cases the crown over the original stopes must be taken out before the pillars are worked, or else this ore would probably be lost by caving, there being no support on the sides of the stope.

Breaking the Ore

Up to date, only large piston drills have been used for breaking ore in the Homestake mine. (Trials are now being run with several makes of stoping, air-hammer drills.) By putting in long holes and picking favorable places, huge masses of rock ore slabbed down. It is this tendency of the ore to break large that accounts for the great amount of shoveling necessary. The ore will not run through chutes, and at each gate block-holers with "Jap" plugger drills are kept busy drilling and breaking the ore so that it can be handled into the cars. About 3 lb. of No. 2 dynamite is consumed per ton of ore placed in the mine-car. Labor at the Homestake costs $3 per day for trammers and shovelers, and $3.50 for machine men. The labor union is not recognized.

Conditions at the Homestake are ideal for the operation of shrinkage stopes, the ore being tough enough to present a back under which the men may work with safety and the walls being good and tight. Caving is the only other mining method that might seem applicable for working such an immense low-grade deposit at a profit. This, however, is not feasible as the ore is too tough and hard. On the lower levels the ore is ahornblendic schist containing much fine disseminated iron pyrite and this so increases its specific gravity that only 10 cu. ft. weigh a ton. In places near the surface immense portions of the orebody have been broken away, and after years of crushing and packing, are not yet sufficiently broken up so that the ore will run in chutes. In these places a small square-set stope is run up a few sets, a grizzly put in at the top and rock-blasted down, being run for waste or ore, according to its character.

Discussion of System

The objection to this method is the excessive amount of shoveling necessitated. Every bit of broken ore is mucked into cars by hand, and this alone means a cost of close to 20 cents per ton of rock handled. The cost of labor amounts to almost three-fourths of the total mine-operating cost. Many efforts have been made to overcome this excessive labor consumption for shoveling, but as yet no satisfactory solution has been reached. However, just now over one-half of the ore is recovered without timbering, whereas formerly everything was worked with square sets.

Example 16.—Gratz Lead Mine, Owen County, Kentucky

Sub-vertical Narrow Vein with Strong Walls, Rill Chutes.—The geological formation is the Cincinnati of the lower Silurian period and around the mine comprises a limestone, horizontal and thinly bedded.

The vein dips nearly vertically and appears to be a fissure crack due to folding, and faulting, if it took place, must have been along the strike rather than along the dip. The filling is barite, calcite, and galena: the first occurs in typical white or brownish orthorhombic prisms, the second in translucent white or yellow tablets, while the galena is in cubes, either in thin bands parallel to the walls or in isolated crystals.

The vein walls exhibit no slickensides or other signs of movement, but the filling is frozen to them as when first deposited.

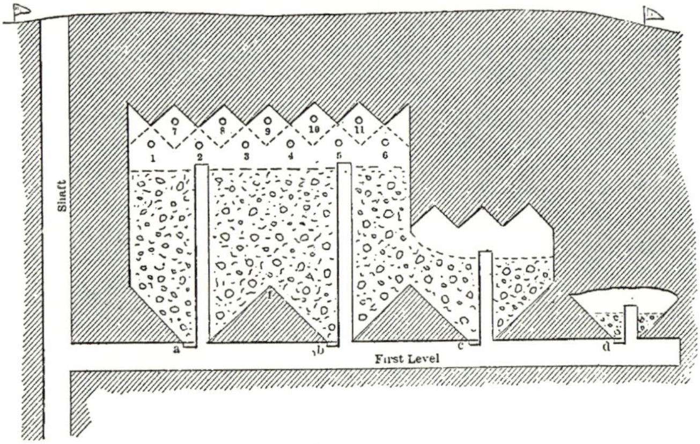

Fig. 52.—Stoping at Gratz mine.

On the first level, where the stoping is almost continuous for 500 ft., the vein thickness varies from 6 in. to 6 ft., with an average of above 15 in. A 3-ft. width of stope, however, must be excavated for machine drilling and all the broken rock is run through the mill.

The mine is opened by three levels, placed 100 ft. apart on the 5x8-ft. vertical shaft, that is only timbered through the surface soil. For drilling the stopes there are three 2 3/4-in. Rand drills and the stoping system is shown in section along the vein in Fig. 52.

Wooden chutes are placed at *a*, *b*, *c*, and *d*, from 40 to 50 ft. apart, and the stope bottom carried up hopper-shape, as shown, so as to leave pillars or "rills" like *a b f* to protect the drift and to avoid the use of timber. These chute pillars can be finally recovered by underhand stoping. In order to save set-ups, the stope back is attacked by the sawtooth system. In stope 1–6, a 3-ft. drill bar would be set up for a certain round only at points 1, 2, and 6, and the drill at 2 would bore two holes above and two holes below the bar in direction 2–7, and then four more in direction 2–8, or eight in all. Holes from points 3, 4, etc., would be put in similarly.

For the next upward round the bar would be set up at points 7, 8, 9, 10, and 11, with eight holes to be bored from each. In this way all the holes are self-cleaning uppers and a 6-ft. depth can be reached without sticking. With the flat and down holes of usual overhand benches, the calcite cubes, chipped out, would tend to wedge the bit and cause delay.

Enough broken ore is left in the stope to support the men at the back, for whose ingress plank mainways *a*-2, *b*-5, etc., are carried up, with entrances next the chute gates. For stoping the 30 to 40 per cent. dynamite, used in development, has been replaced by 15 per cent., as the latter is slower and makes less galena fines for the mill. The mine only runs a 10-hour day shift and the drillmen only bore, all loading and firing being done by an extra gang at night.

This system is well adapted to narrow, steep veins with hard walls where no sorting of waste need be done in the stopes. The rills at the chutes save timber and the breaking of the back in sawtooth profile is designed for boring the most holes with the fewest set-ups of the air-drill bar where upward-pointing holes are the most advantageous.

Example 17.—Alaska Treadwell Mine, Douglas Island, Alaska

(*See also Example* 9.)

Sub-vertical Wide Vein with Strong Walls. Rill Chutes.—For the underground work the main crosscut *C* (Fig. 38) and the drifts *D* are arranged as for the opencut system. In addition, at the ends of crosscuts *C* and 200-ft. to 500-ft. intervals along the deposit, the different levels are connected by raises for manway and ventilation purposes. The drifts and crosscuts are 10x7 ft. and the raises are 6x8 ft. in the clear, no timber being needed for either in this hard formation, which is so free from seams and so difficult to break that cut-holes for the drives must be pulled with 70 per cent. dynamite.

Stoping System.—As the value of the ore does not permit of timbering or filling, the present successful system dispenses with both. The object of the intermediate drift, *E*, is to open communication with the ore-chutes and to furnish a large facial area for the machine-drills to work upon, in cutting out or under-cutting the ground-floor for the stopes. When the intermediate has advanced about 50 ft. the work of cutting out the stope is started. This consists of mining out a chamber 7 ft. high, from 150 to 300 ft. long, and with a width varying with the width of the orebody. In the past it has been customary to cut the stopes with a level floor, but experience has shown that it is more economical to cut the floor so that it slopes up from drifts *D* at an angle of about 30 deg.

This does away with a large amount of shoveling, and the sawtooth stope-floor of ore thus left at *H* is ultimately obtained through the stopes from the next lower level.

When the stope-floor has been cut out, the work of stoping upon the ore is immediately begun. The roof of the stope is arched across from wall to wall of the lode, thus serving the double purpose of supporting the back and offering a better surface for the attack of the machine-drills. The ore is shot down in large, thin slabs, so that the shock of falling, combined with that of the blasting, breaks it up as much as possible. The pieces of rock too large to pass through the ore-chutes are broken by hand and "bull-dozed" with powder to the required size. When starting from the floor, the machine-drills begin in the drift midway between the lode walls and cut out a trench along the center of the back to form the apex of the arch, its height varying with the character of the rock. Two sizes of machine-drills are used: the 3 1/4-in. and 3 5/8-in. Ingersoll-Sergeant, and the holes are drilled to an average depth of 8 ft. A machine-stoping will drill an average of 28.69 ft. per shift of 10 hours and break 34.96 tons of ore with the consumption of 12.53 lb. of No. 2 dynamite. The cost of blasting up the rock after it has been blasted down is a large item in the expense of stoping. One rock-breaker is usually required to each machine, and it takes 0.85 lb. of powder in "bulldozing" for each ton of rock broken.

As no timber is used, it is compulsory that a sufficient quantity of broken ore be left in the stopes to form a solid working-floor for the miners. It has been found that one-third of the broken ore can be drawn off while the stope is being worked, and the surface of the broken ore kept within working distance of the back. In other words, by the above methods, two-thirds of the ore broken must be left in the stope, and cannot be drawn off until the stope is worked up to the next higher level and finished. In the Treadwell mine the slate-horse forms a natural division between the stopes of the north and the south orebodies. The walls of the orebody are supported by vertical pillars, or ribs, 15 ft. thick, and from 200 to 300 ft. apart. For means of communication and ventilation, man-way raises are put in these pillars and connected with the levels. At intervals of 25 ft., short drifts are run in opposite directions from the man-way raise; so that, as the working-floor of the stope advances, each of them is used successively when the workings connect with the main raise, and in turn abandoned and closed up as connection is made with the next higher one. The levels are protected by horizontal pillars from 20 to 30 ft. thick. Heretofore, these pillars have been left in place; yet, even with this saving, fully 20 per cent. of the ore must remain in the mine in the shape of pillars and ribs to support the ground and to prevent caving.

Samples and Maps.—Close attention is paid to sampling and recording the assay value of the ore. As a drift, raise, cross-cut, or other development-work is in progress, a sample is taken after each round has been blasted. These samples are taken either by the shift-boss or the foreman, and their description and location are recorded on a special tag, enclosed with the sample in the sack.

At intervals of 15 ft., and closer if there is any doubt as to the value of the ore, lines of samples—each sample being 10 ft. long and varying with the nature of the ore—are taken across the back of the stopes at right-angles to the strike. These samples are taken by cutting trenches, usually 10 ft. long, 4 in. wide across the strike of the ore, and 5 ft. apart, for the entire length of the new work. A hand-sample is taken from each car at the ore-bins, and again at the crushers a grab-sample is taken by means of large dippers, before the ore goes to the mill. When the mine-sample reaches the assay office, it weighs from 50 to 150 lb.

A complete set of maps is kept, showing in detail the underground and surface workings of the mines, also the value and position of each sample taken and the quantity of broken ore and reserves.

Labor.—On account of the system adopted for working the mines, due to the character of the walls and vein-material, it is necessary to employ only skilled labor in the shafts, drift, raises, etc.

About 60 per cent. of the machine-men and helpers on the island came as laborers and have learned their trade here. They are preferred by the foremen and seem on the average to break more rock than miners who have learnt the trade elsewhere. Machine-men get from $2.50 to $3.00 and muckers $2.00 per 10-hour day with board and lodging.

This mining system is suitable for wide, steep veins with strong back and walls where no waste need be sorted out in the stopes.

Example 18.—Veta Grande Mine, Cananea, Sonora, Mexico

(*See also Examples* 6, 34 and 45.)

Sub-vertical Lenses in Porphyry (Panel Cores with Pyramidal Rill Chutes.—This system of stoping combines square setting and overhand stoping on ore. On the main level the ore is first blocked out with a series of drifts at right angles to each other, one way the drifts being 40 ft. apart, and the other way 50 ft. apart, center to center. The general appearance resembles a checkerboard. All the drifts are timbered with regular sill-floor stope square sets. Chutes are put in every other set. On the next floor above the drift regular stope square sets are put in and the square-set chutes are carried up one floor. On the third floor, that is 16 ft. above the rail, the square sets are put in above the row in the drifts only and the included rectangle is mined out on this floor. From here up this continues with the square sets and the chutes carried up slightly in advance of the central portion of the rectangle.

Enough ore is drawn off through the chutes to give the miners sufficient head-room to work on the ore. In this way these different rectangles outlined by square sets are carried up to the limits of the orebody. There are several kinds of chutes that can be used, and it is not necessary to carry up a regular square-set chute. A simple beveled plank chute is just as good and uses less timber. In mining one of these rectangles the back is filled with holes and all fired together. If there is a horse of waste in the ore it can be easily removed and dropped into the chutes and trammed away. A large amount of waste is left in pillars. The rows of square sets are lagged on the outside, holding the ore in the center of the rectangle until the drawing commences.

Drawing the Core

After the whole body has been worked out in this way, the ore is drawn from the chutes. A certain amount has to be blasted again as it packs. The ore broken in the stopes before the recent shut-down was not drawn for nearly two years after it was mined. In this case a considerable amount of powder had to be used to loosen the packed ore, on which account only a few of the square-set timbers could be saved. However, if the ore could be drawn soon after being broken, the amount

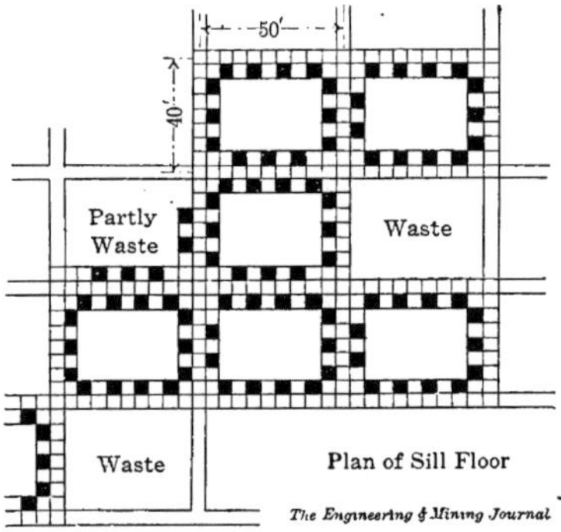

Fig. 53.—Plan Panel-core stopes, Veta Grande mine.

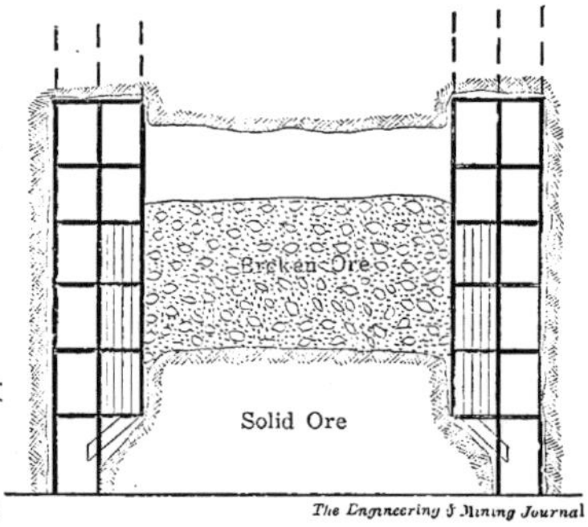

Fig. 54.—Vertical section Panel-core stope, Veta Grande mine.

of the powder needed would be less and a large percentage of the timbers could be saved.

After all the ore that can be drawn from the chutes is removed, there will still remain a pyramid-like mass in the center of each rectangle which cannot be removed in this way. It was from this fact that the system received its name. The pyramid of ore is later drawn by driving a drift into the center of the block and with a raise one set above the sill the remaining ore is drawn. The stoping proper does not commence until 16 ft. above the level, the object in this being to preserve the level drifts with 16 ft. of solid ore above the rails which would be mined from the level below.

The method of mining the block of ore on the level directly below would depend entirely upon the condition of the waste roof to which the first section had been mined. If the roof were treacherous and unsafe, it would be caved and the remaining ore could be mined by the slicing system. Fig. 53 shows the actual method of blocking out the orebody. As shown, chutes are put in every other set with no two chutes opposite each other, as this would obstruct the drift. The chutes are merely small openings cut in the solid ore with a couple of chute jaws and a door attached to the timbers. Fig. 54 shows a section across one of the rectangles. One after the other of these rectangular blocks is carried all the way up to the waste roof and the drawing of the ore does not commence until all have been mined out.

This system takes much timber, and where so large a mass of broken ore stands before being drawn, it takes considerable labor and powder to loosen and draw it. This is its chief disadvantage. It requires solid ore and a strong roof which will stand over the core. The cost for the labor and timber used in placing ore in the chutes is 80 to 90 cents a ton.

CHAPTER X

OVERHAND STOPING ON WASTE IN THE UNITED STATES

EXAMPLE 19.—SOUTH RANGE MINES, HOUGHTON COUNTY, MICH

(*See also Example* 14.)

Dry-walled Drifts. Sub-vertical Amygdaloid Beds with Weak Hanging Wall.—The Quincy mine practice will first be reviewed, as the prototype of that of the South Range. The Quincy amygdaloid, proper, is overlaid by a shaly seam and, to avoid this, the main drifts follow the footwall and leave the lode rock overhead. In stoping out all the lode, the weak hanging wall must be supported, and this has been done, with little use of timber, by a system of dry-walling.

The present South Range mining system was first developed at the original mine, the Baltic. The first system at the Baltic was that of its

FIG. 55.—Dry-walled tunnel, Champion mine.

neighbor, the Atlantic mine, by which the main drifts were roofed over with long stulls, closely lagged, and enough of the ore, broken in the overhand stope above, was left on these stulls to hold the drillmen close to the stoping face. Enough room was left between the hanging wall and the stulls (which were inclined at about 70 deg., with one end on the sill floor, and the other in a hitch on the hanging wall), for the track, where cars were filled from chute-gates above. When the stope was completed to the 15-ft. longitudinal rib, to the left under the next higher level, the withdrawal of its content of broken ore was begun. During this last process, stulls were placed between the walls by the timbermen at dangerous places; but, nevertheless, considerable hanging wall would peel off and contaminate the ore. The net result was that 20 per cent. of the ore was waste and, as all sorting had to be done at the surface, this caused decreased hoisting capacity for mill-ore. Also, the stull system required a regular width of stope and, in the irregular Baltic lode, this meant either an unnecessary breaking down of waste or the missing of bulges of ore.

The Champion.—This mine was first opened by "arching," by which a stone arch 8 ft. thick was left above each main drift, with chutes cut through it at 25-ft. intervals. This arch corresponded to the lagged-stull roof of the Altantic system; and enough of the broken ore was left

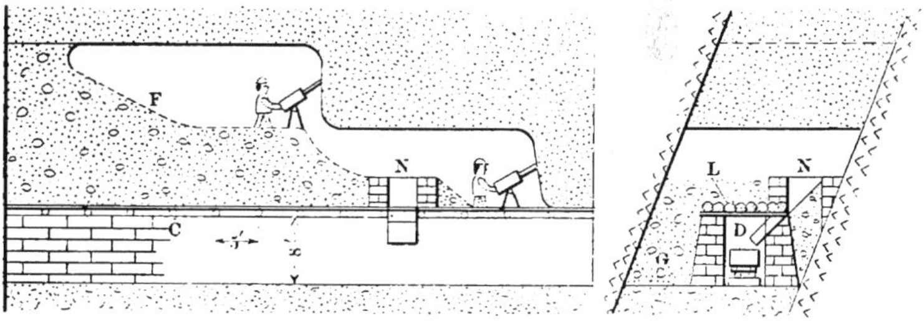

Fig. 56.—Stoping at Champion mine.

in it to hold the drillers up to the face of the overhand stope above. "Arching' had the same disadvantages at the Champion as "long-stulling" had at the Baltic, and has now been superseded by "dry-walling" as adapted by the Baltic in 1900. The Champion mine's "dry-walling" will be described as typical of the present South Range system.

The main levels are driven 100 ft. apart on the 70 deg. footwall, and are cut out the whole width of the ore, whether 10 or 60 ft. The rock is sorted where broken, the ore being hoisted, and the waste used for walling, as shown in Figs. 57 and 58. The side walls of drift *D*. (Fig. 56) are laid dry and 4 ft. thick at the base. The walls are topped with 2-in. plank, on which are laid (at 5-ft. centers) the unbarked, round caps *C*, of 8 to 10-in. diameter, which support the lagging *L* of 3-in. poles. The drift is located near the footwall, but alloys space for the ore chute *N*, set every 25 to 60 ft. along the drift, with hinged, steel troughs for gates.

As soon as the drift-lining is well advanced and well backed by waste *G*, the overhand benches are begun behind it. In Fig. 56 two machines are stoping, while the sorters are handling the freshly broken lode, throwing the ore into chute *N*, and piling the waste at *G* and *F*. The chute *N* is kept just above the stowing, and is built (4 ft. inside diameter) of waste rubble, which has proved superior to the poles and old railroad ties formerly used. Should the waste prove insufficient for filling the stope, some can be blasted from the walls, or a raise can be put through to the next level, and waste run down from the old filling above. When the exhausted upper level is reached, the floor-pillar can be extracted by caving, if the level need no longer be kept open for tramming. A small self-dumping car is often used in a stope to facilitate the stowage of waste.

By this system, a total of 1100 to 1200 men (above and below ground) produce 2500 to 2800 tons in two shifts. Owing to the irregular stope-width, miners work by day's pay, earning about $2.50, while muckers get $52 to $54 a month. Only development is let on contract.

The advantages of the South Range system are, the complete exhaustion of the ore, the saving in timber, and the decreased use of shafts for lowering timber and hoisting waste, all accompanied by safety and good ventilation. A little copper is lost in the stored waste and many men are needed for dry-walling; though with the large output, this only means an expense of about 8 cents per ton of ore hoisted.

TRAMMING

It is only in the South Range mines, with their steep footwalls, that the ore broken in the stopes can all be drawn direct from chutes; in the rest of the district, it is either shoveled off a sollar, placed on the drift-floor, or on a platform at one side, so that the car can pass to stopes

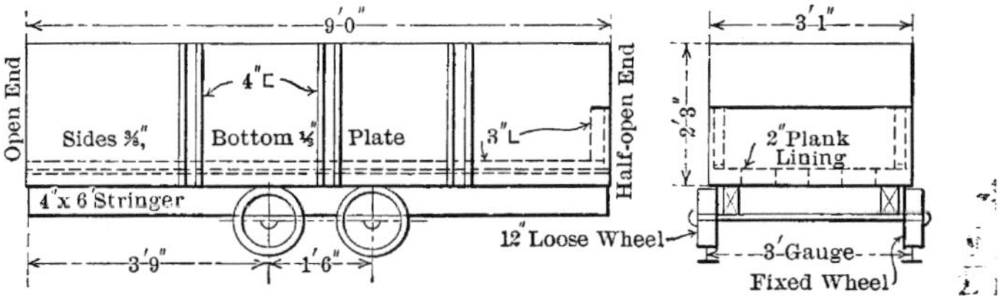

FIG. 57.—Car at Champion mine.

beyond. In capacity, the cars vary from 2 to 3 tons; the latter size is unusually large for man-power, but there seems to be no difficulty in handling it with two men on a track of 3-ft. gauge at the Champion mine; but for longer hauls electric trains are cheaper, and are used at the Quincy with 3-ton locomotives.

In such cars as the Wolverine and the Champion (Fig. 57), the ideal design for capacity and ease of loading seems to have been attained. The car body is placed low (just above the rim of the 12-in. wheels), and is only 2 ft. high, for easy loading from the track level, while capacity is had by the extreme length of 7 ft. for the first and 9 ft. for the second. In order to get a low car-body, a truck with a turn-plate is barred; so dumping is achieved for the Wolverine car by rotating its body around the front axle, while the larger Champion car must be run on a tipple to be emptied. The Wolverine car has loose wheels, but the Champion has one loose and one tight wheel on each axle, which gives better lubrication and less wear.

The copper region has been fortunate in having a bedded formation of great continuity, uniformity and strength; if it had been much faulted or generally brittle, the system of great open stopes could not have been pursued, and consequently the poorer lodes could not have been profitably worked. Nevertheless, there is a limit to even the stability of these formations, and this fact has been forcibly emphasized by recent events.

For many years some of the older mines, notably the Quincy, had been bothered with subterranean disturbances due to settling of the hanging wall in the old stopes; but it was only recently that a calamity resulted. In May, 1906, the main workings of the Atlantic Co. on the Ash-bed lode collapsed, and subsequent movements soon rendered the 5000-ft. shafts useless for further hoisting. The company has not tried to reopen the shafts, but has been fortunate in finding a new mine on its holdings along the Baltic lode.

One explanation of such general caves is that the adherence of the hanging wall to the stope-pillars is so lessened by the great area of nearly continuous excavation that it begins to slip along its sustaining pillars, and thereby so crushes and distorts them that they no longer offer sufficient support to the superincumbent weight. The hanging wall of an isolated stope can be considered as a beam fixed around its circumferential supports; but when many stopes are connected, the hanging wall is then only like a beam resting on its supports, and has consequently diminished stability. Also, a pillar, that is strong enough to sustain the roof of the single original stope, is not necessarily able to sustain the increased strain of an added line of contiguous stopes.

Example 20.—Minnesota Mine, Soudan, Vermilion Iron Range, Minnesota

(*See also Example* 36.)

Overhand Stoping on Waste with Filling from a Descending Hangwall. Sub-vertical Wide Vein with Weak Hangwall.—These iron ore deposits occur in lenses 200 ft. to 1000 ft. long and 5 to 80 ft. wide, and stand at an angle of 65 to 75 deg., with a vertical height of 250 ft. to 500 ft., other lenses occurring below. A number of the deposits were first worked as open pits, which in some cases were carried to depths of 150 ft., when, owing to the weakness of the walls, underground mining was adopted. While the ore was being removed from the open pit, shafts were in several instances sunk into the foot-wall, the intention being to mine the ore with breast-stopes of an approximate height of 20 ft., followed by underhand stopes of the same hight, leaving pillars between of the necessary thickness to support the walls. As work progressed, however, it was found that the chlorite walls were too weak to permit the working of breast-stopes 20 ft. high, there being frequent heavy falls of ground from the hanging wall, and sometimes from the foot. The plan of following breast-stopes with underhand stopes was therefore abandoned, for by working breast-stopes only, but little more than one-half of the ore could be removed, and that only at an excessive cost, the ore being one of the hardest known, to drill.

The Rand 3 1/8-in. piston type is used, with 60 lb. of air, which drills 6 ft. per shift as a yearly average, but in certain places will only make 1 ft. in 10 hours. During a shift's work each drill dulls 45 to 50 bits, and even then powder must be used as an aid by exploding a half stick of 50 per cent. in the hole for every few inches of advance, to enlarge the bottom and prevent the bit sticking. Often 10 sticks of powder are used in boring a hole and, in addition, there are 5 to 10 sticks more necessary to chamber the bottom for the breaking charge of 20 to 50 sticks. Fuse and cap firing is in vogue. The holes are 6 ft. to 10 ft. deep. For a time diamond drilling was employed to bore 20-ft. to 40-ft. holes for breaking ground, but the present high price of diamonds has made this method unprofitable.

The greenstone walls are easy to drill, and as much as 130 ft. per month has been drifted in them with two drill shifts daily. But in the jaspilite "horses" encountered the same men could only make one-third this distance. In the sinking of the main shaft below the 1,150-ft. level, with three machines, the monthly advance in jaspilite with three eight-hour shifts was only 12 ft. to 18 ft.

Mining.—The present system here may be called hangwall filling. To develop it an incline was sunk in the foot wall, and from it, at about 100-ft. vertical intervals, were driven crosscuts in the vein. The ore is then attacked in all directions by overhand stoping until the excavation is 16 ft. by 20 ft. high the whole length and width of the ore body (see Fig. 58).

Next drift sets *d* are set up and spiked together the whole length of the stope with 9-ft. caps and posts, but no sills. At 25-ft. intervals on the footwall side chutes *c*, 5 ft. square inside, are built along the drift sets, and opposite every third chute is placed a similarly constructed manway. The chutes rest on the floor, so they must be filled with waste up to their false bottom of rails, which is high enough to deliver through a hinged, steel spout into the ore car. Above the false bottoms the chutes are lined with 2-in. planks placed vertically.

Meanwhile, waste raises *w* have been driven at 75-ft. intervals in the foot wall, just under the ore, from the stope to the level above. These waste raises extend from the open pit to all the opened levels and are cribbed up in front so that waste can be thrown at any point by removing some cribbing and leaving the part below full, or making a temporary false bottom of wood. Raises *f* are also put through between levels for use as extra manways in case of fire.

When the timbering and waste raises are completed to the lowest floor the filling of excavation *E* is begun. The waste for the first level below the open cut is obtained from the hangwall either by its

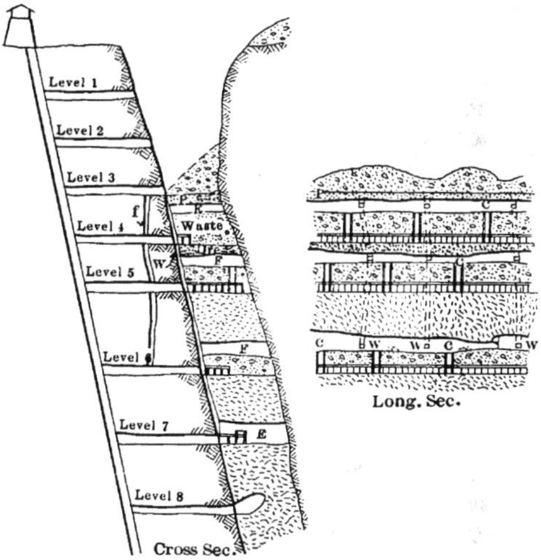

FIG. 58.—Stoping at Minnesota mine.

natural caving or by blasting. If waste for a lower level is not needed till the first level is all stoped, much of it can be drawn from the filling of the latter, as in winter the overlying waste on pillar *P* is frozen hard enough to prevent its breaking through.

Wherever obtained, the waste is drawn down from raises *w* and handled in wheelbarrows to fill the whole stope to a depth of 14 ft., the drift sets being lagged and the chutes and manways being extended to the same height. The drills are then set up on the filling and a 14-ft. to 16-ft. cut made to the back to start a new breast stope, for which the drill is to be worked from a scaffold. Filling can begin behind the breaking, the ore being thrown into the chutes ahead and the waste brought from the raise *w* behind and piled to within 6 ft. of the new back, while the chutes and manways are correspondingly heightened.

This overhead slicing is continued till the next level is nearly reached, under which a pillar *P* is left from 6 ft. to 10 ft. high, according to ground. During the ascent any weak parts of the back are supported by cribs on top the filling, which can be removed on stoping the next slice. When all ore above is removed and the filling in the open cut has settled on pillar P, it can be removed by beginning at each end of the ore body and holing through by blasting. The balance of the pillar is then cut off while retreating, and though the filling follows the broken ore down, the latter comes first, and can thus be recovered and thrown into the chutes. During the recovery of *P* it is liable to cave from the pressure of the waste above, but as *P* is always supported by a number of cribs on the filling, and gives plenty of advance warning of independing disaster, the men are in little danger.

In case pillar *P* does cave it is allowed to settle and its ores recovered by drifting. For this it is necessary to drive spiling ahead, which are of sharpened poles, 4 in. by 16 ft., driven over three-quarter drift sets placed 3 ft. apart. To advance the spiling it often requires considerable blasting of the many boulders encountered.

Application of Hangwall Filling System.—The following are favorable conditions: Highly inclined, hard, wide veins, which will stand without timbers when excavated across their whole width, but whose hangwalls are weak and friable.

It allows good ventilation and requires little timber and no hoisting of waste. Several levels can be worked simultaneously and, development being confined to the soft walls, the hard ore can be broken in wide stopes with minimum expense for drilling and explosives, as most of the filling comes from the natural shelling of the hangwall, much of which can be reused, there is no expense for mining or freight, the only cost being in keeping the waste raises clear and in wheeling and stowing the filling in place. For narrower veins it is sometimes cheaper to break part of the filling from the hangwall of the stope.

Example 21.—Superior and Boston Copper Mine, Globe District, Arizona

(*See also Examples* 32 *and* 42.)

Sub-vertical Wide Vein with Weak Hanging-wall; Rill Chutes.—The present mine production is made mainly from the stopes of the 550-foot level. Here the vein dips about 58 deg. and varies in thickness from 7 to 15 ft., but averages from 9 to 10 ft. The foot-wall is hard and smooth, but on the hanging wall the ore is "frozen" and there is no defined wall. This method of mining was devised by Supt. John D. Wanviz, and is similar to one in use at Zaruma, Ecuador.*

*Trans. A. I. M. E., Vol. XXX, p. 248.

Until recently the ore has been mined by the common square-set system, but in the few months since the first trial of the new system, both the timber and labor, not to mention loss of ore fines, have been materially reduced, and the reduction in costs has been 60 per cent.

Reference to Fig. 59 will make clear the following data:

Drifts 5x7 ft. are carried 100 ft. apart vertically or about 117 ft. along the 58-deg. dip of the vein.

Two-compartment raises are put up at 100-ft. intervals to connect the drifts; the chute compartment is 4x4 ft., and the ladderway 3x4 ft. These raises are on the foot-wall and are timbered only by two lines of stulls with head-boards and one set of lagging.

The division between the two compartments is formed by the first line of stulls which is plank lagged. The second line of stulls is carried

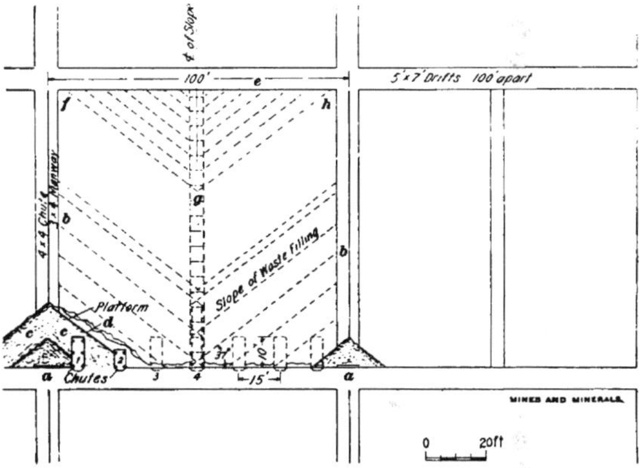

FIG. 59.—Stoping at Superior and Boston mine.

on the other side of the ladderway compartment, but this is unlagged. The outer wall of the chute is thus constituted by a rock wall unstulled and unlagged and the outer wall of the ladderway is likewise constituted by a rock wall, but this is stulled. The main drifts are all timbered with drift sets spaced on 5-foot centers and heavily lagged. Chutes 4 ft. wide are spaced on 15-ft. centers, the chute gates all being placed at once, but the chutes themselves are put up one by one as stoping proceeds above. The chute mouths afford one of the ways of access to a stope as it progresses, since there is always an irregular space 1 to 3 ft. high between the ore back and the top of the drift lagging.

Stoping is started at the lower corners of two adjacent blocks formed by the intersection of a raise with a drift. These two blocks are simultaneously stoped up along the raise and retreating from it horizontally as shown. At such a corner the back is drilled with stoping drills and broken down on to the drift lagging for a distance of 8 to 10 ft. The broken ore is at first discharged into mine cars standing in the drift below by pulling out the lagging and letting it run down.

The first chute, 1, is then placed. This chute is formed by carrying up two lines of stulls 4 ft. apart along the strike of the vein and lagging them on the outside with plank.

The stulls are placed inside the plank lagging instead of outside, as in ordinary practice. The reason for this is found in the fact that when the chutes are abandoned they are not filled with waste, so that empty they have to stand the pressure of the waste filling on all sides.

The entire width of vein from foot to hanging wall is broken out between the rows of stulls. With the first chutes in place, the timber in the lower part of the raise *b* is taken out and waste *c* for filling dumped down the chute from the drift above. This waste assumes a natural slope of about 37 deg. from the horizontal, and it accordingly spreads out from the foot of the raise into the stopes on each side of it at that angle, running farther and farther horizontally into the stopes as they are advanced and the pile of waste increases in height. In the first stage of the work, however, advancing toward chute No. 1, the waste is dumped down till the foot of its slope almost reaches the chute. Dumping waste then ceases for the time being. With sloping floor of the waste now brought within 5 to 7 ft. of the ore back, a sloping wooden floor *d* is laid down on top of the waste extending from the top of the chute to the manway of the raise. Its purpose is to receive the ore broken from the back and discharge it into the chute, thus serving the double purpose of separating the ore from the waste and eliminating all shoveling and tramming in the slope. It will be evident that since the waste will run at an angle of 37 deg., the ore will certainly run on the wooden planks which are laid over the waste on the same angle.

The width of the plank floor *d* is everywhere made that of the vein from wall to wall, and its length, of course, increases up to a maximum of about 60 ft. in extending from the center chute of the block up to the raises on either side of it. The floor consists of 2-in. plank and 15-ft. lengths nailed together by cross-pieces to form sections about 2 ft. wide. In laying down these sections over the waste, the ends are made to form a butt joint in order to insure a clean run-off of the ore. With the floor laid down, the stope drills are put at work to break down the new back, stulls and planks being placed where necessary to afford a good secure footing next to the back. When drilling operations are completed, the stulls are pulled, thus recovering them for further use. The chute at the foot of the sloping floor is always kept nearly full of ore so that wear upon it may be kept at a minimum. After drilling and shooting the back, the ore is at once drawn down to the top of the chute, thus clearing the floor. A number of stulls are next put in as close to the back as possible and the sections of the platform are then raised and one of each placed see-saw fashion over the stulls so that the weight of a larger end of a given section causes the smaller end to press up against the ore back and thus be firmly held. With the platform sections thus disposed conveniently at hand for their next period of use, the top of the chute is now built up and more waste dumped down the raise till the foot of its slope has nearly reached the level of the chute top. The laying of the floor and the mining of another diagonal slice now proceeds as before.

A given chute can be used at the foot of the sloping platform only till the angle of slope carries down the ore in line with the center of its top. After that the top is lagged over, the chute emptied of its ore but not filled with waste, and it is abandoned. Fig. 59 shows chute No. 1 just abandoned and buried in waste with the use of chute No. 2 just begun. The illustration also makes is evident that of the seven chutes raised in a 100-ft. block, all but the one in the center are limited to a height of about 10 ft., as at that height the angle of waste slope carries the flow past to the bottom of the next adjacent chute, which is likewise built up. By working from the two opposite lower corners of the block simultaneously, and thus advancing the diagonal slices toward the center line of the block, they finally intersect at the bottom of the centrally placed chute No. 4. From this point on, chute No. 4 receives all the ore of the block. It is built up from time to time just as the other chutes were, with the exception that it is more heavily and completely lagged by placing 2-in. lagging outside the stulls and 4-in. plank lagging inside the stulls which bear the wear of the chutes. Its height, as shown in dotted lines, is limited only by the floor of the drift above. It will likewise be evident from the illustration that when the top of the waste filling has reached the approximate position of the dotted line *f*, *g*, *h*, the drift *e* must be abandoned for through traffic, or its floor supported on stulls. From this point on, the raises cannot be utilized for throwing down filling, so that waste is dumped from points along the drift itself advancing toward the center line of the block till the central chute reaches the drift and the block is entirely mined and filled.

During the mining of the lower portion of the block, access is had to the stope from the manways of the raises or from the chute mouths in the lower drift roof. After the diagonal slices have connected at the central chute, No. 4, however, access is had only from the drift *e* above by descending through the raise manways into the stope. No objection to this limitation of accessibility has yet been found in several months of work. If found desirable, however, the floor on one side or the other could always be kept down and over it the men could pass in and out to avoid danger of starting the loose waste.

From the loose nature of the hanging wall, it will be evident that working the mine by "shrinkage stoping" would be inadvisable, because caves from the hanging wall would seriously dilute the ore with waste.

The system adopted, as described above, therefore, seems the next best method of reducing the use of timber to a minimum, eliminating tramming and shoveling in stopes, and recovering all the fine ore. The system has the advantage over "shrinkage stoping" of making all the ore broken available at once instead of only about 25 to 40 per cent. until such time as the stope is finished. Another advantage over "shrinkage stoping" is that the drift timber sets and chutes once placed require no reinforcement during the drawing of ore as is frequently the case in drawing a stope completed by the shrinkage method.

Fig. 60.—Metcalf mine and ore bins.

Example 22.—Metcalf Mine, Graham County, Ariz.

(*See also Examples* 29, 30 *and* 40.)

Irregular Lenses in Porphyry; Auxiliary Milling and Square Setting.—At this mine, bodies of oxidized ore form a conspicuous outcrop on the hilltop, (Fig. 60) and were the source of the high-grade ore first mined. The hill is a mass of granite porphyry, capped by the lower members of the sedimentary rock series. Development has demonstrated the existence of four parallel vein systems or stockworks in the granite porphyry, along which oreshoots of varied magnitude are found. The vein systems have suffered severe cross faulting, subsequent to the formation of the primary ore, but prior to the surface enrichment that has occurred. The surface presents a chaotic mass of blocks of quartzite, shale and granitized limestone lying on and in other places completely imbedded in the intruded porphyry.

The oreshoots are generally found at the junctions of the cross faulting with the vein systems. Although, as before mentioned, some of the oreshoots outcropped, the majority of them are found beneath an overburden of barren rock of varying thickness. The horizon on which these shoots occur is variable and their discovery necessitates extensive prospecting from levels of not more than 40 to 50 ft. apart.

Where an oreshoot has been proved in depth and is covered by a heavy overburden of waste, underground mining is employed. The ground in most cases stands well without timbering; stopes up to 75 ft. in width having been worked without difficulty. In ore of moderate hardness, the mode of working is as follows:

On the lowest level on which the ore is exposed, the oreshoot is opened to its full width and length. When the shape of the orebody has been determined, raises are made from the roof of the stope to the surface, the number depending on the dimensions of the stope. The ore is then mined to a height of 20 ft. above the level and a timbered roadway with the necessary chutes and ladderways erected. The overburden of waste is now milled down the raises and leveled off, filling the stope and forming a compact working floor, 15. ft. above the level.

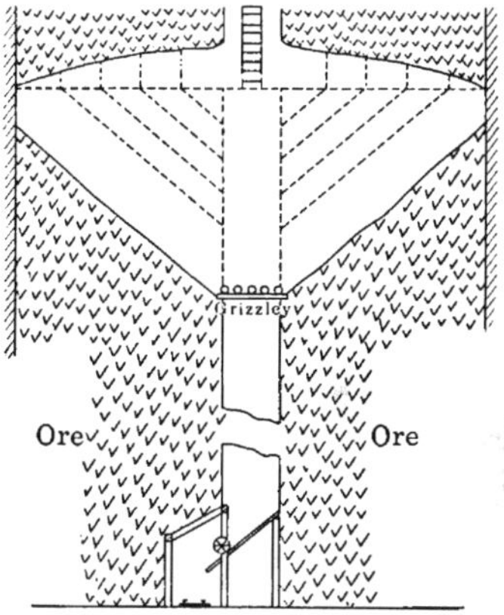

FIG. 61.—Underground milling at Metcalf mine.

The ore is afterward broken by overhand stoping in ascending slices 15 to 25 ft. high; depending on the condition of the roof. As each slice is removed, the overburden is milled down for filling material and to provide the next working floor. When all the overburden has been utilized in this manner, the back of ore remaining is gained by open-cut or milling method. The cost of breaking the ore in stoping is necessarily higher than in open-cut work, but the more economical removal of overburden compensates for this increase.

When the ore is hard and stands exceptionally well the method of extraction is shown in Fig. 61. A raise is made to the top of the ore and extended to the surface or to an upper working for air. At a height of 50 ft. above the level, mining is started from the raise outward, the floor being always left sloping so that the ore will run directly into the chute. When the extremities of the ore have been reached or the roof is as wide as will stand with safety, the bench forming the floor of the stope is now mined. Deep holes, charged with black powder, are used and the ore is broken as freely as in open-cut work. The slower action of the black powder does not jar and weaken the roof of the stope to the same degree as the rapid action of dynamite.

When the first bench, 50 ft. in height, has been worked out, the chute and ladderway are timbered to within 5 ft. of the roof and the stope is filled with waste from the surface; the filling material is leveled and the next block of ore above is attacked in a similar manner.

Stoping with Square Sets

In some parts of the mine, the ore is too soft and friable to permit of any system of stoping without the use of timbers. Ordinary square-set timbering is employed in such cases to support the roof and walls. The stopes are kept full of waste to within one set of the back of ore, timbered chutes and ladderways alone being left open. The waste is obtained from the surface workings and is distributed from a small chute placed in one corner of the stope. This position of the chute is advisable as by commencing the mining of each floor from this point, the filling can be kept close to the working breast without interfering with the shovelers.

Every floor is worked as rapidly as the working faces allow, to avoid excessive weight settling on the timbers. The timbering is arched to compensate for the sinking of the sets in the center of the stope due to the greater weight of roof. This is accomplished by introducing one floor of sets, the posts of which step upward 2 in. to the center of the stope, and descending in like manner to the opposite wall. Should this arch effect at any time be lost by excessive weight, causing the timbers to settle, or by the loosening of the side blocking, another floor of special posts is put in to restore it. Further details are given in example 23.

Example 23.—Copper Queen Mines, Bisbee, Ariz.

(*See also Example* 12.)

Irregular Clayey Lenses in Limestone; Panel *System and Square Sets.*—Of the gangue minerals, the silica is not vein quartz, but fine grained aggregates, and in some stopes it runs like pulverulent sand; while the silicates occur interspersed throughout the altered limestones around the ore. Clay is prevalent, it may be white, gray, or yellow and red from limonite admixture, and it forms the chief gangue of the oxidized stopes and makes their support difficult.

The ore horizon is the Escabrosa limestone with occasional offshoots into the Naco above. The ore occurs in great tables and lenses that follow the bedding planes of the beds. Within the limestone the bodies pinch and swell and many of them, particularly in the central belt, are connected by seams and pipes.

Definite walls are exceptional; the oxidized ores grade into clay and the sulphides often into oxides within a hard limestone casing. A stope outline depends upon the price of copper as compared with the cost of ore extraction and in proportion to the clay matrix the volume of ore may be only a fraction. The ore masses are rarely greater than 200 ft. square horizontally by 100 ft. thick, but occasionally they are larger, as in the Holbrook big stope, which is 600 ft. wide by 800 ft. long on the dip. The ore lenses are not distributed haphazard in the Escabrosa limestone, but favor the line of certain faults and the porphyry contact.

Mining

Much of the ground is soft enough to be removed by a pick alone, but it has been found quicker to loosen it by auger bores and blasting. The auger is of the common earth type with a fixed wooden handle and 4 ft. long. The cutting diameter is the same as that of hand drills, so that, if a boulder is struck, the hole can be finished by single-jacking. For harder ground single jacks are usual, but experiments are now in progress with the hand-hammer air drills. For hard rock 3-in. air drills requiring two men are in vogue. The blasting is always done with 35 or 40 per cent. dynamite.

With the gentle topography and flat ore horizons there are no tunnel sites, so vertical shafts are compulsory. To discover the ore, the ground at the 100-ft. levels is exposed by drifts, following faults or ore stringers where they exist, otherwise the area is divided into more or less rectangular blocks. The volume of ground to be searched is much greater than in vertical veins, as is shown by the annual driving of around 5 miles apiece by the Copper Queen and Calumet & Arizona companies, and in like proportion by the others, besides considerable diamond drilling.

Drifting.—By arching the roof, many of the drifts will stand without timbering, but in bad ground full sets with inclined posts 8 to 12 in. square and 2 in. per foot batter are put in. In the Calumet & Arizona mines, when the drifts penetrate ore, their sets have vertical posts that will join with the stope square sets; but in the Copper Queen the inclined drift sets are kept in, until replaced by the square sets of stoping.

In the Calumet & Pittsburg mine 50 to 60 ft. advance in 7 days is made by using a 3-in. drill to get in a round of 9 to 12 holes, in the

8-hour shift from 7 A. M. to 3.30 P. M., which is then blasted. Muckers then come on and begin to throw the muck back, so as to have the face ready for a new set up, when the night shift drillers arrive at 6 P. M. Tramming the mucking can then continue without interrupting the drillers and when they leave at 2.30 A. M., the clearing of the face again begins as before.

Stoping.—Except for the Mitchell system and the stulls used in the thin beds of the Pittsburg & Duluth, the square-set system is universal here for timbering stopes. For the soft, irregular, and patchy stopes of Bisbee, the square-set system has many advantages. It permits of the easy omission of barren spots, gives the best ventilation and allows the re-entering of an old stope at any time for the removal of filling, now

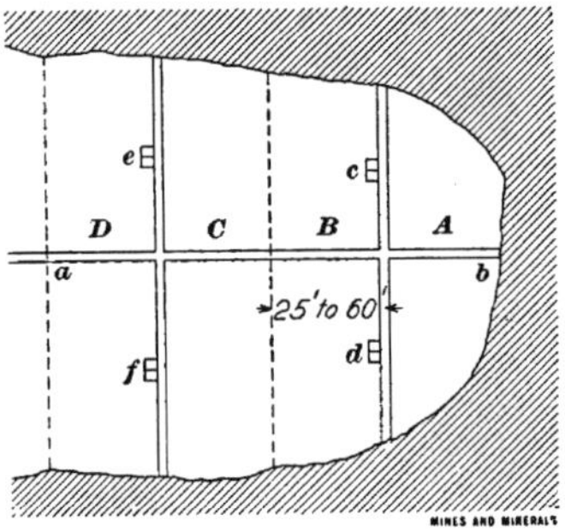

FIG. 62.—Panel-stoping at Copper Queen mine.

valuable but not previously, or for the extension of the sets into lower-grade ground. A junction with an old stope's timbers can readily be affected from any direction and this is especially useful, when approaching from below, for the timbers tend to prevent that sliding of filling, which will occur with untimbered systems. When worked in small panels and closely filled, square sets are safe even in the soft ground of Bisbee.

The Copper Queen Mines have 14 posts between the 100-ft. levels. The upper floor square sets are framed 7 feet vertically by 5 feet square, but the sill floor sets are 9 ft. high to allow space for extra caps and for settling. The stope floors are kept level in spite of ground movement by cutting the posts of the exact length necessary to keep their tops even with that of the corresponding raise post. The width of a panel across an ore body depends on the hardness of ore, but varies from 5 to 12 sets.

The method of mining is illustrated by Fig. 62. A cross-cut *c d* is driven from the sill floor drift *a b* and two-compartment vertical raises *c* and *d* are driven to the top of the ore body. One compartment is used for an ore chute, the other a man and timberway. Panel *A* is taken out floor by floor from the top down, the ore being thrown down the raise to be loaded from chutes at its base into cars on the sill floor, the extra waste for filling is received from the level above through the raise before mentioned. The ore descending from a floor is prevented from mixing with the waste coming to the floor from above by a partition in the raise, or by the use of an adjoining vertical line of square sets for the ore descent. Only such sets are kept unfilled as are necessary for the free movement of the miners. Where there are no levels already existing above, to which a raise can be joined, a cage is put in the manway and operated by an electric hoist set to one side on the level below. A car of waste can then be raised on this cage to the desired floor.

A number of floors of panel *A* can be worked simultaneously in benches by overhand stoping; when *A* is extracted and filled, panel B can be attacked similarly. Cross-cut *e f* and raises *e* and *f* should then be ready for beginning panel *C*. About 25 board feet of timber are lost per ton of ore extracted; the lagging is 2-in. plank, which is largely recovered. The timber comes sawn, from the Pacific Northwest, and costs around $25 per M.

Most of the cars, of 16 cubic ft. capacity, are of the vertical shaft type having a hinge and turn plate to enable them to dump from one end in any direction. The Copper Queen Co. employs simple turn sheets in the drifts; it has installed electric traction and hoists most of its output through one shaft.

CHAPTER XI

OVERHAND STOPING ON WASTE IN MEXICO AND AUSTRALIA

Example 24.—Los Pilares Mine, Nacozari, Sonora, Mexico

(*See also Example* 31.)

Irregular Lenses in Porphyry. Two Methods of Sill Flooring.—The mine is at Porvenir, five miles from Nacozari, where there is an immense pear-shaped "horse" of rock 2000 ft. long and having a maximum width of 800 ft. The horse *a*, as shown in Fig. 65, is surrounded by ore deposits *b*, which vary in thickness from 0 to 200 ft. The interior of the horse has been shattered and at the core there are mineralized areas *c*, one of which is 300x300 ft.

The horse is capped by a brecciated, rhyolitic, iron-stained gossan, varying in thickness from 20 to 75 ft. and carrying little or no copper.

Below the gossan is the enriched mineral zone, also of variable thickness, but averaging about 100 ft. In this zone, the copper mineral changes from pseudomorphic chalcocite after pyrite, to pyrite and chalcopyrite with a slight coating of chalcocite. It is seldom, however, that a complete replacement of pyrite or chalcopyrite by chalcocite has taken place.

Below the enriched zone are found the primary sulphides consisting entirely of chalcopyrite and pyrite. Throughout the entire ore body, from the surface to the lowest workings, the character of the deposit is the same; a shattered rock with the ores existing as the cementing material. Both the brecciated rhyolite near the surface, and the brecciated andesite below it contain copper minerals.

There is a dike *x*, *y*, *z*, Fig. 63, approximately bounding the southeastern portion of the ore body, but more and more approximately traveling the center of the ore deposit as it is followed to the northwest This dike of disintegrated diabase has a width varying in different places from a knife edge to about 30 ft.

Near the surface this dike has a slight dip to the eastward, making it a hanging wall of the ore body, but at the 300-ft. level it changes its direction and dips to the west with increasing flatness from about 70 deg. on the 300-ft. level to about 50 deg. on the 600-ft. level. In the northern portion of the deposit, large spurs and splits from the main dike, varying in width from 1 to 25 ft. and in length from small to great distances are found running into the eastern country rock. This dike has caused great difficulty in the mining operations, owing to its vagaries of direction and its tendency to slough and cave from above.

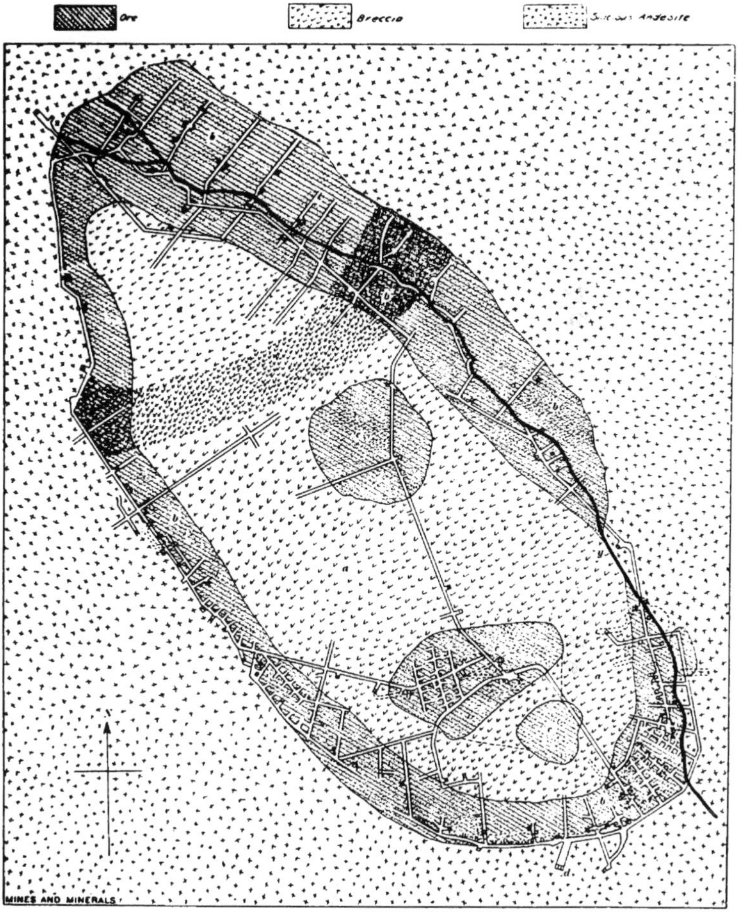

FIG. 63.—Plan of ore body, Los Pilares mine.

From the foregoing it will be understood that the ore body has a definite and fairly regular exterior boundary. The richest ore is apt to lie very close to this boundary, the grade of the ore getting poorer toward the center of the pear-shaped horse until the limit of commercial ore is reached. In what follows, the term "width" refers to the distance from the exterior boundary along a line at right angles to the same to the point at which the commercial ore ceases. This width of ore varies in different parts of the mine, and on different levels, from a few feet to more than 200 ft.

Mining Methods.—Two main working shafts, *d* and *e*, have been sunk in country rock about 50 ft. outside the deposit from which it has been developed. The width and length of the orebody, the kind of ground found in the deposit, the excessive cost of timber at the mine, the cheapness of common labor, and several smaller items, were the considerations which determined the method of ore extraction. The pillar-and-stope method is in use throughout the workings. The whole deposit, or rather the commercial ore area, is divided up into a series of stopes and pillars, the widths of which vary according to the width of the ore and the character of the ground to be worked.

Pillars and Pillar Lines.—The pillars are bounded by imaginary vertical planes extending from the surface to the bottom of the workable ore. Separate maps like Fig. 63 are kept up to date for each level. On each of these maps, the pillars are accurately plotted, thereby showing the location of every stope and pillar, its dimensions, and also the courses of the several pillar lines.

When a stope is to be "sill-floored," the engineer will set pillar plugs on each side of the stope that calls for a pillar. The position of these plugs will be calculated with reference to their distances from their corresponding pillar lines. The distance from the pillar plugs to the pillar will then be given to the stope boss, and it is his duty to see that the pillar line in question is carried forward and the plugs cared for. For a height of two or three slices, these plugs will be changed by the eye by the stope boss; they will then be checked by the engineer and carried on as before. These pillars vary in width from 25 to 60 ft. and are placed approximately at right angles to the country wall. Thus, each stope is bounded on two sides by pillars, while the wall rock on one end and the end of the commercial ore on the other constitute the respective third and fourth sides. When the ore zone is narrow, say from 40 to 100 ft., the above statement applies and the stope will vary from 100 to 150 ft. in length along the length of the ore zone, and pillars vary from 25 to 60 ft. in length also, measured along the length of the zone, depending on the nature of the ground. When the commercial ore zone is much wider, say 250 ft., the length of the stope is either lessened and that of the pillar increased, making a stope 50x250 ft. and the pillar the same size, or, in case of excessive width of ore, a third pillar is introduced running at right angles to the other two pillars and practically making two stopes between one set of pillars, whereas if the ore had not been so wide, only one stope would have been excavated.

Sill Flooring.—The stopes may be "sill-floored" by two methods, *A* and *B*. By method *A* the entire stope area is cut out on the level floor, while by method *B* a floor arch 15 ft. thick is left above the floor level, and from the top of this arch the full stope area is carried up. Method *A* necessitates either a permanent drift in the pillar with cross-cuts running to the stopes and ending in shovelways or chutes, or necessitates the driving of permanent drifts outside the ore with cross-cuts run to chutes or shovelways in the stope; or both classes of drifts may be used for the same stope. Method *A* is generally used in bodies of high-grade ore, or in weak ground. In method *B* a permanent drift must be maintained through the stope and for that reason the drift is protected by the floor arch. This is the method used in the case of wide and long stopes, where the main development drift has been driven along the country wall and must be maintained in order to extract the ore. With method *B* when the ore is 75 or 100 ft. wide, an auxiliary permanent level is driven from the main level through the length of the stope and protected by the floor arch.

Three different methods of stoping are in vogue in this mine. First, square setting; second, overhand stoping on waste of this example, and third, the overhand stoping with shrinkage and delayed filling of Example 31.

Square-set timbering and stoping is well known and also but little used here, so it will not be taken up in detail. It is adopted in soft ground that is liable to cave. After extracting the ore and timbering with square sets, permanent levels may be either driven through the pillar or maintained through the center of the stope by lagging over the sets through the center line of the stope. The stope is then filled with waste up to the top floor of the square sets. Chutes and manways are carried up by lining a given timber set all the way up with 3x12-in. plank and dividing it into chute and manway compartments as in Example 23.

Overhand stoping on waste may be used after a stope has been started by either of the two sub-methods (*A* and *B*) of sill flooring.

Sub-method A.—As an illustration of this system with method A assume a stope *a*, Fig. 64, 50 ft. along the orebody and 200 ft. wide. Corresponding pillars *b* will also be 50 ft. in length. At 50-ft. intervals, cross-cuts *c* will be driven into the stope from the pillar drifts *d* on both sides of the stope in question. Working from the ends of these pillar drifts the entire area of ore within the pillar lines is removed by blasting to a height of 15 feet. This broken ore is trammed over tracks laid through the several cross-cuts from the pillar drifts and by extending temporary tracks from them into the center of the stope. The broken ore is shoveled into cars and taken out via these tracks. After cleaning cut all of the ore, the temporary tracks will be removed. Fifteen feet from the pillar *b* inside the stope, and adjacent to the extension of the pillar, cross-cuts *c*, are built up the chute and manways *e*. This 15 ft. of cross-cut is timbered after extracting the ore. The stope will then be filled with waste from the mill holes *f* to within a distance of 5 ft. from

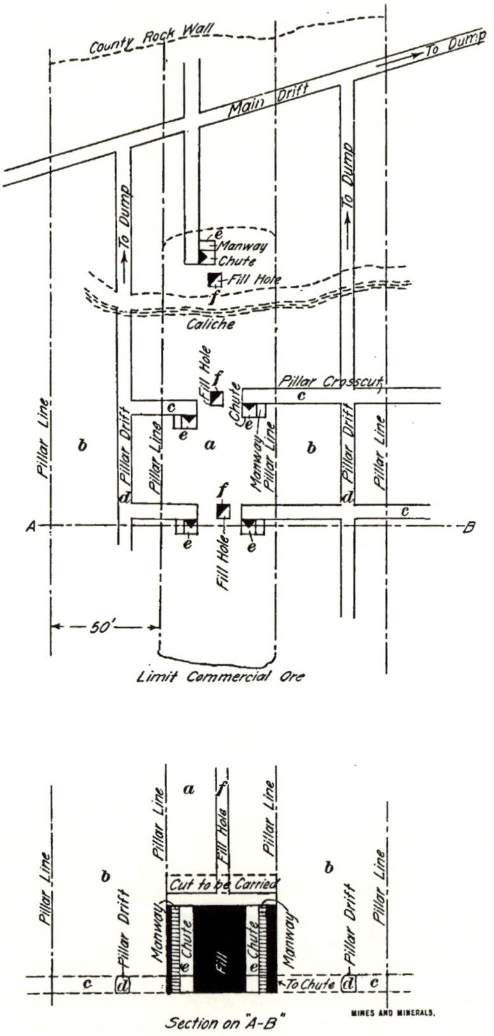

FIG. 64.—Stoping by submethod "A", Los Pilares mine.

the roof, the work of leveling off the waste from the different mill holes being done by shoveling and wheelbarrow work.

These mill holes *f* (see Fig. 64) are adjacent to the ends of the pillar cross-cuts. It is, therefore, obvious that they are raised to the level above simultaneously with blasting out the first slice. They thus serve not only for dumping down waste for filling, but to ventilate the stope.

Chutes and manways are kept built up sufficiently above the waste packing so as to prevent the filling from running into them. The chutes are generally cribbed with 6x6-in. timbers with a 3/4-in. notch on each end. This leaves a 4 1/2-in. opening in the crib between each corresponding pair of timbers. Large rock from the packing is built up around the outside of the chute to keep it in place, and the inside is lined with 3x12-in. plank. The manway, usually about 2 1/2x6 ft., is carried up along one side of the chute and is built out of 2x12-in. plank. A manway is sometimes deemed unnecessary for a chute, in which case the latter alone is carried through the fill. In the manway, a 3-in. diameter pipe is placed with its top level with the top of the manway and its bottom about 7 or 8 ft. above the floor. This serves the purpose of letting drill steel down without cutting the manway lining and breaking ladders which would result from throwing down the steel. After the filling has been leveled off to its proper height, another slice varying from 6 to 12 ft. in height at the breast is blasted from the stope. This slice is started by driving blind raises to the proper height and enlarging these till they intersect, this method enabling the drilling of flat water holes. It is often the case that the same slice will be started in three or four different parts of the stope from blind raises. The ore broken in this method of slicing is put into the nearest chutes by shoveling and wheelbarrow tramming. After the slice has been finished, the stope is cleaned of its ore and waste filling is run in again and the same procedure followed as before.

Sub-method B.—When the ore shoot is narrow, say 30 to 50 ft. in width, for quite a length along the country wall, and the ore is of comparatively low grade, the same method of extraction is used, but with the sill-flooring of method B. For this the main drift is usually protected by a floor arch, as shown at *i*, Fig. 65. The main drift is generally run along the country wall or close to it as at *b*. Cross-cuts *c* are driven every 30 or 40 ft. at right angles to the main drift in the ore, the two opposite term nal ones being on the pillar lines. If the drift to be maintained is in the ore, then these cross-cuts may be driven at the stated intervals on either side of the drift, generally alternating from the left to the right side. Offset from each cross-cut and set out 8 ft. from the center of the main-drift track a 6 ft.-sq. raise is driven to a height of 22 ft. After raising in each cross-cut the stated height, intermediate drifts and cross-cuts 7 ft. high are run from each raise making the floor of each drift 15 ft. above the floor level. These intermediate drifts are connected together and the first floor of the stope is then excavated by enlarging them by b!asting both ways till they intersect. All the ore between pillar lines is extracted to the country rock wall in one direction and to the commercial ore limit in the other. Should the ore prove to extend for 20 ft. or more in width from the center of the main drift, an 8-ft. pillar is left between the main drift and the ore, and an auxiliary drift *d* from the cross-cuts is driven parallel to the main drift. From this auxiliary drift the ore is sliced back, the broken material being trammed out of the various

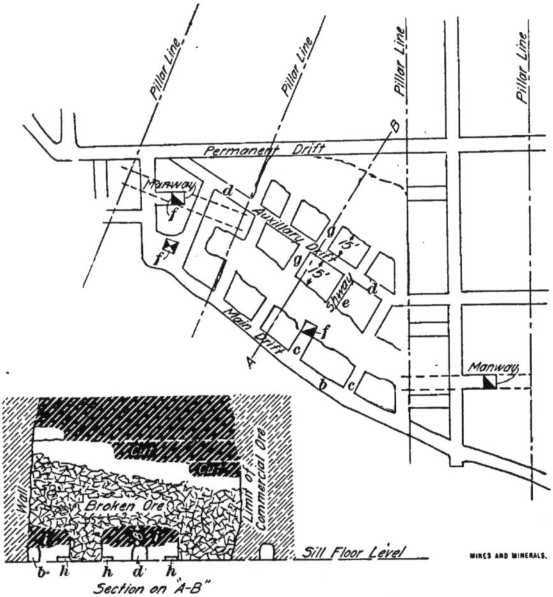

FIG. 65.—Stoping by submethod "B", Los Pilares mine.

cross-cuts first driven from the main haulage drift. This work leaves a solid arch of rock 8 ft. wide on each side and 8 ft. thick over the main haulage drift to protect it. When this method is followed chutes are cribbed up from convenient points about 50 ft. apart. For such small stopes two manways *f* usually suffice, which are placed adjacent to the two chutes at opposite ends of the stope.

After the stope is completed and cleaned of ore, waste is let into the excavation and operations proceed as in the previous case. If the width of the ore is too narrow to admit leaving the ore pillars to protect the drift, it is, of course, obvious that the first system must be followed, after which the drift is timbered and waste filled in over it.

For the filling of stopes one main fill hole is raised to the surface from a level located 100 to 200 ft. above the stopes to be filled. This main fill hole then serves for from three to five stopes or more. From the stopes to be filled, up to the level on which this main fill chute is located, from two to four fill holes are driven for each stope. These fill holes are placed either beneath the center of the track or inclined over to the center from one side. Bearer timbers are laid over these holes and the track run over the bearers. A 5-horsepower electric motor, with a train of 10 cars of 1 ton capacity pulls the fill rock from the main fill chute to the particular stope and fill hole where waste is called for. In this main fill hole at a distance of 75 to 125 ft. below surface (distance varying according to the point of approach or entrance) a grizzly station is cut and a grizzly put in over the fill hole. This grizzly is variable in size, ranging from 8x20 ft. to 10x10 ft. The mouth of the fill hole is chambered outto a convenient size on a side away from the hole leading to surface. Bearers of 12x12-in. timber spaced about 3 ft. apart are placed across this hole. On them are bolted 40-lb. rails so as to form openings of 15 in. square. All rock falling on the grizzly must be broken fine enough to pass through these openings. When boulders come down larger than 15x15 in., they catch on the grizzly and are broken up. This prevents the hole from "hanging up" between this grizzly point and the chute below, and also eliminates any trouble at the chute in the loading of the cars.

Ore Transportation.—On the main tunnel level, and located so as to reach all stopes advantageously, six ore pockets or bins, similar to Fig. 66, have been cut out of the rock within the ore zone, each having a capacity of from 1,000 to 10,000 tons. Each one of these bins is provided with from one to three sets of two chutes each, one set of chutes filling a 30-ton Ingoldsby bottom-dump ore car. Two 10-ton General Electric traction motors running in tandem pull a train of from six to eight of these cars to the main tunnel mouth, Porvenir, which is the terminal of the mine railroad. From here a 60-ton Baldwin locomotive takes a train of 14 cars to the concentrator at Nacozari. On each of the succeeding levels above the main tunnel level as far up as the 200-ft. level, continuous connections have been made to each ore bin as shown in Fig. 66. With dump stations provided for each bin on each mine level, the ore from each working finds its way into the nearest dump. This does away with the hoisting of the ore, a costly item.

Compensating Mexican Labor.—All work underground is done by contract. All development work such as sinking, raising, drifting, and cross-cutting, is contracted to the native Mexicans at so much per foot driven; the company furnishing steel, powder, fuse, caps, etc., the contractor only having to keep his working up to regulation size and in some cases running his dirt to the chute. In case of encountering waste in a working, it will be dumped into a fill stope. Prices of the work per foot vary with the kind of rock and also depend on whether the drilling is preformed by machine or hand. In the stopes both machine and hand drills are used, the miner being paid so much per foot drilled with a machine, or so much per foot with hand steel. In the stopes the supervision of the holes is not limited to the number drilled but they must be drilled as "pointed" by the stope boss to the stipulated depth. Car men receive so much per car trammed, the price varying with distance traveled and whether the ore is shoveled from sheet iron, or a rough

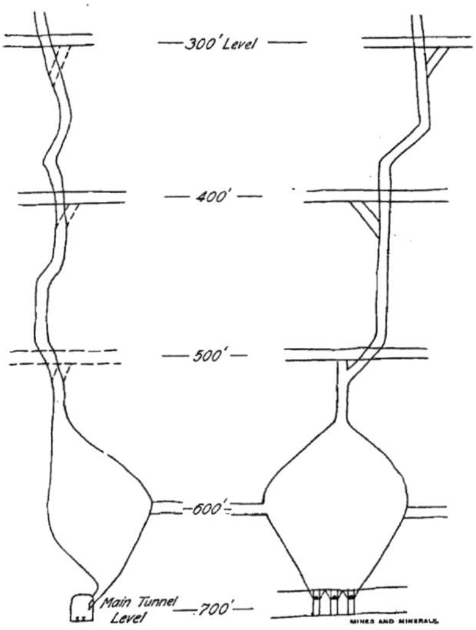

FIG. 66.—Ore bins and chutes, Los Pilares mine.

bottom, or drawn from a chute. In the stopes, shovelers dump wheelbarrows into chutes and are paid by the number of cars drawn from the chutes, which are counted up at the end of the day. As a check, the height of the ore in the chute is taken before starting to work and after finishing; the number of inches in the chute to the car being known. In filling stopes men are given a task of so many wheelbarrow loads per day for a certain wage. All wheelbarrow loads over or under this number are figured and paid for in proportion of the task to the wage. About 1,200 men are usually on the monthly pay roll with an average daily working force of 800 men. From 1500 to 2000 tons of ore is sent to the mill daily from the whole mine.

Example 25.—West Australia

Sub-vertical Veins in Crystalline Schists; Rill Chutes.—When working under ideal conditions, the ore body, if continuous, is divided into blocks by equidistant levels, about 200 ft. apart, and by equidistant winzes on the hangwall side of the lode at 150 ft. to 200 ft. apart. The reason for having the winzes on the hangwall side is that the stopes are thus filled with the least handling. The diagram of Fig. 67 shows how a mine is thus blocked out. It is assumed in it that only one winze is connected with the surface as a pass for filling, but the number of through passes is purely a matter of convenience.

The filling usually consists of fresh residue, which may be of sand, or of roasted "slimed" ore or of raw "slimed" ore, and which may contain up to 25 per cent. of moisture. If the residue contains too much moisture there is a danger of it clogging the passes, so that sometimes it is necessary to stack it on the surface for a short time previous to delivery to the mine, and by this means also much of its residual cyanide content is destroyed. No chemical treatment whatever of the residue to destroy its cyanide contents is practised in Western Australia. The residue received from the surface can be distributed to the various winzes by means of a belt conveyor system along a disused level above the stopes.

The methods of stoping and filling on the rill system are as follows:

On any particular level a leading stope is taken out below the ore to be mined, when the drive is timbered usually, either by single stulls, or when the lode is over 14 ft. wide, by saddlebacks, at intervals of 5 ft. The latter consist of pairs of stulls sloping toward one another like the rafters of a roof, and bearing upon a longitudinal ridging of sawn timber 2 in. thick. The stulls are lagged with poles about 4 in. in diameter of a local wood called gimlet wood, or with old iron pipes. The lagging is in turn covered with old filter cloths, or the sides and linings of cyanide cases or any other inexpensive material which serves to prevent the residues from falling through.

Two alternative methods of stoping the ore are shown at *A* and *B* on the diagram of Fig. 67. In the former all the holes are "down" holes and can be drilled wet, which is an important consideration in view of the necessity of reducing dust production. In this method the benches are taken out at an inclination slightly flatter than that of the natural slope of the filling which in the case of residues is about 45 deg.

In the less common method at *B* the ore is mined by a series of horizontal cuts, and some of the holes drilled must be "uppers" and drilled dry.

Usually during the timbering of a drive ore chutes are put in at distances of about 50 ft., one (P) midway between the Winzes (*W*), the others (*Q*) being intermediate. As stoping proceeds passes about 4x4 ft. in the clear are built above these chutes, usually of 7-in. logs, but sometimes of 9x3 in. sawn timber. Each winze is also "cribbed" up except when it will not be required later on for passing "filling" to lower workings or for ventilation. In such a case the timbering of the level below the winze can be closed up and the winze filled up.

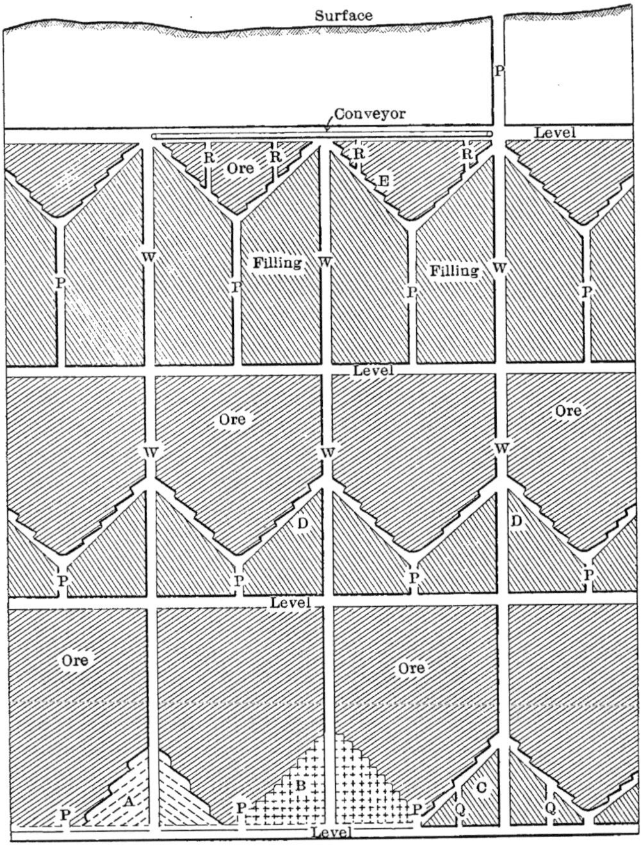

FIG. 67.—Stoping system, West. Australia.

The breaking of the ore and the filling of the stopes with residues succeed one another alternately. Before the benches of ore are blasted eucalyptus saplings or slabs are laid on the sloping surface of the residue filling. These serve to keep separate to a great extent the broken ore from the residue, and assist in its "rilling" into the passes, very little labor being then required. The passes are then built up close to the working faces and covered over to prevent residues from entering them, and the poles or slabs are removed. More residue is then dumped down the winzes into the stopes, filling them up to a convenient distance from the faces. When a stope has assumed the appearance shown at *C*, when all ore can be rilled to the passes P, the intermediate passes *Q* are no longer entirely necessary. As stoping proceeds the appearance of the stopes becomes similar to that shown at D. It is usual to stope a series of blocks on the same level simultaneously, so that the filling of the stopes with residue on both sides of the winzes and the building up of the passes can be carried on symmetrically. When the stopes are nearly beaten out, as shown at *E*, it is usual to sink subsidiary winzes *R* in the triangular blocks of ore left below the level above, through which the residues for filling can be dropped. The lodes in Western Australia are usually very steep, and this rill system is generally particularly applicable to them. Generally, as the dip of the lodes decreases below 45 deg., more and more shoveling is necessary to assist the rilling, and the method becomes inapplicable when the dip is less than about 35 deg.

Filling with residue in Western Australia has been in use for about 13 years, and its present cost is about 10d. per short ton of ore mined.

Example 26.—British and Other Mines. Broken Hill District, N. S. W.

(*See also Examples* 27 *and* 28.)

Sub-vertical Vein in Crystalline Schist. Rill Chutes. Cribs under a Weak Back. *Sampson System.*—The Broken Hill district is situated in a desert country about 300 miles west of Sydney. The great silver lead lode occurs in the Barrier range which runs north and south through the Tertiary plane north of the Murray river and its outcrop forms a narrow rocky line of hills about 1 1/2 miles long. The main lode is 270 ft. thick in places (average about 60 ft.), and it stands nearly vertical and forks in depth into two slightly diverging branches separated by a horse of wall rock. The present primary ore, now mined in depth, consists of argentiferous galena and zinc blende in a gangue of garnet, rhodonite and fluorite of varying hardness and texture and lies between walls of garnetiferous gneiss. The surface oxidized ores have been extracted mostly by huge open pits. The debris from these is now allowed to descend the raises to fill the stopes in the present underground workings. Mill tailing and sand are also used for filling.

The square-set system, using timber from Puget Sound, was in vogue for extracting the enriched oxidized ores, which extended to a depth of 300 to 400 ft., but for the leaner sulphides below methods of stoping which are cheaper in timber and less liable to conflagrations have been introduced. At present, square-setting is only used in the ore too friable to stand by itself and as an auxiliary to the systems of Examples 25, 26, and 27 to be described. The Sampson system of the British mine is worked in the following way: After the cross-cut from the shaft cuts the lode, it is driven across to the further wall and then opened out on each side till the ore is all taken out on the sill floor from wall to wall to a height of about 11 or 12 ft. The face is carried along in this operation by taking out the ore in two stages, the bottom 5 or 6 feet by drilling from the sill, then rigging the machine on a low bulk, or crib, to drill the upper 6 feet. As the face progresses, the drive timbers of the level are put along the footwall side, the leading set always far enough from the face to be unaffected by the blasting. With the exception of cribs here

FIG. 68.—Cribbing back of stope, British mine.

and there the back will stand unsupported. The cribs, built up from the floor and also from the drive timbers, as shown in Fig. 68, are built up with 10x10-in. Oregon timbers, 6 ft. or 7 ft. long. When within a foot or so of the back, long stringers of 10x6 in. or 10x10 in. are cantilevered out (if necessary on every side) to take in any bad ground near top of the crib, the top is blocked and wedged tightly against the rock back, and the whole crib finally tightened by driving in wedges between its lower timbers, with a spalling hammer.

The drive timbers are shown at the right in Fig. 68, and by the drawing Fig. 69. The sills and struts are 15 ft. long, taking three sets spaced on 5-foot centers. The sills butt midway between the sets, while the struts meet over a leg. The leg is tennoned into the sill and strut, and below the sill at each leg a good block is put into the solid bottom. The cap is jointed into the struts and legs. Along each strut a 15-ft. 10x10-in. stringer is laid, thus giving 20x10 in. of strut timber; these stringers being properly spread by a 10x2 in. spiked to the top of the cap while 10x6-in. top laths, laid close together from stringer to stringer as shown, give support to the filling placed on them. It will be seen that there is an 8-in. space below the top laths, this being intended to prevent the weight of the filling from coming on the cap, throwing it instead on to the struts and legs. The weight is thus kept to the sides of the drive, the cap merely taking part of the side pressure. On the outside of the legs a vertical 10x4 in. takes the 10x12-in. horizontal lagging, 15 ft. in length. The reason for this 10x4 in. is, that when the stope is high the weight of the filling becomes so great that it is necessary to put in extra legs to support the strut between the sets, and as the side pressure causes the lagging to bulge inward, the 10x4 will allow it 4 in. play (this has been found no more than sufficient) before the lagging trespasses the plane of the back of the legs. Thus, extra legs can be put in without cutting away the bulged-in lagging or coming within the drive space. The lower 10x4-in. spreader butts 2 in. on the leg. and 2 in. on the sill, is well packed underneath, and its top side forms, with 10x2-in. pieces laid from sill to sill, the floor on which the rails are laid. Outside the sets, over the whole of the sill floor are laid, parallel to drive, 10x4-in. pieces, 15 or 16 ft. long, butting against one another, on which the filling is laid. It will thus be seen that when coming up with a stope from below, the miners will have no timbers shorter than 15 ft. long to catch up, when the last bit of ground is being taken out.

At intervals of from 80 to 100 ft. cross-cuts are put out from the main drives along the foot to the hanging wall. At 30-ft. intervals along the lode drives, the two-compartment timber sets, Fig. 69, are put in butting on the drives. A set forms the bottom for the ladderway and ore chute, built up from the sill as the height of the stope increases, the top set being kept level with the filling. The sets of 10x10-in. pieces are laid one upon the other and held together by the filling. When a stope has been carried up about 50 ft., the ore chute is by then very much worn, and so the ladderway and chute are interchanged.

When all these timbers are in on the sill floor, the filling is run in all around them and the stope filled to within a few feet of the rock back, then starting from the winzes (sunk about 100 ft. apart), another horizontal strip is taken off in the same way for 10 or 12 ft. in height, the back being supported where necessary with cribs. The waste filling sent down the winzes from the level above is built out around the timbering by trucking from the winze chutes in cars running on temporary tracks extending from the winzes. When the filling comes to a crib, another low crib is built on the new filling as near as possible to the old one and the latter then pulled down, and the crib timbers are used over and over again. The chutes and ladderways are built up level with the filling as the stope progresses as before stated. In this way the ore is taken out from wall to wall till the back is 60 ft. above the sill.

It is then not considered safe to carry the back further up by horizontal stripping, but instead, starting from the winzes, the ore is taken out in diagonal strips or "rills," illustrated by Fig. 70, the slope of the back being approximately that of the filling's slope of rest. By this method of stoping the only weak place is the part where the top of the stope meets the level above. This point has to be well supported by an extra number of cribs, also the points over the edge of the filling which will by

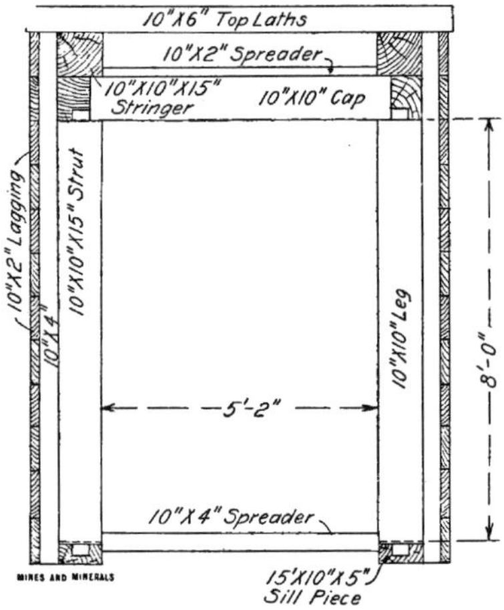

FIG. 69.—Drift timbering, British mine.

this have reached within 6 or 7 ft. of the level. The old 10x4-in bottom timbers are caught up by a line of cribs spaced on about 10-ft. centers extending from wall to wall, the bottom timbers being directly supported by long 10x10-in. stringers laid across the tops of these cribs. In this way the old bottoms are all supported by the cribs, which are left in, and the filling shoveled up around them as close as possible. Thus, the ore is all taken out, even the last pyramid-shaped piece of ground practically unsupported by the walls not presenting any special difficulty. In the British mine this method has been very successful. The lode there does not attain as great a width as in some of the other mines, its greatest width being 130 ft., while the average width is not more than 70 or 80 ft. The walls are both firm and the ore throughout of a consistently compact character, so that no other method has been required. The filling used which was formerly mill tailings is now zinc plant residues. Placed in a damp condition and having greater mobility than waste, it sets better and finds its way into the far corners, so that shrinkage is reduced to a minimum. On these accounts the British mine has mined cheaper than any of the others, the underground costs for some years going about 9 shillings per ton.

In all the other mines open stoping with temporary cribs has been adopted, but the other details usually have been modified, In some cases, struts in the drive sets are short, only reaching from leg to leg, and the sills are put across the drive instead of parallel with it. In the Block 10 mine, when coming up under old bottoms, the whole of the last 40 ft. is taken out by square sets started off the filling, the old bottoms on level above being caught up from the top sets.

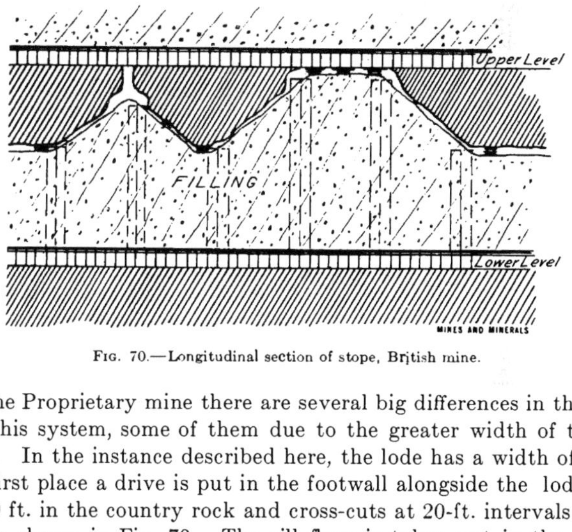

FIG. 70.—Longitudinal section of stope, British mine.

In the Proprietary mine there are several big differences in the working of this system, some of them due to the greater width of the lode worked. In the instance described here, the lode has a width of 200 ft. In the first place a drive is put in the footwall alongside the lode about 20 or 30 ft. in the country rock and cross-cuts at 20-ft. intervals off this drive, as shown in Fig. 70. The sill floor is taken out in the manner already described, and eventually these cross-cuts tap the main timbered drive which runs down the middle of the lode. The drive in the country is an advantage in case the timbered drive collapses at any point.

The chutes and ladderways, formed by a double square set, are placed about 30 ft. apart along the drives. The double set is close lagged all around outside with horizontal 10x2-in. pieces, and the inside of the ore-chute set is lined with vertical 10x4-in. hard wood (stringy bark) butting and spiked at the caps and struts of the sets. This is the only case where other than Oregon timber is used below ground. The hardwood makes an excellent chute lining, as the wear is very slight and the polished surface soon acquired is of benefit, whereas the Oregon timber is soon cut away and the chips give trouble in the concentrating mill. If one of the lining pieces is knocked away, the outside 10x2-in. lagging prevents filling from running into chute. It should be mentioned that during the last few years the Proprietary has used for the major part of its mine filling the sand residues from the zinc plant. These necessitate close lagging; where spaced lagging is mentioned, the filling is waste.

On account of the width of the lode, part of it, in section somewhat less than half the width as shown in Fig. 71 is left as a solid pillar except where cross-cuts, etc., cut it on the sill floors and only the lode on the

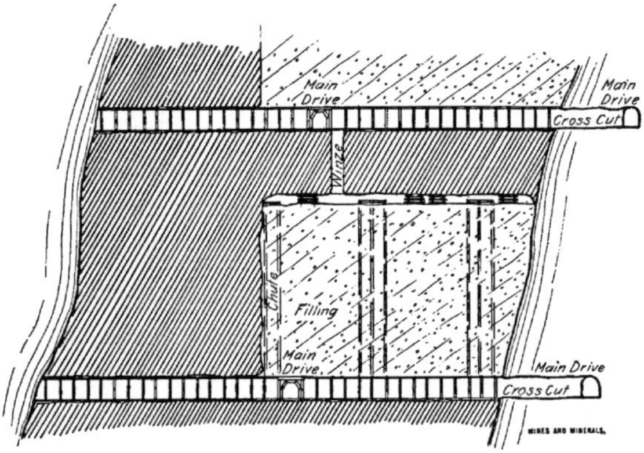

Fig. 71.—Cross-section of stope, Proprietary mine.

footwall side is worked out for the time being. A line of chutes forms the boundary of the stope at the pillar. The stope is taken up in the usual way by cribs, etc., till the back is 75 ft. above the sill. When the stope is being filled, vertical 10x4-in. timbers are placed 6 ft. apart against the pillar and across these as the filling rises are placed 5x2-in. "paddock laths" a few inches apart, the intervening spaces being covered with waste pieces of candle boxes and saw-mill "flitches."

The remaining 25 ft., up to the level above, is removed in the manner illustrated by Fig. 72.

At each of the waste winzes, which are about every 100 ft. apart, a cross-cut, 16 ft. wide by 8 ft. high, untimbered except for cribs, is taken out across the block from wall to pillar and is then tightly filled from the winzes. On this filling another cross-cut is driven about 8 ft. high and 7 ft. wide and also filled, the sides being first chamfered down to the sides of the 16x8-ft cross-cut. Above this cross-cut another one 8x7ft. is driven, this time timbered with "clap-me-down sets," which catch up and support on their tops the bottom of the level above as shown in Fig. 72. These sets are now filled up as close to the under side of level as possible, only the set alongside the pillar being always left unfilled. In filling the sets, a 10x2-in. board is put along each side of the drive against the bottom of the legs, and the struts take the ends of vertical paddock laths, as before explained, the intervening spaces between filled as before with candle-box pieces, which serve to hold the filling. Stoping across the block now ceases, but is resumed in the direction of the lode's strike, by blasting down with a sloping breast (Fig. 72) from the side of this top run of sets to the under side of the 25-ft. deep block. The breasts, which extend right across the block

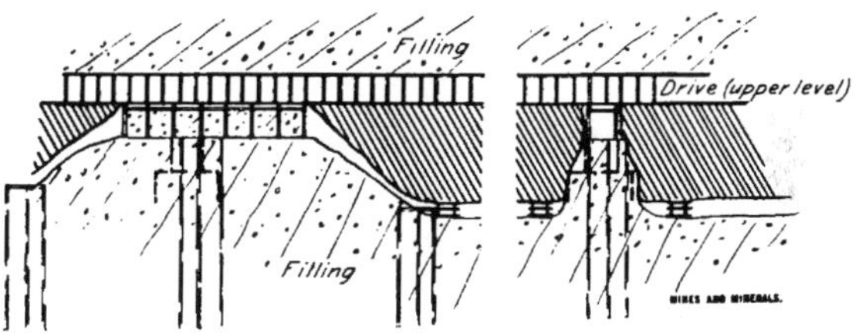

FIG. 72.—Long. section of stope, Proprietary mine.

and are worked from chute to wall, are carried along from one cross-stope to another till the block is worked out. As in the case at the British mine, the top of the stope is the weak part, and has to be well supported by the sets, cribs also being put in under the back whenever necessary. When the breast has advanced sufficiently, another cross-run of timber sets is put in alongside the first. The filling for the first sets now has to be dumped down where the drive above crosses the top of the stope. The set next the pillar, as before, is always left open, so that when the footwall block is worked out, there is just below the level alongside the pillar an open drive of sets which connects all the middle ladderways and chutes. By this means, openings are left for attacking the pillar which will probably be removed in a somewhat similar manner, depending on conditions existing when the time (not yet arrived) comes for working them. If, at any place, the breast is too broken to be worked safely by the open method, the lower face of the block is caught up on cross-cuts timbered with "clap-me-down sets," stepping them up and backward to the top run of sets.

In some cases, instead of working out the ore to a certain height by horizontal slices, it is worked out by the sloping breasts entirely. This method has two great advantages in that nearly all the drill holes required are down holes and the waste being on the slope, the filling is easily and cheaply done. Where the ore lies in flat layers, the back will probably support itself better when worked by slopes than when it is mined in horizontal slices. On the other hand, it is more difficult to support the sloping back by cribs. A place has first to be prepared on the waste sill for laying the first timbers, and the sloping back does not come squarely on the top of the crib. The waste probably does not set so tightly on the sill as in other cases where much trucking has to be done, but still where the walls are firm and the back is good and does not require much support, this method can be advantageously adopted. It has been used extensively in the South Mine where conditions are favor able. Here an extra number of ore chutes are put in, and the cribs are often built up from the tops of some of these.

EXAMPLE 27.—PROPRIETARY MINE, BROKEN HILL, N. S. W.

(*See also Examples* 26 *and* 28.)

Sub-vertical Broken Vein in Crystalline Schist. Cross-cutting in Panels.—In the Proprietary Mine a class of ground was met with in which the hard sulphide ore was found to be broken up into big and little boulders. It was impossible to work this by the open method, and if square sets were used, the sudden loosening of a big boulder would be likely to knock over a dozen or more sets, with disastrous results. A system was therefore adopted in which as small an area of the back as possible was left unsupported while working out the ore. The piece of ground worked in this way was approximately 60x60 ft. To illustrate this method we will consider that one floor shown in Fig. 73, is just finished, *A*, B, and *C* are the jump-offs of ladderway and chute, while *M* is the waste winze. *A* is first raised one floor and a drive started off toward B connecting on its way with *C*. When drive *A* B is complete a cross-cut is put out to the waste chute. Two men are then started at say *a* to cross-cut toward the wall, and when about a third of the way in, another pair start at say *b*. When *a* is two-thirds of the way in, a third party starts at *c*. When the cross-cut *a* reaches the wall, the miners are withdrawn to start another cross-cut and the filling gang is put in. This consists of two truckers and one shoveler, the three working on contract, together with one man on wages to look after the packing. The cross-cut is filled in a manner similar to that previously described. The filling is shoveled up as tightly against the rock back as possible, and by the time this cross-cut *a* is filled, cross-cut *b* will be ready and so on. Toward the end the corss-cuts will be put in alongside the filled ones. Usually

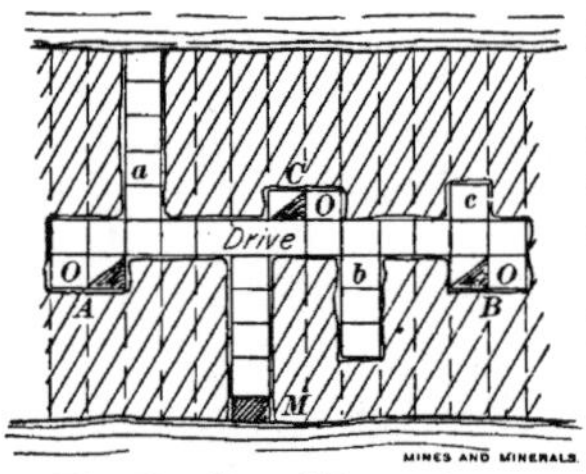

FIG. 73.—Plan of Cross-cut stope, Proprietary mine.

the cross-cuts on the *A* side of the waste chute are finished before those on the other side, and while those on the *B* side are being finished, the drive on the *A* side is filled up, the jump-off *A* is raised a set and the drive again started on floor above toward *B* as before. When the cross-cuts on the *B* side are all filled, the drive is filled and lastly the waste cross-cut to *M* is filled and thus a floor about 8 feet thick is completely taken out.

The drive sets used in this method are shown in Fig. 74. The legs of the set are about 7 feet long, are tapered, being 10x6 in. at top, and 8x4

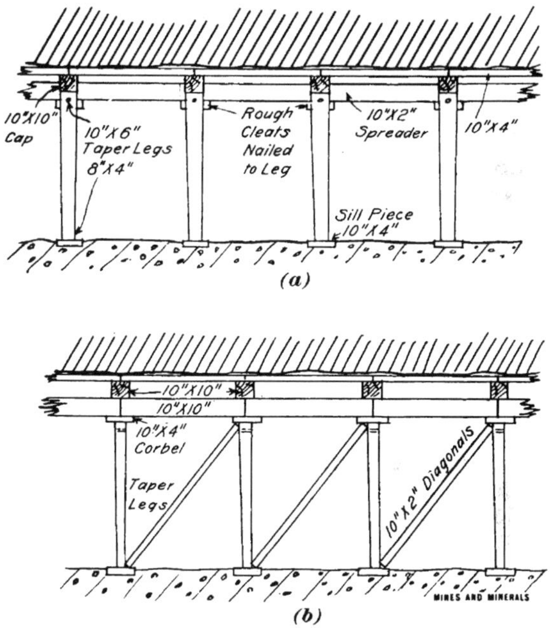

FIG. 74.—Timbering of cross-cut stope, Proprietary mine.

in. at bottom and have a hole of 1-in. diameter about 6 in. from the top. These legs foot on to a 10x4 in. scrap piece about 18 in. long. The cap is 10x10 in., its ends simply laid on the tops of the legs, while across the caps is laid whatever timber is necessary to support the rock back (this at times is a fairly high crib). A 10x2 in. spreader butting half on the cap and leg is held up by rough cleats nailed to the top of the legs. When the cross-cut timbers are put in, the drive caps are extended along the sides of the cross-cut, becoming struts, as shown in Fig. 74 (*b*), and butt over a 10x4 in. corbel on top of the leg. Across the struts, 10x10 in. caps are laid with timbers on these to catch up the back. No spreader is used in the cross-cut, and where the pressure at end of one is considerable, diagonals are put in between sets as shown. The sets are approximately 6 ft. wide and 6 ft. apart, but there dimensions are not strictly adhered to, as the sets are only of a temporary character. If a boulder bulges into the drive the set may be made narrower to suit, or if available timbers are longer or shorter than necessary, the sets are built to suit, the size of the sets being always subordinate to the ground and timber. When opening out a new floor, some of the timber is recovered from the floor below usually all the caps and some of the legs and other parts being saved, the hole in the tapered leg being used to put in a drill or bar to assist in withdrawing. About two-thirds of all the timber used is recovered. In this system it is desirable to drill by hand labor, using shallow holes and light charges, on account of the unstability of the ground. By the above means, this difficult ground is not only safely worked, but at a cost little more than for ordinary ground with square-setting.

CHAPTER XII

OVERHAND STOPING WITH SHRINKAGE AND DELAYED FILLING

Example 28.—Central Mine, Broken Hill, N. S. W.

(*See also Examples* 26 *and* 27.)

Sub-vertical Veins in Crystalline Schists.—Auxiliary Cribbing and Square-setting.—In this mine the lode reaches its greatest and most constant width, and it was not considered feasible to work the immense orebody, 2 per cent. solid ore, by the ordinary systems in vogue at Broken Hill. The method adopted, as illustrated in Fig. 75, was to divide the lode up into 50-ft. sections by vertical planes running across the lode and working out every alternate section as a stope, leaving the others as pillars ti.l the stopes are worked out and filled. These nlled stopes act then as pillars while the former pillars are worked out. This is carried out in the following way: A main drive, timbered by 8x6x6ft. sets is first driven approximately down the center of the orebody to certain section lines and this drive afterward connected by cross-cuts to a waste drive in the footwall.

At every stope block the whole of the sill floor is taken out, using square sets from wall to wall. A winze is sunk from level above about in the middle of the side of each stope, half the winze in the pillar and half in stope, as shown at *W*, Fig. 75. The cross-cut, or gangway, timber sets next the pillars are left open and the row of sets joining the ends of these are also left unfilled. All the other inside sets are filled with waste excepting the chute sets which are started in the rows next the gangways. The stope is then taken up from the top of these sets by the overhand stope and crib method of Example 26 with the difference that the gangway row of sets is carried up on each side of the stope and left unfilled, acting as a barricade to keep the waste clear of the pillar, and also as a place from which the pillar may be attacked. The rock pack of the stope is given a slight slope downward from the winze side to the opposite side. When the back has a height of 60 or 70 ft. above the sill, square sets are started on the waste, and the rest of the stope is taken out in this way. So great are the ore resources of this mine that only a few of the stopes have been worked out and no necessity has yet arisen for working the pillars on a large scale.

The intended method for working them out, however, is indicated in Fig. 75, at *M*. Starting from the hanging wall, the part furthest from the country drives and the shaft, a row of sets is put in against the wall from one filled stope to next, and this row is carried up from one level to the one above. This is done from every level and at every pillar simultaneously as nearly as possible. These sets are all filled except the last set of the row on the winze side. This one is made the waste chute from which to fill the next row taken out. In this way the pillars are gradually sliced away from top to bottom by vertical strips parallel to the drives, working from hanging to footwall till only worked-out ground is left behind. Enough work has been done in this way to

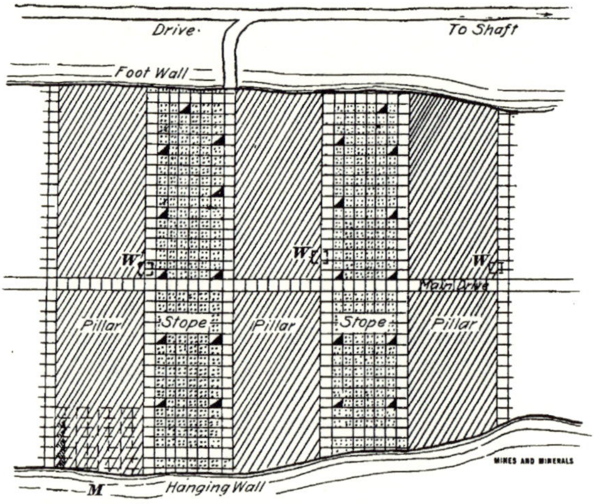

FIG. 75.—Plan of stoping, Central mine.

demonstrate that under ordinary circumstances the pillars can be successfully worked in this way, but on account of the heavy moving ground characteristic of the lode at this part of it, it is probable that the system when worked on a large scale will have to be modified. The room-cavieg system of Example 46 seems to the author more applicable for mining these pillars, as slicing horizontally could be more easily controlled than slicing parallel to a steep and heavy hangwall.

EXAMPLE 29.—KING MINE, GRAHAM COUNTY, ARIZ.

(*See also Examples* 22, 30 *and* 40.)

Irregular Lenses in Porphyry: Auxiliary Back-caving and Underhand Stoping.—The two lenticular oreshoots of the mine are 700 and 500 ft. long respectively and fill a fault-fissure in a granite porphyry hill. The faulting has been severe, but in the absence of any sedimentary rocks, the amount of displacement cannot be determined. The ore is chalcocite and chalcopyrite in a gangue of brecciated granite porphyry and varies in width up to 30 ft. The vein dips at an angle of 70 deg. and the walls are strong and well defined. The steep slope of the mountain permits of the vein being worked from adit levels, the lowest of which gives a vertical depth of 600 ft. below the outcrop.

Haulage Roads

Main haulage roads are driven in the foot- and hanging-walls, parallel with the vein, but at a distance of from 15 to 20 ft. from it. From these roads, cross-cuts are made at intervals of 25 ft., those in the hanging-wall being staggered or spaced midway between those in the foot-wall as in Fig. 76.

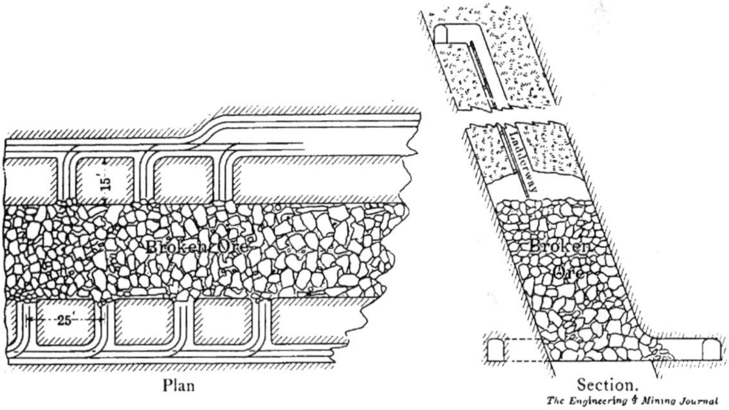

FIG. 76.—Stoping at King mine.

The ore is then broken from wall to wall for the whole length of the ore-shoot. The broken ore is at first shoveled out, but as stoping progresses it is allowed to accumulate, sufficient being removed to allow a working space of 6 ft. between the broken ore and the roof. Two-thirds of the broken ore is left till the stope is worked out, the ore serving as a working floor for the miners and also prevents caving of the walls.

Overhand Stoping

Access to the stope is obtained from raises made in the roof at intervals of 100 ft. and connected to an upper level. From these raises, the roof is broken in horizontal slices of from 10 to 15 ft. in thickness. As the miners work outward from the raises, the sag or belly of ore between generally breaks off, leaving the roof sufficiently arched to allow the block to be broken from on top. Large horses of hard, barren ground frequently occur in the vein and these are left in as pillars, to support the walls.

Occasionally, parts of the vein are too soft to be mined safely by overhand stoping and the mode of attack is changed. From the two raises between which the ore is softer than usual, a drift is made 20 to 30 ft. above the back of the stope and connecting the raises. Midway in this drift down-holes are drilled in the floor and sides. As these holes are blasted and break down the shell of ore between the floor of the drift and the stope, mining is continued back to the raise until the whole of the shell has thus been broken by underhand stoping. In using this method, the roof of the drift which connects the raises must be sufficiently high to allow the handling of the long jumper drills needed in breaking down the floor. When approaching an upper level, the ore is always broken by underhand stoping.

When the top of the orebody is reached in stoping, the remainder of the broken ore is drawn off through the cross-cuts. A certain admixture of wall rock and ore is unavoidable when the last portion of the ore is drawn, but this is easily removed on the sorting platform over which the ore is passed. The switches at each cross-cut on the road-way allow the shovelers to load their cars without interfering with the haulage. The ore is sledged and loaded by contract, and when a car is full it is pushed into a side track, where the mule train is made up. A few miners are required to block-hole the larger pieces of ore as they appear at the shoveling openings. The advantages of this method are obvious; one worthy of special notice is the security in which the shoveler works.

Example 30.—Coronado Mine, Graham County, Ariz.

(*See also Examples* 22, 29 *and* 40.)

Irregular Lenses in Porphyry; Auxiliary Back-caving of Rooms, and Subsequent Pillar-caving.—This mine, one of the most important holdings of the Arizona Copper Company, lies on the southern slope of the Coronado mountain, a granite *massif*, whose precipitous sides form a conspicuous landmark in the district. The great Coronado fault is at the base of these granite bluffs. It strikes east and west and can be traced for a length of two miles. Its movement has been downward and westerly at its eastern extremity, resulting in a vertical displacement of 1200 ft. between the basal quartzite on the south of the fault and of that resting upon the granite on the north.

The vein, which fills the fault-fissure, is followed on the south side by an intrusion of fine-grained green diabase, varying in width up to 70 ft.

The Coronado oreshoot is approximately 2000 ft. long and will average 35 ft. in width. It has been opened by a three-compartment shaft to a depth of 700 ft. The vein is practically vertical. The north or foot-wall is of slightly altered granite and the south or hanging-wall is of quartzite to a depth of 150 ft., below which the vein enters the granite fissure. The zone of sulphide ore is reached at a depth of 250 to 300 ft.; above this level, small bodies of oxidized ore have been found.

The ore of the sulphide zone is chalcocite, in some places entirely replacing and in others forming a coating on pyrite and chalcopyrite. The gangue consists of crushed and altered granite and diabase; in this respect the vein differs from most of the others of the district. Horses of granite are occasionally found in the vein, the outer shells of which will be typical ore, gradually merging to an interior of slightly altered granite, showing no line of demarcation.

Mining Methods

The orebody is contained between walls of granite, the foot-wall is exceedingly hard and the hanging wall is hard, though liable to slab off in large pieces. The greater par tof the ore is of medium hardness and, not being frozen to either wall, parts readily from them. A back of ore will generally stand well without support if properly arched; it is advisable, however, to work it out rapidly to prevent "air slaking." No sudden change in the width of the orebody has been found, and no sulphides occur in the walls as is often the case in the porphyry deposits.

The system of Example 22 was formerly employed in hard ore, when sufficiently close to surface to allow of waste filling being easily obtained.

The expense of breaking and leveling waste for each slice and the exposing of unskilled laborers under a high roof, were the vital objections to the continuance of this system. These defects were overcome in the present system in which the shovelers work outside the stope, the miner is kept so close to the back of ore as to allow constant scrutiny and the handling of waste is reduced to a minimum.

In preparing a level for stoping by the new system, two methods have been employed. In the first method shown in Figs. 77 and 78, all of the ore between the walls is removed for a length of 75 ft. and to a height of 20 ft. This space after being floored with 2-in. plank is filled with waste from the old stopes above to within 5 ft. of the back. New roadways are now driven in the foot- and hanging-walls, paralleling the vein at a distance of 15 ft.

Chute raises are carried up at intervals of 25 ft. along these roadways and connected to the stope by crosscuts. The broken ore from the stope runs through the cross-cut; the grizzly allows the finer material to pass into the chute and the larger pieces are broken by a laborer stationed on the grizzly. The chutes are all connected on the level of th grizzlies by small drifts, from which a ladderway extends to the level.

In the second method shown in Figs. 77, 79, and 80, the sill floor of th

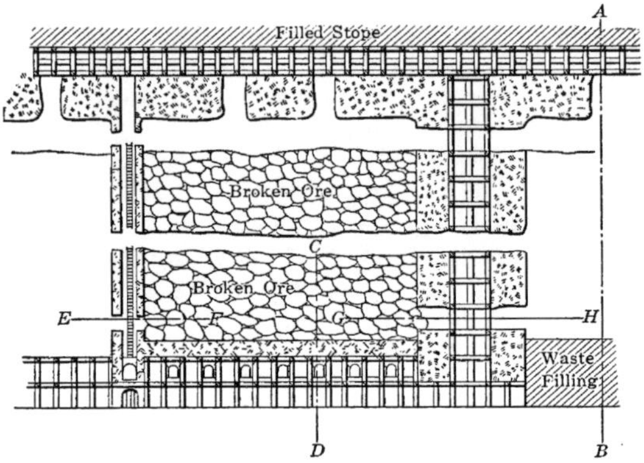

Fig. 77.—Vertical section.

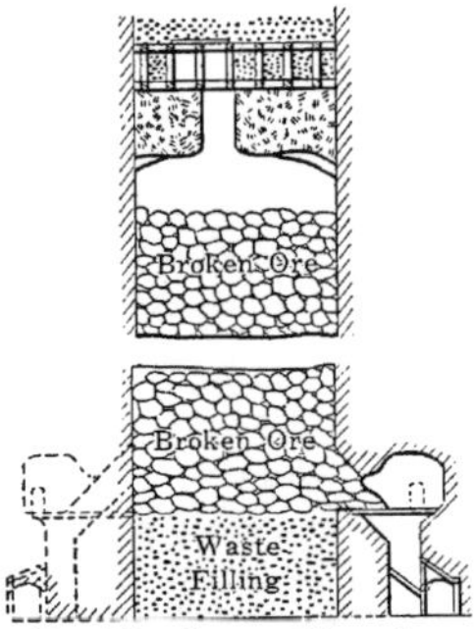

Fig. 78.—Cross-section of stope at *A-B*.

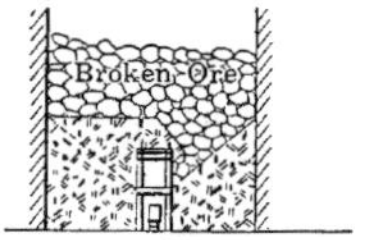

Fig. 79.—Cross section of stope at *C-D*

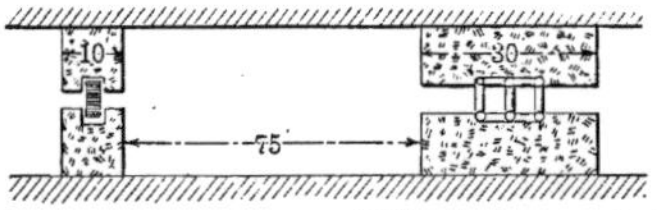

Temporary pillar at section *E-F*. Fig. 80. Pillar to be removed by slicing section *G-H*.

Stoping System at Coronado Mine.

stope is started 15 ft. above the tramming level. This level is in center of the vein and is timbered two sets high for the whole len of the stope, leaving a shell of ore between the top of the timber and floor of the stope above. On each alternate side of the upper sets, inclined funnel-shaped raises communicate with the stope, the floor of which, viewed from inside the stope, consists of two rows of hoppers. The broken ore passing through these openings falls upon the 6x8-in. lagging with which the floor of the upper sets is lagged. By opening the center lagging, the ore is raked into the cars placed beneath.

The cost of the preparatory work is less in this system than in Fig. 78, the loading of the cars being direct is cheaper, but the rapid cleaning out of the stope, which is an important matter, is subject to more delays.

Pillars

Two classes of pillars are employed to support the roof and walls, small temporary stoping pillars and larger pillars to be removed later by top slicing. A section along the vein in Fig. 77, shows a pillar of ore 30 ft. long, a stope of 75 ft., a temporary pillar of 10 ft. in length, another stope of 75 ft., and again a pillar 30 ft. in length. The 30-ft. pillar is provided with a chute and ladderway, from which drifts at intervals of 15 ft. give entrance and ventilation to the stopes. The smaller pillar is 10 ft. long by the width of the vein and contains a ladderway with small drifts, as in the larger pillar.

Overhand Stoping

The ore is broken by overhand stoping, Waugh drills being used. The stope is kept full of broken ore, sufficient only being drawn to leave a working space between the floor of broken ore and the back of the stope. Work is confined almost entirely to the ends of the stope adjacent to the pillars with the purpose of leaving a sag or belly of ore hanging between. This eventually breaks down by its own weight and is block-holed from on top. In an eight-hour shift, of which two hours are consumed in blasting and picking down, each machine will drill from 90 to 110 ft. of holes.

Should the back of ore turn soft and render it inadvisable to work beneath, the ore can be broken down underhand by connecting drifts from the raises and breaking the floor as described in connection with the King Mine of Example 29.

The levels are 200 ft. apart, and when a stope is within 15 ft. of an upper level, the breaking of ore ceases. Two raises are then made from the roof of the stope beneath the waste filling of the level above. The small temporary pillar is broken by first undercutting and then blasting from inside the ladderway. The roof of the stope is now carefully dressed and the stope emptied of its broken ore as rapidly as possible.

Filling and Extracting Pillars

The waste filling from the level above is now allowed to run into the empty stope and, when full, a working floor is leveled off. To extract the shell of ore left, square-set timbering is employed. As the ore is removed by retreating to the chutes in the pillars, the sets are caved and the waste allowed to follow.

To extract the pillar left between the stopes, work is commenced beneath the upper level. Square sets are employed and a mat laid, upon which the waste is caved. The ore is then removed from beneath the mat by descending slices 11 ft. thick, using posts to support the overhanging mat, as in Example 43.

When the broken ore is drawn from the stope by chutes in the walls, there must necessarily be a "hog back" of broken ore left along the center of the stope. This is removed by spiling a timbered roadway through it and withdrawing the broken ore. As soon as the waste appears, another set of spiling is blasted out, retreating in this manner to each end of the stope.

Where the sill floor is above the roadway as in Figs. 79 and 80, the stope which next ascends from below is carried up to this level and the shell of ore removed as before described. The system has been satisfactory and has resulted in a substantial reduction in the cost of mining.

Example 31.—Los Pilares Mine, Nacozari, Mexico

(*See also Example* 24.)

Irregular Lenses in Porphyry; Slicing and Delayed Filling.—Where this system is employed the sill flooring of a stope is practically the same as method *B* in Example 24. The ore extracted by this system of stoping has a better grade than that mined in Example 24, and for that reason no chance is taken of mixing it with the waste filling. The ground is also much firmer, allowing a large stoping area without danger of caving and thus losing the stope. At intervals of from 12 to 20 ft. cross-cuts from the main drift (which is generally on the wall) are driven into the ore for a distance of 20 ft. from the center of the main drift. If the orebody is not wide, these cross-cuts will suffice to draw the broken ore of the stope, as there will not be too much space between the far side of the stope and the end of such a cross-cut which, because of its purpose, is termed a shovel-way. If the orebody is wide, as in Fig. 65, an auxiliary drift *d* is driven through the ore at approximately right angles to the pillars and located about two-thirds the distance between the country wall and the limit of the ore. Cross-cuts *g* are then driven at intervals of from 12 to 20 ft. from this auxiliary drift to both the left and the right, 15 ft. long, and after leaving a 15-ft. pillar between the side of the drift and the end of the cross-cuts, the remaining area of the stope is "silled" on the level floor. The pillar thus left forms the base of the floor arch over the drift, and is pierced by the cross-cuts which serve as shovelways. Turn sheets are then placed in the main drift in front of each cross-cut and from 6 ft. to 8 ft. of track laid in each cross-cut. A platform *h* raised 2 ft. above the rails is then placed in the cross-cut at the far end of the track and a flat iron sheet placed on top of it, at the same time raising the roof above the platform about 1 ft. This track arrangement allows the car to be turned into the shovelway, leaving the main track open. The broken ore runs down on to the iron sheet over the platform from which it is shovbled, thus giving the car man an easier task in filling his car.

Fifteen feet above the level floor, the stope is cut out back over each drift, leaving it protected by the pillars and floor arch *i* as seen in the section shown in Fig. 65, but thus acquiring the whole area between the

Fig. 81.—Town of Los Pilares.

pillars overhead for stoping. From the top of the arch up, the stope is worked by two and three slices being carried forward at the same time. The broken ore accumulates in the stope and only enough is drawn from below through the shovelways to allow the miners to drill by standing on ore. Manways *f* are carried up through the center of the pillars limiting the stope on each side. From one manway 20 ft. above the level floor an intermediate cross-cut is driven to the stope and at each succeeding 20 ft. similar cross-cuts are driven. From the pillar manway on the other side, the first intermediate cross-cut to the stope is driven 30 feet above the level floor, and others at intervals of 20 ft. above it. These manways and their connecting cross-cuts give an inlet and outlet to the stope for every 10 ft. By this method, a stope may be carried up 100 ft., or even 200 or 300 ft., before being drawn and filled. Again a stope may be worked from two or three levels at the same time, each level being driven up till only a thickness of 12 to 15 ft. of solid ground separates two stopes.

At this point the uppermost stope is drawn through its shovelways, after which the floor arches, etc., protecting the haulage drift are shot down. The solid ground separating the two stopes, one over the other, is then drilled with a large number of holes, say 50 to 60, from the top of the broken ore, the holes heavily loaded and shot down, thus making the two stopes one. By this method all the ore left for floor arches, etc., is eventually recovered except that on the lowest level worked.

These great stopes may or may not be filled with waste rock soon after drawing off the ore. When the mine was first opened they were sometimes left standing empty for months without accident. In recent years, however, the tendency is to fill as soon as possible after drawing the stopes. As illustrative of the standing qualities of the rock, no better example can be cited than that of the old No. 1 stope worked out during the first years of mining. This was located near the Pilares shaft (Fig. 81), was 100x100 ft. in floor plan and was worked up from the 400-ft. level clear to the oxidized capping, a vertical distance of 280 ft. Although the capping was here only 25 to 30 ft. thick, the empty stope stood for 18 months without caving before it was filled.

Filling the Stopes.—In filling stopes mined by this system, fill holes are run direct to the surface over the stope where they are widened out to a size of 12x12 ft. On the surface this 12x12-ft. hole is then further enlarged to a roughly funnel shape by churning holes from 12 to 30 ft. in depth above the edge of the fill hole. These churn drill holes are sprung with dynamite, and shot with black powder, the rock breaking from such shots falling directly into the stope. By having the fill hole of this large size, it is seldom choked with large rock.

CHAPTER XIII

OVERHAND STOPING WITH SHRINKAGE AND SIMULTANEOUS PILLAR-CAVING

Example 32.—Miami Mine, Globe District, Arizona

(*See also Examples* 21 *and* 42.)

Irregular Lenses in Porphyry.—Rill Chutes and Slicing of Pillars.—At present the main prospecting shaft, No. 1, has been sunk to a depth of 720 ft. and, as ore was encountered at 220 ft., its thickness at the shaft is at least 500 ft. The area of the orebody (see Fig. 84) is about 10 acres and it is covered by 60 to 250 ft. of porphyry capping.

Below the 220-ft. level sub-levels have been driven at 25-ft. vertical intervals down to the 370-ft. level, and on these sub-levels the orebody has been almost completely blocked out into 50-ft. squares by drifts 7 ft. high and from 4 1/2 to 5 ft. wide.

The 50-ft. interval of ore between the 370-ft. level and the 420-foot, or main-haulage, level has been left solid to protect the haulage level while mining the ore above it. Before detailing the method of mining, a word as to the character of the ore.

Superintendent N. O. Lawton, the deviser of this mining system, says: "The enriched schist of the spheroidal-shaped orebody has been greatly fractured, crushed, and later softened or altered by percolating water, so that in its present form it is quite easily drilled and broken. The fractured condition will facilitate mining by causing the ore to break into pieces easily handled by the shovel or run in chutes. Little of the ore will break in so large pieces as to necessitate rebreaking or block-holing to avoid choking the chutes. Most of the ore caves easily, so that to avoid timbering except at soft places, the levels are driven the narrow width of 4 1/2 to 5 ft. by 7 ft. high, with the roof carefully arched."

System of Mining.—It is apparent from the above description of the ore that, because of its soft crushed nature, which causes it to quickly cave wherever excavations of any width are left untimbered, a system of caving, practical where the ore is hard and stands well, would here be unsuitable.

The accompanying illustrations, Figs. 82 to 85, will make clear the following description. Fig. 82 is a plan of the orebody showing method of rectangular system for tramming levels located 50 ft. below floor of stopes. Fig. 83 is a cross-section through rooms and pillars. Fig. 84

is a plan of first mining level, showing method of cutting-out room preparatory to stoping. Fig. 85 is a detail plan of sub-levels, showing method of stoping rooms.

On the first main-haulage level, the 420-ft. haulage drifts have been driven spaced on 50-ft. centers, as shown in Fig. 82. The ground is to be excavated by a series of rooms 60 ft. wide alternating with 40-ft. pillars as best understood by reference to Fig. 83. Every alternate haulage drift is to be provided with a mill hole to draw the broken ore from the rooms while the other drifts will come beneath the

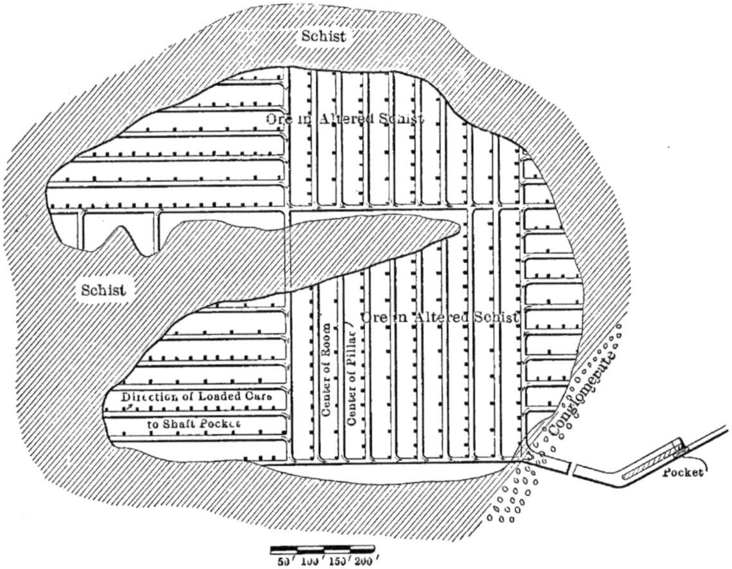

FIG. 82. First haulage level, Miami mine.

pillars. From these haulage drifts 4x5 ft. raises *e* have been put up so as to come in the center of both the rooms and pillars. This spacing arrangement has necessitated the carrying of the pillar 25 ft. thick on one side of a pillar drift, and only 15 ft. on the other, as shown in the plan Fig. 84. The haulage drifts beneath the rooms have chutes spaced on 25-ft. centers, while the drifts beneath the pillars have chutes every 50 ft. It will be noted on Fig. 83 that the raises to the rooms are branched from a point 25 ft. above the haulage level in order to make them effective in drawing the broken ore all the way across the room. As previously stated, the orebody has been already completely blocked out preparatory to starting caving in the near future.

Operations will start on the first mining level, Fig. 83, 50 ft. above the haulage level. The top line of Figs. 84 or 85, represents the limit of commercial ore. Mining will commence at this limiting line and retreat from it. Starting then at the end of the drift cross-cuts will be driven both ways along the limiting line to the pillar lines. These cross-cuts will be carried as wide and high as found practical, say 7 ft. high by 8 ft. wide. The roof of these cross-cuts will then be drilled with upper holes, using hammer-drill stopers, and shot down to a height of 7 ft. more, making the total height of the cross-cuts 14 ft. Then starting again at the drift a slice or drivage will be taken each way in a similar manner, breaking to the previous excavation and then shooting to obtain the height of 14 ft. In

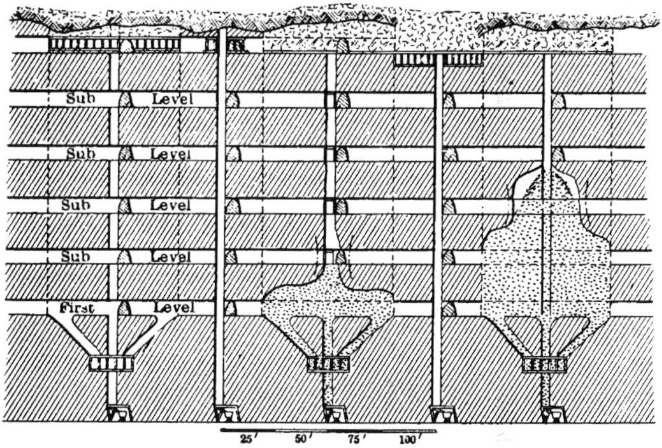

FIG. 83.—Cross-section of stope, Miami mine.

this way, retreating slice by slice from the limiting line the ore above the floor of the room is broken to a height of 14 ft., although at no point do the miners work under a roof higher than 7 ft. The ore thus broken will fill the branched raises of Fig. 83, previously driven; but only as much ore as necessary will be drawn, as it is desirable that it shall pack under the solid back as closely as possible, thus supporting its weight.

The miners are then transferred to the first submining level located 25 ft. above. As 14 ft. of ore has been broken below, the thickness of solid ground underfoot is now but 11 ft. Reference to Fig. 83 will make clear the method of procedure to break up this 11 ft. Starting, at a central raise down holes are placed all around it and shot through to the broken ore in the room beneath. Simultaneously work is started at the next raise and the floor likewise broken down. Drill men then drill the floor and roof of the drift which connects the two raises blasting them both simultaneously in 4- to 5-foot sections, retreating, from the holes started at the raises, toward each other till they meet about in the

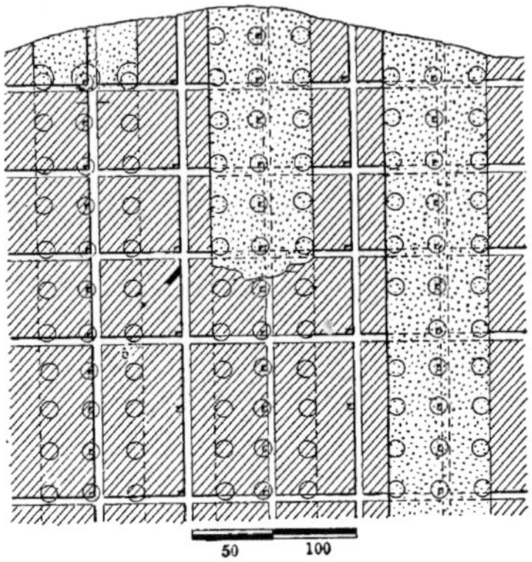

FIG. 84.—Plan of first mining level, showing chutes, Miami mine.

center of the drift. This work breaks the floor of the drift through into the room full of broken ore below, and breaks up its roof to a height of 14 ft. The drill men now start again at the raises, and leaving in the solid floor to work on, slice back the walls for a width of about 7 ft. but

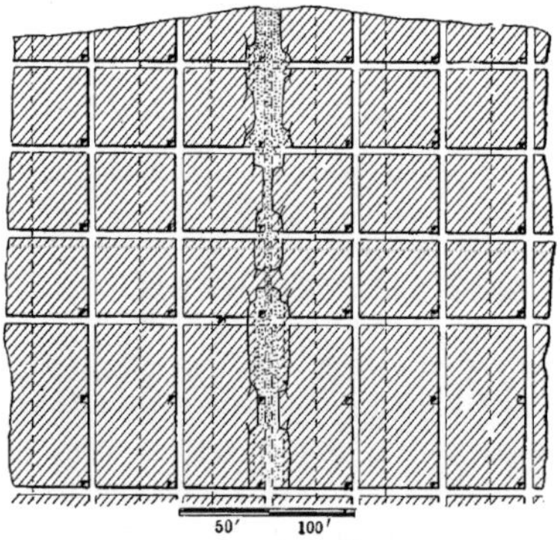

FIG. 85.—Plan showing rooms during stoping, Miami mine.

only 7 ft. high, thus paralleling the slot broken through to the room below and again meeting in the middle of the block as before. The floor and roof are now simultaneously drilled and shot, breaking through into the room below, and raising the roof to a height of 14 ft. In this manner slice after slice is worked back parallel with the original drift until the pillar-line limits are reached, thus breaking up the entire 25-ft. block below the first submining level, and likewise demolishing the lower 14 ft. of the block below the second submining level. The miners are then transferred to the second submining level, and the breaking process repeated from sublevel to sublevel until the leached capping is reached.

In spite of the fact that the soft, crushed nature of the rock makes it a bad roof to work under, it will be noted that the method always provides safety by keeping the height of the roof under which miners work at only 7 ft. and affording ample ways of ingress and egress. Owing to the shattered condition of the rock there should be no difficulty in breaking it fine enough to avoid choking of the chutes.

For placing the side and down holes, light 2 1/4-in. one-man piston drills will probably be used, and for uppers the air-hammer drill.

It will be readily apparent that quite a number of men can be worked in a room when it is once started from several points, and to that end the compressed-air pipe system now being installed has been designed to furnish air to a large number of drills per level without reducing the desired pressure. The main air pipe is 10 in. in diameter down the shaft, and from it two 8-in. diameter pipes are branched off at the main haulage levels, the 420- and 570-ft. On each haulage level the 8-in. pipe is hooped completely about the rectangle of Fig. 82. Each of the haulage drifts l branching from the rectangle is to have a 4-in. main into which 2 1/2-in. pipes are tapped and carried up the pillar raises to the various sublevels where 1 1/2-in. pipes connect with the drills. By this system of piping, an ample supply of air at the proper pressure is provided.

Drawing the Broken Ore.—As the ore runs about 12 cubic feet to the ton solid and about 16 to 17 cubic ft. broken, it will be understood that as the breaking up of the rooms proceeds, enough of the ore is drawn through the chutes to allow a sufficient space before blasting for a free breaking. Yet at all times the top of the broken material is kept close under the solid back in order to prevent the falling of the large masses into it. Such would have to be block holed under a roof of dangerous height to prevent choking of the chutes. After two adjoining rooms have been broken the length of the ore body, or, say, 200 to 300 ft. back from the limiting line of Fig. 82, the intermediate pillar may be mined by the method of slicing of Example 43, as shown at the top of Fig. 76 (*b*). It will, of course, be understood that the irregular dome-shaped masses of comparatively small horizontal area are best mined by this same slicing method, before the rooms are broken up to the level where the width of the ore is fairly uniform across the whole ore body. This slicing system breaks the capping to the surface and gets its weight on the ore body before the rooms reach the top submining level shown in Fig. 82.

With the rooms on each side of the pillar broken up to the top submining level, the first slice of the pillar is mined until the broken ore is encountered. The level of the broken ore in the rooms is then lowered to that of the top of the solid pillar, thus allowing the mat of timbers and broken materials *n* to come down and bringing the weight of the roof on the timbers above the pillar, thus crushing them flat to form the mat. A second slice of the pillar is then mined, the ore in the rooms drawn down to the new top of the pillar, and so on. Eventually the ore is all broken and drawn down to the first mining level. The main haulage system of the mine, which up to this time has been on the 420-ft. level, will then be transferred to the 570-ft. level, and mining will proceed above it, starting on the first mining level 50 ft. above, or the 520-ft. level. In this way the 50-ft. block between the 420-ft. and the 370-ft. level will ultimately be mined.

By getting a thick mat of timbers on the top submining level before the drawing of the rooms commences, and as the capping comes down keeping it at as uniform a level as possible by drawing the chutes uniformly, it is expected that little ore will be lost by mixing with waste.

Tramming.—The drifts on the 420-ft. haulage level have been laid with 30-lb rails set to a 24-in. gauge. The ore will be drawn from the chutes into cars of 2 1/2 tons capacity and hauled in trains to the shaft by electric locomotives. At the shaft the cars will discharge into a 700-ton pocket, below which will be a skip filler holding exactly a skip load. The skips will be of 7 1/2 tons capacity. With a skip placed beneath the filler, a lever will discharge the filler into the skip.

Summary of Mining Conditions.—Thanks to the compact nature of the Miami ore body and the character of the ground, ore can be mined with the use of little or no timber and a minimum of explosive. A force of but comparatively few men can break and handle enough ore to maintain a steady output of 2000 tons per day. With the ground thoroughly blocked out and with the present equipment as it is, the doubling of the output would be simply a matter of increasing the number of miners.

The management has established the cost of mining the present ore body at $1.25 per ton, but the average cost should eventually closely approximate $1 per ton, including hoisting to the surface. The method of mining seems well adapted to insure both the safety of the miners and the extraction of at least 85 per cent. of the ore body. Exclusive of mining losses, therefore, the present development justifies an expectation of hoisting at least 14,000,000 tons, which is said to average 2.75 per cent. copper as determined by systematic churn-drill prospecting.

As the development plan for this system is similar to that of several others, like the room-caving of Example 46, or the block-caving of

Example 42, the stoping could be altered to follow one of these similar systems at any time that it might seem desirable.

Example 33.—Boston Consolidated Mine, Bingham, Utah

(See also Examples 3, 37, 41 and 43.)

Irregular Lenses in Porphyry. Block Caving of Pillars.—A caving system has been devised and adopted here. Briefly, this consists in weakening the block of ore by means of a series of ore-filled rooms, and then, when the remaining pillars are shattered, the ore is drawn evenly from under a large area of capping and the surface allowed to settle gradually.

The mine is opened up by two levels, one about 200 ft. below the steam-shovel workings near the top of the hill and the other, the main haulage level, 150 ft. below the first. On this lower level there are two main haulage drifts. From these a system of parallel side drifts are turnd off at intervals of 120 ft. At distances of 200 ft. along these drifts, raises 5 ft. square, inclined at an angle of 60 deg., are driven to connect with drifts on the level above, and from these main raises, or chutes, as they become later, a series of branch raises or chutes are driven so that the collecting of ore may be concentrated as much as possible. These raises are equipped with air-operated sector gates, which have 24-in. openings, and work admirably as a train of ten 8-ton cars can be loaded in 4 min.

The tops of these raises and branch raises hole into the center of the drifts on the level above, so that they have to be fenced off by means of guard rails nailed to four upright sprags. A series of knotted ropes, hanging from the guard rails like the low-bridge signals on railroad tracks, permit the placing of the guard rails high enough not to interfere with the dumping of a car. This arrangement prevents any one from walking into the chutes.

Blocking out the Rooms

On the upper level (as in Fig. 86) the orebody is blocked out by a series of drifts 60 ft. apart and a series of cross drifts 400 ft. apart, since that is the length of the room used wherever the size of the orebody permits. These drifts are 6 1/2x7 1/2 ft. in the clear and are driven on contract at \$2 to \$2.25 per foot, making their total cost \$3 to \$3.25 per foot. In driving these drifts a 5-ft. round is drilled each shift by one man using a 2 1/2-in. Sullivan piston drill, but rarely does the round break over 4 ft. in the clear. The powder consumption is about 2 lb. to the foot. These drifts in the future pillars are laid with double track, so as to permit rapid tramming.

At intervals of 30 ft. along the pillar drifts, drifts 30 ft. long are driven in both directions. The mouths of the drifts are staggered, so as not to come opposite to each other, thus possibly weakening the pillar too much. In addition the tracks in these short drifts are arranged so that the cars are run out to the nearer cross drift. The ends of these short drifts are then connected with each other by means of drifts parallel to the pillar drifts. The short drifts at the end of each pillar are not driven within 15 ft. of the main cross drifts, in order not to weaken their pillars.

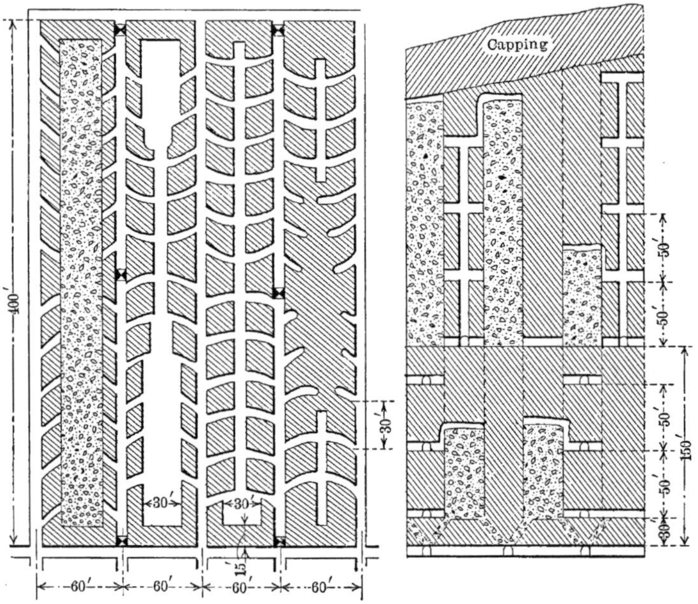

Fig. 86.—Stoping system, Boston Con. mine.

As soon as these short drifts are all connected, this central drift, which marks the middle of the future room, is slabbed out by means of two flat holes, one near the floor and the other near the roof. These holes are drilled from 16 to 18 ft. deep and are parallel to the drift. By blasting two rows of holes on each side of the room drift, the room is enlarged to a width of 30 ft. Model 9 Water Leyner drills are used, and in drilling these long holes a 2 1/2-in. starter and a 1 1/2-in. finisher bit is employed. A 1-in. water pipe is carried along the pillar drift to water these drills.

In alternate pillars, at the same time that the room drifts are being driven, raises are put up at the middle of the pillar drift and also near each end where the pillar drift meets the main cross drifts. These serve as future manways in mining the rooms. They are driven merely large enough for a ladder and an air pipe.

As soon as the room has been enlarged to a width of 30 ft. and the ore mucked out, a chute having a mouth 24 in. wide is built across the end of each short cross drift. These chutes are used in drawing off the excess of broken ore from the room. Miners then begin to drill the roof of the room full of uppers, using Leyner air-hammer drills. These holes are drilled so as to look forward about 75 deg. and toeing out to the sides so as to keep the walls of the room vertical. These holes are drilled about 7 ft. deep and in rows of 5 holes across the stope, these rows being 8 to 9 ft. apart. It takes about 5 drills to a hole, the starters having a 1 1/2-in. bit and the finishers a 1-in. bit. These Leyner hammer drills use solid steel and drill from 100 to 120 ft. in a shaft, the air pressure being only 80 lb. per sq. in. at the receiver. This low pressure is used because with a higher pressure the bits cut too fast to permit them to turn freely.

These holes are not loaded until half the length of the stope has been drilled. Then they are loaded with about 4 sticks of 30 per cent. Hercules dynamite and blasted all at once. About 0.45 lb. of dynamite is required per ton of ore broken. One man, by using a nicked fuse, spits about 50 holes, a 7-ft. fuse that burns at about 45 sec. per foot being used. Thus the stope is carried up in alternate halves.

The excess ore, which amounts to about 40 per cent. of the total ore broken, is drawn off before each blast, so that there is 12 ft. of open space below the roof when the holes are spit. The ore at first was drawn off at only one side, but this left the pile slanting toward that side and necessitated leveling it off each time before drilling began. To avoid this the tapping drifts are driven from each side and the ore drawn off on both sides so that it keeps a fairly level surface for the men to work upon.

Three manways, one at each end and one in the center of the side of the pillar having the raise in it, are carried up through the ore. These are made by blasting out a triangular notch in the pillar and cribbing it off on three sides. This manway is about 3x3 ft. in size and is merely large enough to allow a man to pass through it after the air pipe has been put in. The men climb the cribbing of split timber, a vertical pole being nailed to the cribbing for the men to hold to when climbing. These manways are carried up through the ore for only 50 ft., since at every 50 ft. a drift is driven to the stope from the three raises in the pillar.

The ore is drawn off in cars having a capacity of 28 cu. ft. and is run by two men to the main chutes. Only 4 or 5 men work in a stope, and one of these is busy all the time picking down the back of the room and the roofs of the different drifts in the pillar connecting with that room. The rooms are carried up until the ore becomes too poor to pay—at present when it carries about 1.45 per cent. copper. The room is then abandoued; and as soon as the stope on the other side of the pillar is completed, the air pipes and tracks in the raises and drifts in the pillar between the two stopes are removed.

Caving the Ore

As the area undermined by rooms increases, the roof gradually settles, but as the top of the broken ore in the rooms is within six feet of the roof when the rooms are abandoned, the capping cannot drop far. This undercutting throws weight on the pillars, and after an area 200 by 400 ft. has been undermined, they begin to crush without any further weakening. At present five stopes are being worked on the upper, and three on the lower level.

After the orebody on the upper level has been undercut by rooms, mining on the lower level begins under that area. On the lower level the rooms are placed so as to come directly under the pillars on the level above. These rooms are mined in the same manner as on the level above, with the exception that their floors are 30 ft. above the level, so as to leave a pillar 30 ft. thick to protect the main transportation drifts. This necessitates the driving, from the main level, of inclined raises, 30 ft. apart, to tap the rooms on each side so as to draw off the excess ore. The manways in the pillars are also dispensed with in the lower blocks of ground, for it has been found, owing to the weight on the lower pillars, to be cheaper to drive drifts to the different rooms at intervals of 50 ft. vertically from a raise placed in a part of the lower level not being undermined than to maintain the manways in the pillars.

After the whole of the orebody has been cut up by rooms and the lower pillars have been weakened by raises put up to tap them, drawing will begin throughout the whole orebody, the ore being removed evenly under the area so that the capping will settle regularly. Thus the mixing of ore and capping will be prevented as much as possible. When this is completed, the ore below this level and the pillar above it will be mined by a similar method of undercutting by means of rooms.

The company is mining 2600 dry tons of ore a day at its porphyry mine, and employs 351 men underground and on surface. This gives an average of almost 7 tons to a man and from 8 to 8 1/2 tons to a man employed underground. This is remarkable when one considers that only ore broken in the rooms is being taken out at present.

The wages of the men per 8-hr. shift are: Machinemen and timbermen, $3; helpers and muckers, $2.50; trainmen, $3; helpers, $2.50; blacksmiths, $4; toolsharpeners, $3.50; shift boss, $4.

The cost of driving the large 9x9-ft. haulage drifts is about $8 per foot, as the cost of such a drift 1600 ft. long, one-third timbered, was $8.08 per foot, inclusive of the cost of laying track and of putting in the trolley wire. The cost of driving a 6 1/2x7-ft. drift is for labor (contract), $2 to $2.25 per foot, making the total $3 to $3.25. For driving 5x5-ft.

raises the cost is, for labor (contract), $1.75, making the cost per foot for raises 180 to 200 ft. high $2.25 to $2.50 per foot.

At present the cost of mining the ore is 44 cents per ton. The cost of development is 2 cents per ton of ore developed in a room and a pillar, for the company charges this at the rate of 10 cents per ton to the ore, that is drawn off from a room; or to only 40 per cent. of the total ore broken in a room. The rooms and pillars are equal in size. Therefore, disregarding the ore removed from the raises and drifts in the pillars, which work comes under the head of development, there remains, after the whole area is caved, 80 per cent. of the ore that needs only to be drawn through chutes and trammed to surface.

At present it costs 17 cents a ton to draw the ore from the rooms, so that the cost of drawing the 60 per cent. left in the rooms and pillars ought not to cost more than 20 cents per ton. Consequently the cost of mining figures out as follows: Forty tons of ore mined at 44 cents per ton, $17.60; 160 tons of ore mined at 20 cents per ton, $32; development cost for 40 tons at 10 cents per ton, $4. Total cost of extracting 200 tons, $53.60; or 26.8 cents per ton. The cost of superintendence, taxes, etc., when 2600 tons of ore are mined a day, will amount to about 2 cents per ton. Allowing a factor of safety of 25 per cent., or 6 cents per ton, to cover unforeseen difficulties in mining the lower pillars of ore, it appears that by this system the ultimate cost of mining a block of ore will probably approximate 35 cents per ton, the same as steam-shovel mining.

This method of mining is quite bold, but from the results obtained on the small area already caved, it appears that the ore breaks in a fairly perpendicular plane to surface. This will greatly decrease the tendency of the ore and capping to mix when it becomes necessary to cave one section of the orebody before an adjacent block is touched. But the main difficulty from mixing of the ore and capping will come from unequal settling of different portions of the block that is being caved. The Bingham porphyry ore, from the nature of its formation, is much broken up by small fracture seams. Owing to this fractured nature of the orebody on caving, it breaks up into fairly fine ore, and so there will be few large boulders to block the chutes or to cause the ore to hang up above the chute mouth and form a grizzly through which only fine ore could pass, as would be the case if large masses followed down that were crushed but little by the weight thrown upon them by the undermining of the block. It therefore appears that, if a system of recording the approximate tonnage drawn from each chute is used, there should be no great difficulty in drawing the ore evenly from under the capping in each block, and consequently no trouble in causing the capping to follow evenly af er the ore with little intermixing of capping.

The advantages of this method of caving are many: The ore is broken in large rooms; the method of stoping is adapted to the use of air-hammer drills; the blasting is done over large areas so that as little time is consumed by blasting as is possible, and the air vitiated only between shifts; the work is done systematically throughout an area, so that one can tell when all portions of the block have been reached; the ore is only dropped a few feet at a time and over large areas, so as to diminish the amount of ore mixed with capping; the ore is drawn systematically so that there is little danger of leaving ore behind; the amount of drifting and raising required in mining a block of ore is very small; the gathering of the ore is rendered cheap by concentrating it at a few chutes equipped with easily and rapidly operated gates.

Ground to be adapted to this method must be weak enough to cave readily, and yet strong enough to stand in rooms 20 or 30 ft. wide. The ore must break up into small chunks when the weight of the capping causes the pillars to cave, or otherwise difficulty will be experienced in drawing ore from the caved area. Apparently there should not be much difficulty in drawing the ore on the lower level, as the pillar 30 ft. thick ought to be amply strong enough to protect the haulage drifts.

Example 34.—Duluth Mine, Cananea, Mexico

(*See also Examples* 6, 18 *and* 45.)

Irregular Lenses in Porphyry; Block-caving of Pillars.—Pillar-caving is a combination of overhand stoping on ore and a caving system. As is necessary is nearly all caving methods, the first step is to prospect and thoroughly outline the orebody by means of drifts and raises. Fig. 87 shows an orebody on the 200 level which extends above the 100 level.

After sufficient prospecting work has been done, the size of the sections to be mined and the pillars of ore to be left were decided upon. Pillars are usually about 50 ft. wide, with sections from 75 to 100 ft. wide extending across the body. Because of the irregularity of the upper portions of these bodies it is necessary that they be mined by means of square sets in order to follow rich stringers.

At the Cananea-Duluth the orebody is mined by square sets from the 100 level to the top of the ore. These sets are then all removed and the pillar-caving system proper begins. In the meantime the section to be mined is blocked out on the 200 level by means of drifts and regular square-set raises are put in at intervals, as shown in Fig. 87. The sill raise set is 8 ft. 5 in. high and the second set is 7 ft. 4 in., making practically 16 ft. from the rail to the top of the second set. This completes the regular raise sets, for at the top of the second set drifts are run connecting all the raises in the section. These drifts are then widened from 12 to 15 ft., after which they are carried up vertically by means of overhand stoping, the miners working on ore, only enough ore being drawn off so as to permit them to be within easy reach of the back.

These drifts are finally carried up to the level above, cutting out a number of small pillars which have been cut loose from the waste above by the square-set stope and are now partially supported by the ore surrounding them.

Construction of Chutes

Formerly, cribbed chutes of 8x8-in. timbers were carried up in the broken ore with a manway compartment, 2 1/2x5 ft., and a chute, 5x5 ft. It has been found that a 3-in. plank chute is practically as good, with a saving of considerable timber. The inside dimensions of the

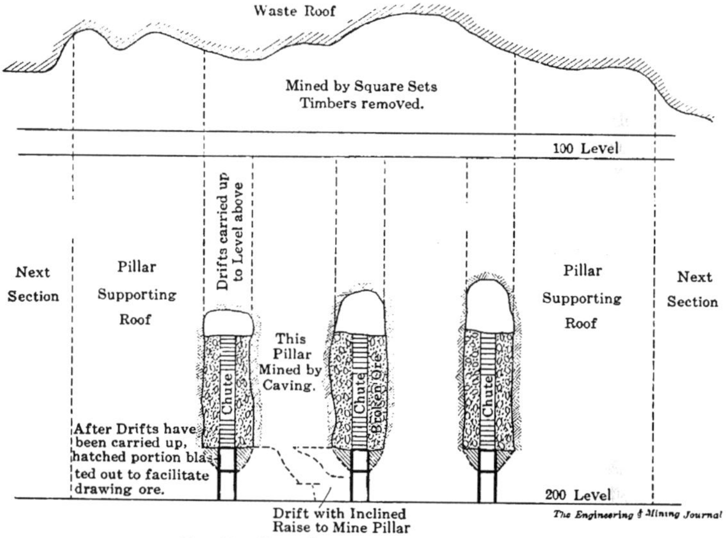

FIG. 87.—Vertical section of stope, Duluth mine.

combined chute and manway are 3 ft. 3 in. by 6 ft. The chute in the clear is 3 ft. 3 in. square, with a manway 2 ft. 6 in. wide. The 3-in. planks are placed on edge, with ends beveled at 45 deg. The dividing partition is a 3-in. plank which fits into a notch cut in the side pieces. As the back advances, the chutes are carried up, surrounded with ore.

Horse of Waste

The matter of handling a horse of waste is not difficult, as it can be broken and easily drawn off through one or more of the raises that are carried up from the level. It has been found possible in mining by this system to place the raises close together, thus almost entirely eliminating the wheelbarrow by shoveling directly into a chute.

Drawing the Ore

The next operation is to draw the ore. This is accomplished by drilling holes in the solid ore which surrounds the second square set in each raise, as shown by the hatched portion in Fig. 89. These holes, after being blasted, form a mill-hole around the raise. In this way the ore is drawn off with the occasional use of a small amount of powder. The chute planking comes out with the ore. The short pieces are usually unbroken, while perhaps 50 per cent. of the side pieces are unbroken and can be used again. By this means all the ore is drawn from the section and the small pillars are left standing.

Mining the Pillars

The pillars crush down and break, due their own weight and a few small slips that usually exist in this class of porphyry ore. In case a pillar does not break down, a drift is run on the level underneath it and

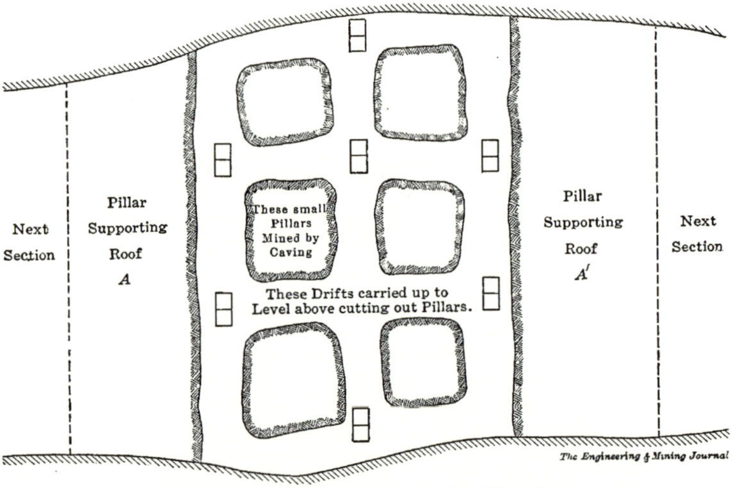

Fig. 88.—Plan of stope, -16 ft. above level, Duluth mine.

a raise is run up a short distance into the bottom of the pillar. One side of this raise is filled with holes, the base of the pillar is blasted out and the pillar falls. From this drift a new set of inclined raises in the bottom 16-ft. block of ore are used to draw off the ore in the pillars. These raises are merely flat sloping floors of heavy timbers, with head room blasted out so that a man can stand up and bar and draw the rock down the chute and into the car. The chute bottom is made almost flat, so that the ore cannot run down it, but piles up at the bottom. Large boulders are easily plugged and blasted at the mouth of the chute without injury to the timbers. Any waste can be sorted before it is loaded into the cars and need not be mixed with the ore. The small boulders are broken with hammers before being loaded into the cars.

Further Developments

The next step is to mine out the sections on the other side of the large supporting pillars A and A', Fig. 88. This is as far as the method has been worked out and therefore future developments will be watched with great interest. There are several courses which can be followed in the subsequent mining. If the back and the pillars supporting it are sufficiently strong, it may be possible to mine out another section directly under the first, from the 300 level to the 200. Again, it may be possible to mine the supporting pillar by caving it, as in the mining of the smaller pillars, provided that the waste roof will stand without any support. If, however, the main pillar could not be mined in this way, the back over the sections on either side of the pillar would be made to cave in and the pillar itself would be mined by the slicing system. If this last were done, the remaining ore below the 200 level would be mined by slicing.

The method has been considerably changed from that first employed. Originally the section was mined without leaving the small pillars. It was then simple overhand stoping on ore. The back then was usually quite unsafe, not because of any great weight, but merely due to large masses of ore breaking away on small fractures, which are common in almost all kinds of porphyry. After one of these stopes caved, burying several men, the system as described was evolved. Since then it has given the greatest satisfaction and as now employed is quite safe, as the men always work near the back and when mining pil.ars are well protected.

Cost of Pillar Caving

The method requires practically no timber and the greater part that is necessary can be used again. In practically every step in this method the breaking of the ore is done with the least possible amount of powder and labor, while the ventilation can easily be kept good and it is comparatively safe. For its application it is absolutely necessary to have a strong, solid ore and a strong roof; and both the ore and the waste roof should require no support with the exception of a few stulls to hold up small slabs and loose boulders. It is also necessary that the ore have definite boundaries and be large enough for advantageous work.

The body should be large so that it can be divided into sections and be blocked out as shown in Fig. 88, preferably extending from one level to the next. The amount of waste in the ore must always be small. A small amount of sorting can be done in stopes and the waste drawn off through chutes, but the proportion of ore to waste must always be large. Again, if the ore were inclined to pack, it could not be economically drawn and would practically have to be mined over again.

This system has such rigid requirements that its application is limited. With the exception of this its disadvantages are few and unimportant, while on the other hand it is the cheapest method of mining at Cananea. The cost at the Duluth mine (which is new and presents favorable conditions) is only 40 to 50 cents, for the labor and timber to place a ton of ore in the chutes, as compared with 75 to 85 cents per ton at the nearby Elisa mine. The Elisa is worked by the system of "overhand stoping on waste" as described in Example 22; its orebodies are also irregular lenses but are in limestone instead of porphyry.

CHAPTER XIV

BACK-CAVING INTO CHUTES OR CHUTE-CAVING

EXAMPLE 35.—HARTFORD MINE, NEGAUNEE, MARQUETTE RANGE, MICHIGAN

(*See also Example* 13.)

Caving into Rill Chutes on Levels, in Sub-vertical Wide Vein.—The Hartford mine (Oliver Co.), lying a short distance northeast of Negaunee, has a jaspillite hangwall and a soft, hematite ore. On the lower levels the ore shoot is 100 ft. wide by 300 ft. long, and its method of development and extraction is shown in Fig. 91.

Drift *d* is first driven in the foot wall, 150 ft. above the next level above, with cross-cuts *b* to b^3 and c to c^3 turned off from it in both directions at 50-ft. intervals. The foot-wall raises r are put up to the stopes as needed. Stoping is started from a 30-ft. raise as *K*, the most advanced stope being at the end of the lense at K^3. In stope K^1 the original raise has been widened by breast stoping at the top, from which it was cut down by underhand benches to a funnel shape. Next, the back of the stope is attacked by driving a raise *n* into it around the periphery, leav-

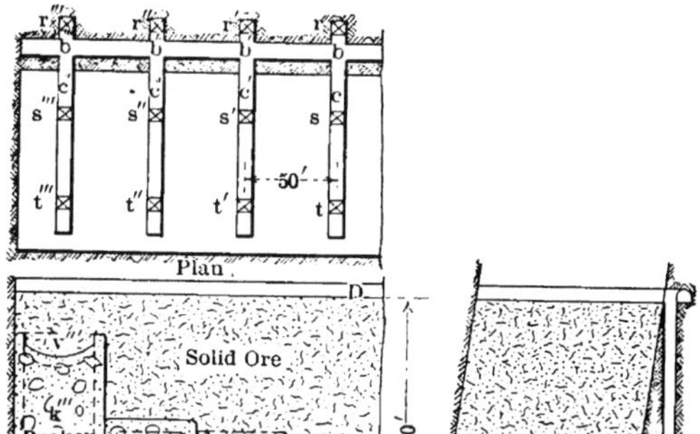

FIG. 89.—Stoping at Hartford mine.

ing enough broken ore in the funnel to form a footing for the tripod of the stoping drill.

This peripheral raise leaves a hanging core v^1 in the back of K^1, which, as the raising continues, will become so heavy as to break off by its own weight with the effect shown in stope K^2. In K^3 height has been gained by several such breaks. Only one drill is worked at peripheral-raising in one stope, and as its two runners are never under the core, they run no danger from its downfall.

Cross-cuts *e* are driven to connect the peripheral raises with a foot-wall raise *r* in order to provide an entrance into a stope after the core has fallen. Stope K^3 should not be holed through to the next level *D* till the ore there, corresponding to the sill-floor pillar and to wedges w, w^1

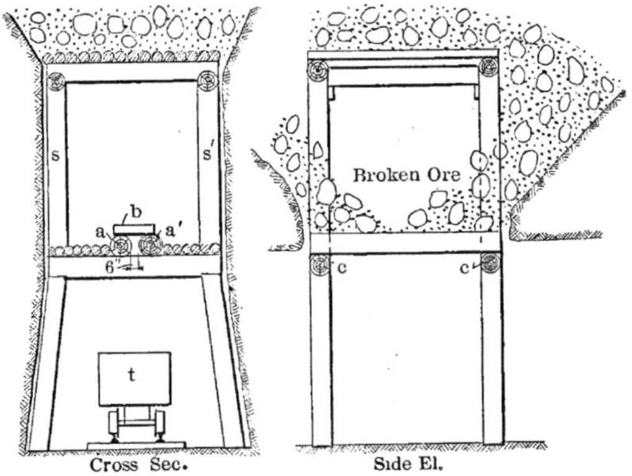

FIG. 90.—Chute at Hartford mine.

and w^2 on level *d* has been removed. This is accomplished by putting up a raise from K^3 to D, and from this attacking these pillars by the room-caving system of Example 45, and throwing the broken ore down into chute K^3.

As stopes K^3, K^2, K^1 and *K* get higher their diameters increase and are merged, so that the miners from there on work the peripheral raise mostly along the foot and hang walls. In wide veins two mills should be placed abreast as *m* and m^1; but these also will merge on ascending. When nearing the upper level *D* the ore is kept near the back so that, when holed through, any débris in *D* from the walls of the emptied stope above will not descend far. By withdrawing the ore gradually from adjoining completed mills the débris can be kept mostly above the ore and a mixture avoided.

The caved ore is not drawn through the usual spout-gates into tram cars, but falls into a car from a central slot in the roof of the haulage drift. To arrange this enough lagging poles, over two adjoining drift caps *c* and *c*' (Fig. 90), are omitted to give space for two 12-in. poles *a* and *a*', the opening between which is covered with 6-in. poles *b*, which can be easily taken up when loading a tram car *t* beneath. Two sets, *s* and *s*', are placed as a screen over this slotgate for the purpose of regulating the passage of the broken ore sliding down from the stope above, as at *x* in Fig. 89.

With this screened slot there is no chute to be choked by the huge masses that often fall off the core above, as these are held back until they can be broken up. The core is often drilled and blasted, when there is danger of its breaking off in chunks too large to be easily shattered after dropping.

This method requires a nearly vertical and a strong hangwall, but the ore is best adapted when somewhat friable, as a dense, tough ore would tend to hang up and cave only in large masses, very difficult to break up below. The vein must be wide enough to permit of cutting a core which is large enough to pay for the raising around it. This method takes no more timber and much less drilling and powder than the underground milling of Example 12, as considerable of the ore is crushed in the caving. Ventilation is good, and wide lenses can be stoped with a minimum of development work. The system does not permit of underground sorting, and some ore is consequently lost by contamination when the filling finally falls into the stope mill.

Labor

The Hartford mine, at the time it was visited, was producing, exclusively by back-caving, 1000 tons in two shifts, employing 24 air drills and 255 men above and below ground. This force included 20 men on dead work, 16 on four diamond drills and 12 men on the stock-pile loading. For each shift there was a general mine foreman and a stope boss on each of the four levels. The development work of sinking, raising and drifting is contracted by the foot and stoping by the ton, one mill hole being let to a relay of four men for each air drill. Stope contracts were being let at 15 cents or 16 cents a ton, the total mining cost being under $1 and often as low as 80 cents.

Example 36.—Pioneer Mine, Ely, Vermilion Range, Minn.

(*See also Example* 20.)

Sub-vertical Wide Vein: Caving into Chutes from Sub-levels.—The Vermilion is the most northerly of the Minnesota ranges and strikes N. 70 deg. E. along the 48th parallel, and, though the ron formation extends here disconnectedly for 80 miles, the only important mines yet located are near the towns of Ely and Soudan, which are 20 miles apart. The local productive zone is the Soudan iron formation of Archaen age, which rests on the Ely greenstone (a basic igneous rock), and is covered by layers of intrusive granite and porphyry. The Soudan contains the typical iron formation rocks, and of these jaspilite is especially abundant.

The ores include very hard, specular hematite and softer reddish hematite, more or less hydrated; but neither of these are "paint" ores in the sense of easily staining the skin red, like those of the paint ranges, the Gogebic and the Marquette. The iron contents vary from 60 to 70 per cent., the phosphorous averages 0.06 per cent. and the silica and moisture about 5 per cent. each. The ore bodies lie near the bottom of the Soudan formation and follow pitching troughs of folded Ely greenstone. A typical deposit is that of the Chandler-Pioneer mine at Ely, which follows an east-west trough and is covered with a thick layer of barren iron formation, except at the west end, where the bare outcrop first revealed it to mankind.

The Pioneer mine produces at full capacity 1,000,000 tons yearly and has two working shafts, *A* north and the newer B south of the ore body. Of these *A* is vertical and 20 1/3 ft. by 7 ft. inside of timbers, the pump compartment being 4 ft. and the two skip and ore-cage compartments each 5 ft. long. The 5-ton skip is of the car type and its four wheels run between two 7-in. by 8-in. wooden guides on each side of a skipway. In the skipways are also central cage-guides for the double-deck cages, which are placed temporarily over the skip when handling men. When idle this auxiliary cage rests on a truck running on a track, supported by the head-frame. The truck can be moved to and fro by an endless rope device worked by a small hand windlass. To enable the cages to be slid into the shaft there is a slot through both decks to pass the skip rope.

Shaft *A* is lined with wood, but shaft B is supported by a steel frame and is inclined at 53 deg. It has three compartments, a pumpway and two skipways. At the stations the skip track is spiked to 12-in. square stringers, but elsewhere to longitudinal plank only.

Tramming.—Two 30 h. p. Goodman electric locomotives handle 1500 tons in 10 hours at shaft B on one level, with an average haul of 400 ft. The track has 24-in. gauge, 30-lb. rails and 0.5 per cent. grade toward the shaft. A locomotive goes out in one, and returns in the other, of the two parallel drifts along the ore body. One motorman for each train and two dump-men at the shaft pocket is the operating force.

Breaking Ground.—The air is conveyed at 65 lb. pressure through an 8-in. pipe for 1 1/2 miles from the central compressor at the Sibley mine. The Little Giant 3 1/8-in. drills are used with + bits made of octagonal steel and operated on counter-weighed tripods. The powder is 35 per cent. dynamite and is fired by fuse and cap.

In the diagrams of Fig. 91, the main levels, *A* and D, are spaced 100 ft. apart vertically with two sub-levels, *B* and *C*, equi-distant between them. On the floor D (see plan, Fig. 91) drifts are run to block out 50-ft. by 75-ft. pillars, made large to prevent a premature squeeze, which might close the haulage-ways. On sub-levels *B* and *C* drifts are run longitudinally near the foot wall, and off from them are turned parallel crosscuts 25 ft. apart. These are connected by vertical raises r, which are put through to level *A* at 25-ft. intervals. Between *C* and D some of

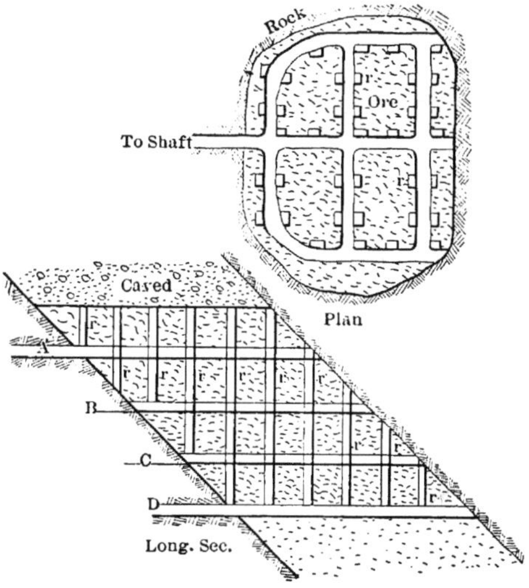

FIG. 91.—Stoping layout at Pioneer mine.

the raises are inclined in order to reach the gates in a fewer number of drifts. Block D is finally removed by breaking it down into the level 100-ft. below with the aid of extra raises.

After blocking out, the caving is begun in slices, and first the slice *A* (above the abandoned level) is attached, next *B* and finally *C*. To be safe, the first cave-panel of level *A* must be kept at least 50 ft. horizontally back from where the men are working at the first panel of sub-level B, and the hang wall is drawn down uniformly from the farther end inward, along the whole ore body before attacking the next panel behind the first. The attack on a slice is begun from a raise by driving at the sub-level *B* a drift for the length of two sets in three directions, one toward the hang wall and two longitudinally. (See Fig. 92.) The face is then attacked beyond each of the three end sets (*n*, *g*, and *h*) of the drifts, and large open spaces, like *f m n* and *h k p*, are blasted out. The domed back is then bored with a pointed bar and blasted to excavate another shell like *f m′ n* and *h k′ p*, and the boring and blasting of the back is continued until the dome becomes too large and dangerous to work beneath.

If the dome caves partly and then hangs up it can often be started again by exploding 8 to 10 sticks of powder, held below the cave on a long stick. Should any uneaved wedges be left above level B they can be recovered when caving *C* by raising through them from below. In drawing down a slice while caving care is taken to draw equally from all the chutes, so that the caved waste above will settle equally and not mix with the ore. When the ore falls from the dome in big masses it has to be blasted to allow it to pass the raises. Much ore runs into the raises by gravity, but at the start of a dome it has to be shoveled.

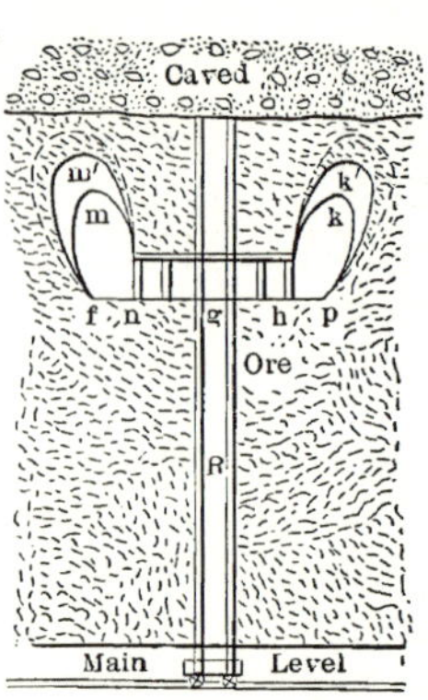

Fig. 92.—Enlarging stope at Pioneer mine.

Timbering.—The greenstone walls need little support, but all the development openings in ore must be closely timbered to resist the heavy pressure of caving. The main haulage ways are supported by three-quarter sets, having caps and posts, 8 ft. long by 1 ft. to 2 ft. diameter, with the latter battered 2 in. per foot. The raises are cribbed closely with round sticks, halved at ends and 6 in. to 10 in. by 5 ft. The caving pressure often crumbles the drift sets, but seldom before their usefulness is about over. Little timber is recovered from the caved ground, and that saved is only good for cribbing.

Application of Sub-level Chute-caving System.—Its success is favored by the following conditions: The deposit should be large, with regular and well-defined boundaries and with a uniform and yielding hang wall on not too steep a pitch. The ore should be soft enough to crush under moderate pressure, and should be all of the same shipping grade, as no sorting is practised underground. The value per ton should not be high, for some ore is unavoidably mixed with the caved waste and lost.

Where applicable it is a cheap system, being easily ventilated, safe, and requiring no t mber and but little powder in stoping. Even the development openings are not proportionately numerous, so the total timber, drilling and powder used per ton of ore is small in amount, counting both surface and underground force. The output per shift is about four tons for each man employed.

Example 37.—Utah Copper Mine, Bingham, Utah

(*See also Examples* 3, 33, 41 *and* 43.)

Irregular Lenses in Porphyry. *Caving into Chutes from Sub-levels.*— The underground workings of the Utah Copper Company are opened by two main levels, one even with the bottom of the gulch and the other 200 ft. above. Formerly the caving method was used on both sides of the creek, Fig. 93 but now underground mining is confined to the ground to the northeast of the creek, it being the intention to use steam shovels for mining all the ore on the other side. The lowest, or transportation level, is cut up into irregular panels about 75 ft. square, although lately the tendency is to increase the size of these panels somewhat. These drifts

Fig. 93.—Caved surface, Utah Copper mine.

and cross drifts are driven 5x7 ft. in the clear for single track, and 5x11 ft. for double-track service. These drifts rarely require timbering as the ground stands well, but where weak, 8x8-in. sets with a cap 5 ft. in the clear, and posts 7 ft. 3 in. long, having a batter of 1/2 in. to the ft., are used for single track, and 10x10-in. timbers, 11 ft. cap and same length post for double track. In driving these drifts 3 1/8 in. drills are used. Back holes are drilled about 4 1/2 ft. deep and cut holes about 5 1/2 ft. It takes from 7 to 10 holes to break the ground. A 4-ft. round is broken in a shift, and the cost is from $3 to $3.50 per foot.

The ore is hauled in trains of eight or nine 2 1/2-ton bottom-dump cars by 5-ton electric locomotives. A direct current of 500 volts is used, but as the trolley wire is placed about 6 1/2 ft. from the rails there is little danger, especially as at chutes and low places where a person might touch the trolley wire it is enclosed in an upside down trough (about 6 in. wide and having 4 to 6 in. sideboards) nailed to the sprags that support the trolley wire. The wire is strung from 4x4-in. sprags placed about 30 ft. apart. A thirty-pound rail is used. The main loading chutes are fitted with double rack doors so as to facilitate loading.

The main raises are driven 5x7 ft. in hard ground, but only 4x6 ft. in soft ground, since in such ground they widen out with use only too quick. In driving the raises a 2 1/4-in. drill run by one man is used, and it takes six to nine holes to break a round 3 1/2 to 4 ft. deep. These main raises are driven on contract at a cost of $2 a foot, the contractor furnishing labor only. These main raises are driven vertical for 30 ft. and then are turned so as to have an inclination of 60 deg. in wet ground and 50 deg. in dry ground in the direction of the surface slope. This enables the raise to serve more ground, while the change in angle breaks the fall of the ore and prevents its packing in the chute. The 30 ft. at the bottom that is vertical has proved quite sufficient to insure an even feed at the mouth of the chute. In extremely soft ground the chutes are cribbed. The branch raises are driven approximately 5 ft. square and on company account, 2 1/4-in. machines being used. These branches are sometimes advanced on a slope as low as 45 deg. it is said. All these raises are driven without a manway, and later, when mining of the ore begins, become blind chutes.

Occasional raises are put up to serve as manways; these are driven about 5x7 ft. in size so as to be large enough for air pipe, ladder and timber-slide These ladders are made with wooden legs and half-inch round iron for rungs so that they will not be broken by drills that may occasionally slip out of the chain when being hoisted or lowered. All timbers and drills are hoisted through these manways by hand.

Sub-level Interval

From the main raises sub-levels are driven, and most of them are connected with the surface so as to provide excellent ventilation. These sub-levels drifts are driven 4x7 ft. in the clear, and, where it is necessary, are timbered with round sets having caps 4 ft. in the clear and posts 7 ft. long. In mucking out the drifts, wheelbarrows are used unless the run is greater than 75 ft. Then tracks are put in and cars used.

The level interval was 17 ft. at first. This was increased to 25 ft.; then to 30, and finally to 33 ft. In a few places an interval of 50 to 60 ft. was tried, but with so high a back of ore it was impossible to control the caving satisfactorily so a great deal of ore was lost. From this experience it seems that 35 ft. is the limit of economic caving by this method in the Bingham porphyry.

When it is desired to begin caving, all the raises in that part of the mine are fitted with ordinary chute mouths, and at the same time a grizzly made of 10x10-in. timbers, spaced to give openings 18 to 20 in. wide, is placed over the top of the raise in the slice below. The bulkheads are necessary not only to prevent large boulders from getting into the main chute and blocking them, but also in order that the chutes can be bulk-headed in case capping begins to run from above.

After the raises have been so equipped, the tops are widened until caving is almost imminent. Then holes are drilled until it is certain that the roof will cave when they are blasted. Adjacent raises are similarly widened out and blasted, and the panel caved. This begins at the boundary and progresses away from it as the different chutes run capping and have to be bulk-headed. All these raises are put up on an angle of about 50 deg. by one man using a cross-bar and a 2 1/4-in. machine. He stages up with a couple of sprags. By driving the raises inclined the work is rendered easier, less dangerous, and cheaper.

The bottoms of these raises are placed within 25 ft. of each other on the sub-level and when caving begins they are almost together at the top. As can be seen, a pyramid of ground with a base 25 ft. square is left standing. In order to start this to caving a chute mouth is built and a raise driven in the block. One hole is kept 2 ft. ahead of the other holes so that the miner will know when he approaches loose ground. So long as this hole does not show solid ground above, the holes are blasted, the broken ore drawn off, and drilling begun again. But whenever this hole shows the ground to be crushed and broken, the sides of the enlarged top of the raise are drilled, and then all are blasted at the same time, caving the raise. This is called by the miners a "general fire or general blast." The drilling of one hole 2 ft. deeper than the others insures against the round's leaving only a shell of a roof to catch the miner when he begins to pick down the back, for it has been learned by experience that a raise with a solid roof 2 ft. thick will not cave.

The "general fire" in this raise usually caves the whole pyramid, but in order to get the ore near the base it is necessary to put in along the drift chute mouths quite near together, so as to draw off all the ore. This is especially the case where, as it sometimes happens, a drift has to be driven to tap a pillar that has caved before all the ore could be drawn off. The ore from these secondary chutes is drawn into 1200-lb. cars running on a 12-lb. track and trammed to the nearest branch chute.

Finally before abandoning that portion of the mine, branches from the branch raises in the block below are driven so that their tops are within 15 ft. of one another. The tops of these are widened out into a funnel shape by means of water holes, and the bases of the pillars drilled so that, when all the ore that can be obtained from the ground above has been drawn, these pillars can be blasted and the stope caved. The ore then runs into the chute in the block below and is drawn off through it. Thus the ground above each sub-level is caved working back from the boundary, care being taken not to undermine any portion that has not yet been caved on the sub-level above.

Toward the end each chute runs mixed ore and capping, easily recognized by its reddish, oxidized appearance. Drawing continues and the capping is sorted out until one man can no longer run 20 to 25 small cars (1200 lb., capacity) in a shift. Then the chute is abandoned. Sometimes the chutes get hung up, but a stick of dynamite soon starts them again. In case the hang-up occurs well up in the chute, a lighted primer fastened to a long pole is placed against the hang-up.

Disadvantages of Utah-copper Caving Method

There are many drawbacks to the caving method used by the Utah Copper Company. The ore is broken mainly in drifts and raises. By this method the amount of development on each sub-level is excessive, for each raise must be met by a drift on each sub-level, so as to block it off whenever capping begins to run into a chute. Indeed, a more expensive system of undercutting a block would be hard to devise. Again, instead of making use of light air-hammer drills, these raises are driven by means of heavy piston drills mounted on a bar. In blocking out the ground, wheelbarrows in some cases are used, although it is known that through all the drifts some ore will have to be trammed in cars.

For successful caving it is necessary to have the surface settle as regularly as possible, and to drop it over as wide areas as feasible, so as to avoid mixing the ore and overburden at those places where the capping slides down past the ore. Even in systems where this is done, the loss of ore is considerable. But with this chute-caving method the ore and capping mingle together throughout the height of ore caved, a distance of about 33 ft.; for, as the ore is drawn off through each chute, the capping follows down, rubbing past the rough sides of the ore whose fracture planes have been opened up by the weight thrown on the pillars in the block. This mixes some ore with the waste.

Drawing continues until the chute is filled with a mixture of ore and waste too low in grade to pay for mining. Then this chute is blocked off, another raise is put up to cave the pillar adjacent to this chute, and the process is repeated. This mixing of ore and capping becomes still greater as the caved ore is drawn down past the capping that fills the chutes that have already been drawn. The percentage of ore lost can only be approximated, but considering that to mine the ore, said to be 310 ft. thick, nine sub-levels are necessary, it would amount to at least 7 per cent., and where only three or four sub-levels are caved 15 per cent. to 20 per cent., if not more.

Another disadvantage of the Utah Copper method is that it is not systematic. A raise is put up to cave a pyramid of ground, the shape of which is not known. Ore is pulled from the chute mouth until, owing to mixture with capping, the grade becomes too low to pay for mining.

Then similar raises are put up to cave other pillars having unknown shapes. To decide upon whether all the ore or even most of the ore has been obtained is impossible. Besides the Utah Copper Company, only the neighboring Ohio Copper Company is using this method of mining.

Cost of the System

At the time observed, the Utah Copper company was mining 1500 to 1600 tons of ore per day by caving and was working 225 men underground. As it was said that as much ground was being prepared for caving as was being mined, this would indicate an average of almost 7 tons to the man underground. The average pay of these men is probably about $2.65 a day, so that labor alone costs 38 cents per ton. Assuming that labor makes up 60 per cent. of the cost, this would indicate a mining expense of 63 cents per ton. This cost represents the cost of caving the ore and does not include the cost of blocking the ore out, which item is very high, owing to the numerous drifts that have to be driven on each sub-level and because of the large number of raises. Besides, it is known that when the ore was mined by the older system, in which the caving was done by enlarging rooms, the cost of mining was at first $1.25 a ton and later it was reduced to $1.10. Considering that considerable timber is necessary in obtaining the last ore in each slice, the total cost of caving probably exceeds 80 cents per ton.

CHAPTER XV

BLOCK-CAVING SYSTEM

EXAMPLE 38.—PEWABIC MINE, MENOMINEE RANGE, MICH.

(*See also Examples* 8 *and* 46.)

Sub-vertical Wide Vein. Blocks Cut Off by Underhand Stoping. No. Chutes.—This mine is located just northeast of Iron Mountain and is operated by the Mineral Mining Company of Milwaukee, Wis. The main orebody is a deep lense, about 2,000 ft. long by 200 ft. thick, with a soft talc-slate footwall and a silicious-slate hangwall. It dips northward 76 deg. to 90 deg. and is overlaid by horizontally bedded Lake Superior sandstone. The bulk is hard, low-grade, silicious hematite, like that of the Traders mine; but within the walls is also a large shoot of high-grade, blue hematite resembling Chapin ore.

Slicing.—The high-grade shoot is soft and is stoped by slicing, but with only five sub-levels between the levels (125 ft. vertically apart). This gives a distance between sub-levels of nearly 21 ft. and a back to cave, above the room cap, of over 12 ft. Since no floors are laid down before caving, as in other mines, it might be thought that the resulting contamination would be disastrous; but, as the high-grade shoot lies entirely within the vein, it is not waste but low-grade ore which follows a caved back onto the 2-in. plank sollar, from which the shovelling is done.

Block-caving.—This system is always used for the low-grade ore, but within the length of the high-grade shoot the former ore is not touched until the latter has first been removed by the slicing of Example 45. Block-caving is diagrammed in Fig. 94.

A block is 250 ft. long on the vein and 125 ft. high between the levels. The block is laid out by driving a wide footwall drift *f* and four cross-cuts *c* to *c'''* about 80 ft. apart to the hangwall, connected by drift *h*. Next raises *r* are put up at 50-ft. intervals along cross-cuts *c* and *c'''* to within 20 ft. of the filling above. Cross-cuts *b* and *b'''* are then driven, over *c* and *c'''*, at the top of these raises and wooden chutes put at their bottoms to allow gravity loading.

From cross-cuts *b* and *b'''* underhand stopes of 8-ft. width are then cut down to *c* and *c'''*. Simultaneously the block has been undercut by breast-stoping from the cross-cuts *c* and *c'''* and from drift *h* until it is only supported by small round ore pillars *p* (dash and dot in the plan,

Fig. 94), except alongside *c* and *c'''* where transversal strips *K* (dash and dot) are left, and these are drilled, to be broken later. As the ore is hard no timber need be used in this development of a block.

The 250-ft. block is now free at the top and practically free at each end, so that as soon as the supporting pillars *p* and *K* are blasted out by sections the caving can begin. But settling is slow, and a month may pass before the block has reached the sill floor and half a year before self-crushing has reduced the ore to first-size ready for removal.

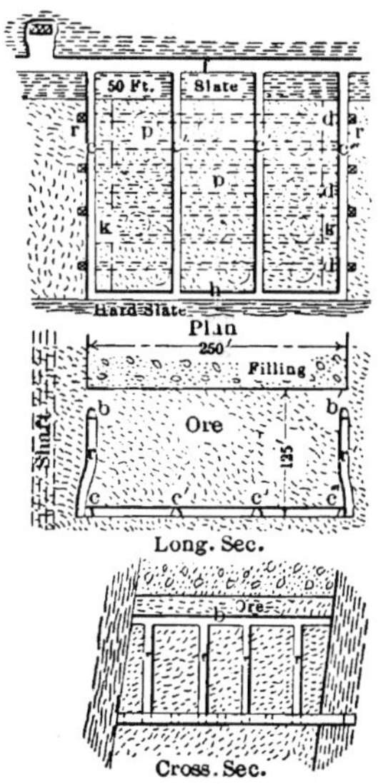

FIG. 94.—Stoping at Pewabic mine.

Drawing of the caved ore is then begun by allowing it to run from the face of the seven drifts *d*, which, like the reopened cross-cuts *c* and *c'''*, have been driven (closely timbered) through the broken ore at 25-ft. intervals. The ore falls onto plank sollars, to be shoveled into tram cars, and when any face ceases to run ore, and shows filling, the next inward set is blasted out to get its superincumbent ore, and the withdrawal is continued until the cross-cuts *c* and *c'''* are reached. To exhaust the block it is now only necessary to drive new drifts from c and *c'''* half way between the caved drifts *d*, and recover their superincumbent ore by the same withdrawing process and finally withdraw the crosscuts themselves.

During this drawing there is at each drift face a crew of one miner and four muckers. The miner keeps the ground open by blasting any scaffolds, and watches the safety of the shovelers. Two muckers load and the others shove the tram car to drift *f*, to be attached to the endless-rope haulage system. The stoping is only conducted during the open season, the winter being occupied in cutting out the blocks. For an output of 1,800 tons daily 500 men and 31 air drills are required; but, as a large portion of the extraction is rich ore (got by room-caving) this gives no average output for block-caving alone. It is estimated, however, that the total cost of mining the lean ore alone by block-caving is only 65 cents to 75 cents a ton.

Recently, in starting a top block just under the sandstone capping, it failed to part well from the sandstone when dropped, so inclines had to be put down from the footwall drift and raises put up from their ends to recover the huge masses adhering to the sandstone. Such mischances can always be avoided by cutting away the ore from any adhering side of a block by a narrow stope before dropping it. Here the ore usually parts easily from its slate walls, so the isolating stopes need only be cut at the block's ends; but in other veins it might be necessary to cut stopes along either or both walls. This system can be also applied to soft ore by the use of some timber during development.

Block-caving is adapted to any long-bedded body of brittle homogeneous ore with tough definite walls if the ore parts easily from the walls; if not, the bed must be of sufficient width to allow the space and expense for cutting an isolating stope on the clinging wall. If a wall is not tough it will peel off and mix with the ore, if such wall is a hangwall or a highly inclined footwall. If a wall is indefinite a contamination will also take place, unless a regular isolating stope is inserted between the block of good ore and the transition zone.

Given a suitable deposit, block-caving takes less powder, drilling and timber than most of its competitors. It is safe, provides good ventilation, and a great and easily varied output from one level, while blocks (if not superimposed) can be worked at the same time on different levels. In ores which are very hard to drill, but yet are brittle enough to break up by block-crushing, this system is very economical. It has been successful at the Pewabic mine for seven years in producing annually about 250,000 tons of the leaner ore.

Example 39.—Mowry Mine, Santa Cruz County, Ariz.

Sub-vertical Pipe in Limestone; Blocks Cut off by and Caved onto Square-set Floors.—The town lies in a small basin bounded on all sides, save one, by low hills formed of the same Archæan granite and schist as the basin itself. On the excepted or north side, the bounding hill is limestone and it is in the contact between the basin granite and this hill that the ore-bodies occur. The contact runs east and west and dips about 80 deg. northerly, while the abutting limestone dips 45 deg. in the same direction.

At the surface the ore outcropped as a series of disconnected shoots, extending along the contact for half a mile.

Much of the ore is spotty and crumbly in nature and is a mixture of clay and $CaCo_3$ with Fe_2O_3, MnO_2, and $PbCo_3$, with most of the silver in the last. The largest chimney is 180x65 ft. in section, and is the richest as the ore appears to become less manganiferous and more siliceous as the dike is receded from. As in Leadville, the lead carbonates occur hard and siliceous, as well as soft, while the accompanying iron and manganese oxides increase the resemblance of the two camps. Most of the ore is so soft as to run freely with moderate pressure and on this fact depends the success of the excellent caving system hereafter described.

The vertical position and soft nature of the largest shoot permits the use of the caving system illustrated in Fig. 95. Two shafts are sunk, one at each end of the shoot, in the walls, and levels established at 150-ft. vertical intervals. On the first level, these shafts are then connected by drift *b*, through the shoot, and cross-cuts *d* turned off from *b* spaced on 25-ft. centers. From this drift and cross-cuts, the balance of the ore on the sill floor can be removed. All the cavity is timbered with square sets of 10-in. Oregon fir, 5 ft. square horizontally and 6 ft. 10 in. vertically to center lines. The first floor above the sill floor is now removed and the same size sets inserted, while chutes *C* are constructed by siding up nearly every alternate set with 2-in. planks. The usual wood gates, for dumping into the 1-ton mine cars running on tracks *b* and *d* are now inserted below each chute.

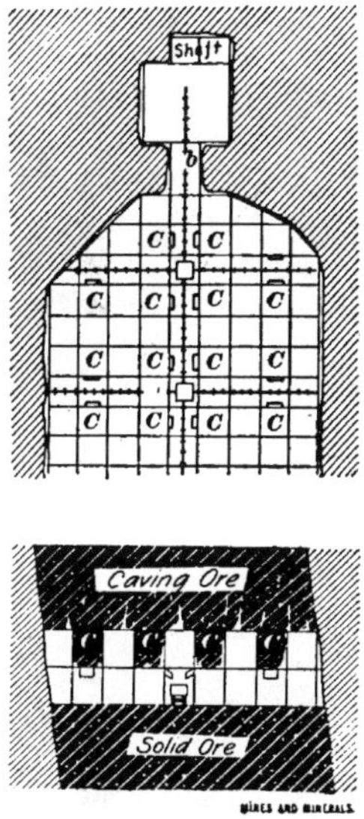

FIG. 95.—Stoping at Mowry mine.

Caving can now begin by removing the roof lagging over the chute sets, so that the ore will sink from the back into the chutes. By drawing from all the chutes uniformly, the ore can be made to settle vertically, so as not to distort the square sets. Occasional angle braces or filled sets may be necessary. Where the back is strong locally and bridges a chute, it can be barred down by miners standing on the first floor below in a vacant set. If the chutes get clogged with boulders, blasting is resorted to. Where the back is too hard to cave, as happens with some of the silicious ore on the granite footwall, square sets are carried up above the first floor, high enough to remove it.

By this system, 80 per cent. of the ore is removed without timber, powder, or filling, and without working under an unsupported back. Its only disadvantage is that but one level can be stoped at a time. But a ower level can be preparing, while the first level is being stoped, and in a body of this size with plenty of cars, the output need be limited only by the hoisting capacity of the shafts. At present from 100 to 125 tons are hoisted daily, of which some 80 per cent. is concentrated.

As the mine is less than 400 ft. deep, the shafts have only one cage and one pump compartment and the hoists are single-drum. Sullivan 2 3/4-in. rock drills are used for development and for the hard silicious stoping. About 100 men work under ground, nearly all being Mexicans except the bosses.

Example 40.—Detroit Copper Mine, Morenci, Graham County, Ariz.

(*See also Examples* 22, 29 *and* 30.)

Irregular Lenses in Porphyry; Blocks Cut off by Cross-cuts and Raises and Caved into Chutes.—Block-caving stopes are laid out in much the same way as the slicing stopes of Example 46, by drifts and cross-cuts on the level and raises to the top of the ore, but mining a top floor with square-sets is unnecessary. While a square-set top floor with timber mat would temporarily afford a better separation between waste and ore, it soon is disarranged and the timber becomes a nuisance, blocking the chutes. It would not be necessary to push the raises to the top of the ore if the waste line could be established and maintained otherwise. If the raises are brought up to the waste, it is a good idea to strip them of timber before the actual caving begins. The operation is as follows. At a certain distance—20 to 35 ft. below the top of the ore—a working floor is started, called the 'sub,' or sub-level. The height of these sub-levels depends on the character of the ore. The easier and more regular the ore breaks, the higher are they taken. On them all raises are connected by drifts and cross-cuts, leaving blocks about 20 ft. square, in case of 5x7-ft. raises, 25 and 30 ft. between centers. These blocks are subdivided again by other intermediate drifts and cross-cuts run parallel to the first series, leaving the block of ore standing on four legs. These are weakened by drilling and blasting as far as safety permits, and again drilled preparatory to the final blast, as are also the backs of all the drifts. This work is done with hammer-drills (Waugh, Shaw, and others). When a part of the stope has been prepared in this way the pillar and back holes are loaded and blasted electrically. The broken ore is shoveled into the chutes from that part of the stope still standing and drawn out below through the original chutes, and through others that are driven up to near the sub-level and then holed through into the broken mass of ore. The original raises are, as a rule, vertical and timbered. The secondary ones can be inclined, and if the ground is solid may not require timbering. The ore is drawn out until mostly waste begins to show at the chute. Occasional blasting to break boulders will usually keep a chute going as long as needed. In principle, this method of stoping makes it possib'e to dispense entirely with timber, except in the original raises, where it can be recovered. In practice stulls will generally be found necessary here and there, but the less timber used the better. Figs. 96 and 97 illustrate the method of working a block-caving stope.

Comparison Between Square-setting, Slicing, and Block-caving.—Square-setting can be done under the best conditions prevailing at Morenci for about 80 cents per ton of ore extracted, at least 20 per cent. of which would be for the timber alone; but unfavorable conditions, such as heavy ground, by necessitating reinforced timbering, careful filling, and small-sized stopes, make the cost run up to $2 and more per ton of ore. Slicing, as in Example 46, would cost, under favorable conditions, perhaps 60 cents per ton. While requiring nearly the same amount of timber, it increases the tonnage mined per man and employs a greater proportion of

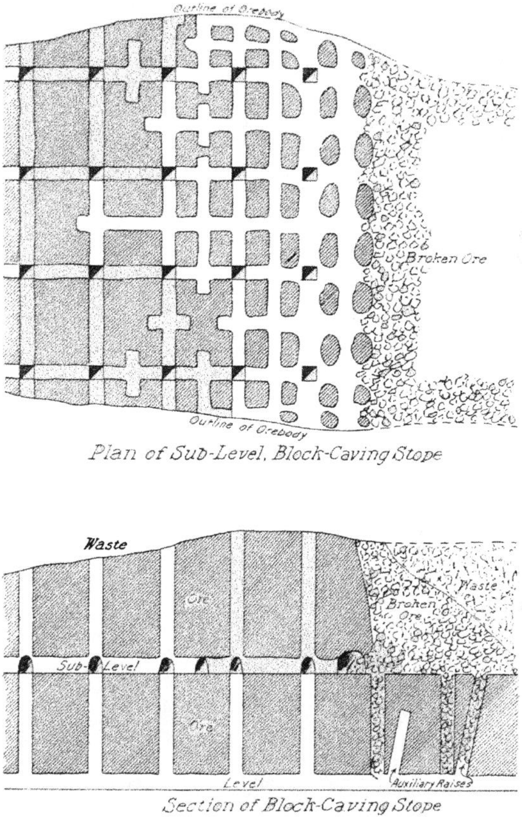

FIG. 96.—Stoping layout at Detroit mine.

common labor. Heavy ground will not affect its operation as readily as it does square-setting. One disadvantage is that first-class ore, from 6 per cent. copper up, used for direct smelting, can not be sorted out easily. An effort has been made to accumulate it in the stope and to run it out through one chute, set aside temporarily for this purpose, but the practice was abandoned. Leaving the leaner parts of the orebody, which can be done easily in square-setting, is also rather difficult and costly in slicing, as it breaks up the continuity of the mat and necessitates new square-setting below, should the orebody change again for the better. These difficulties make it necessary to mine the low-grade ore as well as the other. Block-caving can be done for about 40 cents per ton. It gives the greatest tonnage per man and shift, and reduces the timber bill to almost nothing, but it is apt to result in mining large amounts of low-grade ore that would otherwise not be mined, and which, with the

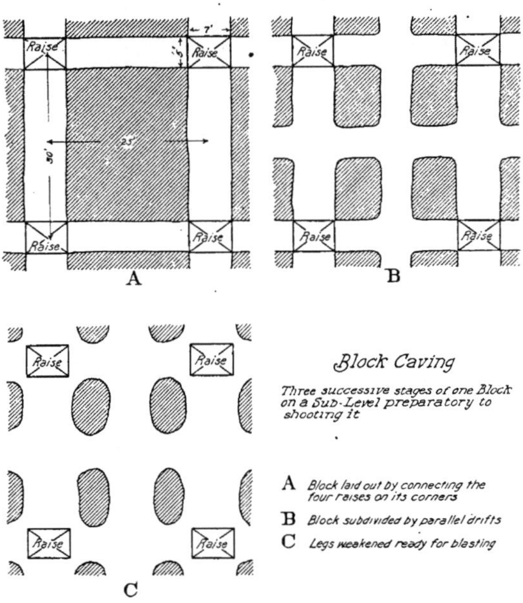

FIG. 97.—Plan of stoping details, Detroit mine.

loss in concentration, can not possibly pay expenses. Heavy ground affects this method but very little. Both slicing and block-caving represent a large initial expense, and take long preparation before a stope is ready for extraction, but once started the production is more centralized and a very large tonnage can be rapidly mined from one stope.

A careful sampling should follow closely the opening of a block of ground to be mined by the caving method, as mine samples as ordinarily taken prove usually to be higher than the ultimate mill sample covering the same block of ground, and in extraction, much of the profit of what should be won by cheap mining may be lost in taking too low a grade of ore. Sometimes more careful sampling might result in abandoning part of a stope already blocked out, as too low in grade to afford a profit, which otherwise might have been mined at a loss. To reap the full benefit of the cheaper working methods, improvements in mine sampling and concentrating are two important factors, and the Morenci companies, especially the D. C., have conducted researches in both branches for several years past. The floor plan of the stope on the different levels should be as simple as possible, and only enough drifts opened to afford the necessary facilities for tramming and starting the raises. The upkeep of the raises and tracks will run into a rather heavy repair bill in any event, and this should be kept to the lowest possible point consistent with the rapidity of working and handling materials and ore. Morenci is somewhat handicapped in reaping the full benefit of the cheaper caving methods. In the first place, its orebodies lack the regularity and the even tenor of value of some other porphyry ore camps. Then, its orebodies are not intact, having been worked in their richer and more accessible parts by square-setting, and as a result, in places there are some badly cut up pieces of ground which now remain to be worked by the new system. Last, but not least, mining in Morenci is not very speedy. The camp is somewhat conservative; hand-drilling is still the prevailing practice, machine-drills having been introduced only these last few years. More rapid mining would certainly reduce the cost, especially of slicing, and make possible the recovery of a good deal of timber that is now blasted and lost, at the same time that it would increase the tonnage output per man. The work is now frequently handicapped by the closing in of the mat due to settling and by the necessity of close stulling. New orebodies that afford the opportunity of close prospecting by boring present many advantages, as in these instances they might be mined by attacking the top by slicing, while raising and driving is going on below, thus minimizing the time required for keeping these workings open. Systematic boring, too, would give at the same time a more reliable sampling than by any other method, even if it is liable to give a slightly higher average, as was pointed out recently by L. D. Ricketts at Cananea. Boring might also help to avoid losses in other ways. For example, at Morenci an orebody was prepared for block-caving, when another orebody was discovered in close proximity. By the time the second orebody was opened the ground of the entire vicinity was breaking and settling, owing to the removal of the first. The drifts could not be kept open by reinforcing the timbers, even with use of angle braces or doubling. The raises were settling, and almost constant timber-changing and easing of ground had to be done.

EXAMPLE 41.—COMMERCIAL MINE, BINGHAM, UTAH

(*See also Examples* 3, 33, 37 and 43.)

Bedded Lenses in Sloping Limestone. Blocks Cut off by Cross-cuts and Raises and Caved into Chutes.—At this mine there are large lenses of low-grade silicious ore, 200 to 500 ft. long and 30 to 70 ft. thick with limestone walls, and for them the caving system is safely and economically applied as follows:

No timber is used, except for chute gates, as the ore will sustain itself over the small excavations opened. The mine is developed by an incline, following the general dip of the limestone, and along it are started drifts at 100-ft. intervals. An ore lens dips with the limestone and it is first blocked out by driving and raising on its footwall, so as to divide it into 50-ft. blocks, as do sub-levels *a* and *b* and raises *s* and *s'* in block *A* of Fig. 98. Extra holes are put in the sides and back of these drives, to be blasted later as will be explained. Caving begins in the top strip and in the blocks at each end of it, the work thus proceeding both ways toward the center of the lens. Only when the top strip is completely caved does the caving begin in the end blocks of the next lower strip, though development work may go on anywhere.

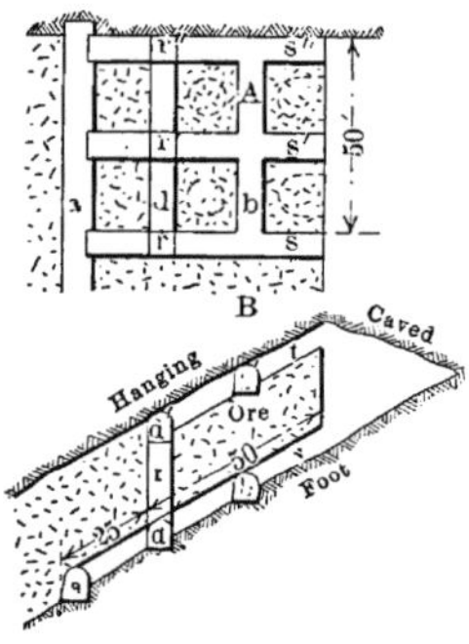

FIG. 98.—Stoping at Commercial mine.

When sub-dividing a 50-ft. block below caved ground, as in Fig. 98, we have as a start the sub-levels *a* and *b*, the footwall raises *s* and *s''* and the raise *s'*, down as far as sub-level *b*. The first work is to put in drift *d* on the footwall and extend raise *s'* from sub-level *a* to *b*. Next a chute gate is placed in sub-level *a* below each of the raises *s*, *s'*, and *s''*, so that the broken ore can be spouted from above into mine cars and proceed to some chute connected with the main level below. Vertical raises, *r*, *r'*, and *r''*, are then carried from drift *d* to the hangwall and there joined by drift *d'*. Along the hangwall three raises are now put in (directly above the footwall raises *s*, *s'*, and *s''*) and finally drift *b'* is opened.

As extra holes have been put also, where necessary, in the sides, back, or floor of these openings in order to save extra set-ups of the machine drill, the further excavation is simplified. Holes are now blasted along the sides of the hangwall drifts and raises until only round pillars (shown dotted at *A* in Fig. 98) remain and these are drilled. The foot-wall openings are then treated likewise to leave only drilled pillars. The round hang-wall pillars are next blasted and then the foot-wall pillars, and the whole blocks should then break off along vertical plane *d d'* and fall on the footwall if the following points have been considered:

The block is a cube of which the upper side was separated at the start; by the given work, the top and bottom are quite free and the lower side cut up by the three vertical raises, leaving only the two sides (that have been rimmed around their edges by the extra holes) to be sheared by the caving. Should the shearing of the merely rimmed sides not take place, one or more intermediate raises must be put in along it, parallel to raises *s* and *s''*. At the next block *B* of this strip, where two sides, nstead of one, are caved at the start, only one side remains to be separated from the matrix by intermediate raises between *s* and *t*. After caving, the material that is too large to run through the gates must be block-holed, by ascending through the chute gates into the cave. After the completion of blocking out and final blast ng of the raises, enough time is allowed for self-crushing of the block before beg nning to draw the ore.

Example 42.—Inspiration Mine, Globe District, Ariz.

(*See also Examlpes* 21 *and* 32.)

Irregular Lenses in Porphyry. Drummond System Blocks Cut off by Overhand Stoping, Ore-mat under Capping. The underlying principle of this system is in having drawing-off stopes from which the ore is raked into nearby drawing-off raises. With this as the nucleus from which the method is expanded, Supt. Drummond has devised a very ingenious method of caving the ore in mass.

On the "Tramming Level" (see Figs. 99 and 100) the mine is cut up into a series of blocks 75x200 ft. The drifts *d* running across the orebody are made large enough for electric haulage, while the other drifts *e* are merely subsidiary drifts for connecting the haulage drifts with one another at convenient intervals. So these drifts are placed 200 ft. apart and driven only the ordinary size, as none are intended for tramming except one used as a main longitudinal entry. Along the haulage ways *d* and on one side of them, raises *r* are put up at intervals of 33 ft. These later become the loading chutes for the trains and so are fitted with chute-mouths. After this raise has been carried up vertically for 12 ft. two branches are driven to the "Mucking Level" 50 ft. above, at such an angle that they hole into that level at such a distance from the mucking openings, which connect with the drawing-off stopes *s*, that the broken ore will run out almost up to the edge of the raise, and yet is prevented by its angle of repose from quite getting far enough to enter the chute that the raise becomes later. During mining, these chutes will be covered with grizzlies having an opening 14 in. wide between bars.

On the Mucking Level, the orebody will be cut up into blocks by a series of drifts *f* and *f'*, and cross-cuts *s*. The drifts *f* (directly over the

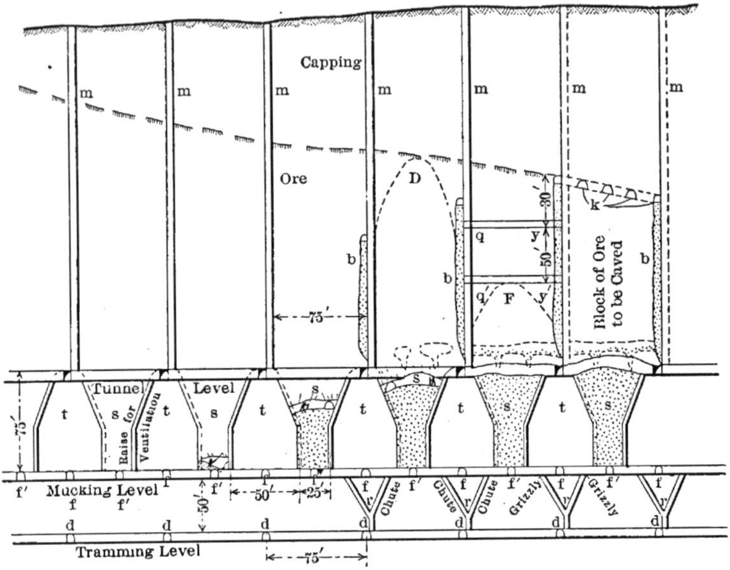

FIG. 99.—Long. section of stope, Inspiration mine.

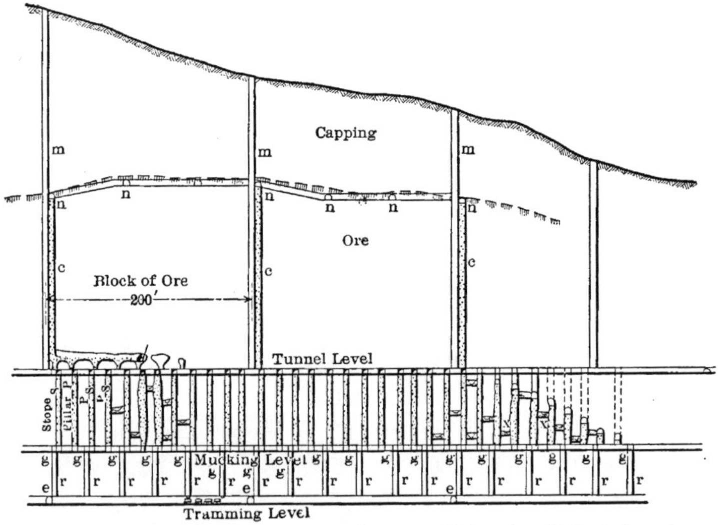

FIG. 100.—Cross-section of stope (drawing off stopes are being enlarged), Inspiration mine.

haulage drifts *d*) will serve later as entries to the mucking places when drawing begins.

The other drifts *f′* will serve for tramming the ore from the chutes that are used in cutting the stopes *s*, and also for entry to the raises and chutes that serve the narrow stopes *b* and *c* that are used to cut the orebody into blocks, so that the ore will be more under control while it is being caved.

The drawing-off stopes *s* will be 25 ft. wide at the bottom, and 50 ft. wide at the top on the "Tunnel Level," 75 ft. above. These stopes will go up vertically for 40 ft., and then the sides will be given a flare. In cutting them, raises will be carried up at the ends and to give good ventilation and for safety during mining the stopes will be joined by cross-cuts like *x*.

As the stopes are carried up, they will be widened in hard ground so that only a narrow pillar is left between them, while in soft ground the stopes will be kept narrow. In this way the 200-ft. block will be undercut by a series of narrow stopes with pillars between so narrow that they will collapse when the ore is drawn off.

In the meantime above the Tunnel Level the orebody has been cut up by a series of narrow stopes *b* and cross stopes *c* into a series of blocks 75 ft. wide and 200 ft. long, that are centered above the drawing-off stopes *s*. To afford ventilation to these blocking-out stopes *b* and *c*, and also to aid in the cutting off of the orebody at the top from the capping, a few raises *m* are put up to surface. These will pay for themselves, as the supplies can be lowered through them, while this work in the upper part of the orebody is being done.

Cutting the Ore off from the Capping

As soon as the blocking-out stopes *b* and *c* have reached the capping, the cutting off of the capping from the ore begins. This is done because it is thought that there will be less mixture of the ore and waste rock if there is a break between the two. Then when the ore caves, it does not pull the capping down with it, but rather the capping will follow down after and on top of the ore. This cutting of the ore free from the cap rock is done by means of a series of parallel drifts *k* (with a pillar 12.5 ft. wide left between them) and cross-cuts *n*. This pillar is drilled as the drifting advances; later, when the boundary of the ore has been reached, the pillars will be blasted retreating, and the capping dropped. These drifts will be driven in the ore immediately under the capping, but not necessarily all on the same level.

When the ore has been cut off from the capping and the orebody is cut up into the 75x200-ft. blocks in the manner already described, the blocks are undercut on the Tunnel Level. This is essential in order to insure that all the ore throughout the area will get in motion when the drawing of the ore begins, for the drawing-off stopes do not touch one another by about 25 ft. This undercutting is effected by opening out a series of rooms between which pillars are left which are small enough so that they will collapse when the ore is drawn out from between them, or when the bottoms of them are blasted.

When all this has been accomplished the drawing of the broken ore in the stopes *s* begins. This causes the pillars *p* to collapse and then the whole mass of ore in the drawing-off stopes under the 75x200-ft. block is set in motion. This, as the ore is drawn out, causes the pillars above to collapse and come down and the whole mass of ore in the block itself is then on the move. Some of it will break coarse and the rest of it small, but whenever a boulder of ore gets down so as to show in the drawing-off holes it is blasted. Then the raking of the ore into the drawing-off chutes is resumed.

The lower part of the orebody will be broken up by its fall, due to the giving way of the pillars *p*, but in time, as mining progresses further ore will begin to come down in large masses. These will come down as far as the drawing-off stopes, and then will temporarily block them. As the drawing of the ore progresses, an open space will form under this immense boulder of ground until there is quite an open space below the boulder roof of the stope, but the weight of the ground above the boulder. or slab, as it might better be called, will finally cause the arch to break, In fact the beauty of the system is that, if the slab is thick, the weight of the unsupported ore will cause the center of the arch to begin to dribble away so as to approach a dome or arch of equilibrium, as is the case in stopes where they begin to cave. This crushing action will cause the ore to break up fine. But, as the drawing-off stopes are large, the arch will give way before the stope is drawn empty of fine ore.

In the Drummond method there will be great wear and tear on the pillars *t* that protect the mucking places. These will require a great deal of support to keep them open. Possibly the mucking drifts *g* will have to be concreted, or else especially large timber sets will have to be used, the first ones of which are protected from the wear of the ore by an iron shield made of steel shapes.

The ore will be drawn as regularly as possible from under the whole area by counting the carloads coming from each chute. But there may still be a great irregularity in the speed with which the capping settles, causing a large loss of ore. To prevent this, C. T. Rice proposes that the block to be caved in mass not only be cut off at the top as described already, but that then the ore be undercut and the block dropped on a level (as *q y*, Fig. 90) 30 ft. below the lowest part of the capping. This would form a protective barrier, or mat, to keep the ore and the waste apart until right at the end of the drawing of the block.

If this were insufficient, another cut-off level (as $y'y'$) could be made below to provide another mat that would float or ride on top of the ore as it was drawn from below. The purpose of the breaks that this undercutting would cause are several. When soft ore is undercut and it begins to cave, it eats up into a dome like D (Fig. 99) having sides with a very steep angle. Consequently it is very possible that domes filled with broken ore would be formed during the drawing of the ore from the caved area that would extend up to the capping and nibble on into that before the ore around these domes got much into motion. But if this dome were broken, as would be the cause if the ore were dropped in two or three cut-off levels before the capping were reached, this would be prevented for the ore around one of these domes (as *F*) would be subjected to a shearing action whenever the dome extended up as high as one of the levels on which the ore had been dropped. This shearing action is just what is desired so as to get the ore broken fine, so that it will pull evenly. Where the orebody is thin, the mat could be made of waste instead of ore if it were thought advisable, but this would be dead work and add proportionally to the cost of stoping. Finally the dropping of the ore in a block on several different levels above the Tunnel Level would insure that the ore would be more thoroughly broken, and so draw more evenly from under the capping.

The cost of blocking out the ore and preparing it for caving should not be more than 35 cents a ton, even when a top mat of ore is provided and the block is cut up by dropping it at vertical intervals of 50 ft. The ore can be mucked into the drawing-off chutes, including the boulders blasted, for not over 10 cents per ton. Consequently the cost, including everything, should not amount to much over 75 cents per ton.

CHAPTER XVI

SLICING UNDER MATS OF TIMBER IN PANELS

Example 43.—Old Jordan Mine, Bingham, Utah

(See also Examples 3, 33, 37 *and* 41)

Irregular Lenses in Limestone. Also Slicing at Low Moor, Va., and *Square Setting at Bingham.*—For this slicing the square sets 6 2/3 ft. high by 5 ft. square in the clear are framed, from 8x8-in. Oregon timber, in a Denver Engineering Works double-ended framer. A lense is first cut, for example, at the 200 level by a drift *d h* (see Fig. 101) and two two-compartment raises *a a′* and *b b′* are put up from it to the top of the orebody and timbered with square sets. Then the whole top floor *e f* is excavated by breast stoping from the raises, and timbered with regular square sets, using 2-inch plank spreaders between the posts instead of sills, and a plank floor. Then holes are bored near the base of half the posts (alternate lines) and loaded with 1/8 to 1/4 lb. of dynamite each. By bunching 12 to 15 of the fuses, it is possible for a dozen men to light the fuses for a whole floor simultaneously.

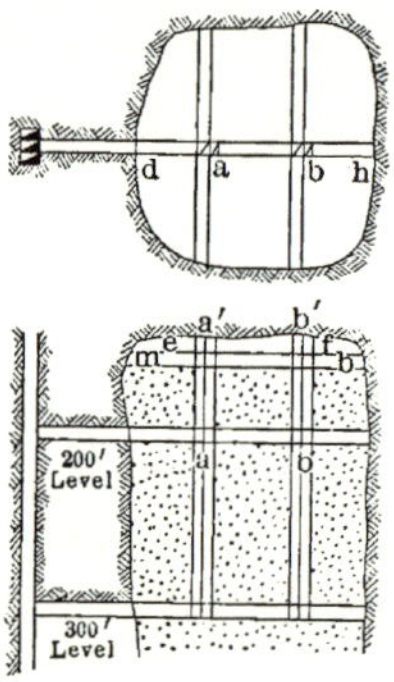

Fig. 101.—Stoping at Old Jordan mine.

The entire back falls on to the plank floor, as its complete caving has been assured by blasting holes in any solid portion of the back at the same time as the post holes. The next move is to start from the raises on the lower floor *m n* and excavate it all, catching up the floor above with square sets during the advance. When the second floor is finished, it is only necessary to drill and blast its posts and bring down the back again before starting another slice. The plank roof and the timbers in the sinking mass seem to render the back cohesive enough, so that no spiling is needed for driving a new slice. The excavation and caving proceed downward, slice by slice, until the 200-ft. level is passed, when excavation must be thereafter begun from raises put up from the 300-ft. level.

Though slicing takes as much timber as regular square setting, its advantage lies in the fact that safety can be attained without the use of extra filling, which may be expensive to obtain, when the whole orebody is shipped. In irregular ore lenses like these, slicing may lose less ore than untimbered caving. The ore-breaking cost more than in square setting, as the latter is overhand stoping and allows one more free face than the breast stoping of slicing, which is broken by a bottom cut off the solid. Only one floor can be sliced at a time, but a large production can be attained by increasing the number of raises. In panel-slicing, waste need not be moved away but can be sorted out and stored where broken.

One of the worst troubles of slicing is poor ventilation, for not only is there no open airway from level to level, but the rock oxidation hastened by brecciation and timber decay of the sinking back creates heat and gas, especially when a slice is left unworked for some months. In one such case in the Old Jordan, the temperature is high, in spite of an artificial circulation maintained by a centrifugal exhauster, belted to a vertical air engine. In case such a delayed floor becomes too heavy to hold as a whole, it may be divided into panels, one of which is extracted and caved before the next one is attacked.

The Old Jordan method is similar to the one used at Low Moor, Va.,* in the eighties to extract a vein of iron ore 30 ft. thick and dipping 60 deg. Not only was the ore so soft as to require close timbering, but the floor would creep if excavations were kept open long. The hangwall was also weak, being a band of broken flint and clay. Nevertheless the slicing system overcame these obstacles. The stopes at Low Moor were higher than the drifts, being 12 to 15 ft. high; and levels were 4 to 6 stopes or 60 to 75 ft. vertically apart.

The slicing system was not in use by the Highland Boy or Boston Con. mines at Bingham, which used square setting for their huge sulphide lenses. The ore is very heavy and goes 20 to 21 tons to a set, 5 ft. square by 6 2/3 ft. high in the clear, so that many sets cannot be left open at once without disastrous caves. No sorting is done, as all the ore broken is shipped, so the filling used is obtained by the exploratory cross-cuts. There is no sharp boundary on the periphery of a lens, so that the pay-ore limit can only be determined by assay. A chute set is left open at 50- to 100-ft. intervals for the passage of the ore to gates in the adits below. At the Boston Con., the mining cost is given at only $1.25 per ton, but such a low figure is largely due to the high specific gravity of the ore. The timber for the sets comes from Oregon already framed and this effects a saving in freight of $4 per M.

Example 44. Cumberland-Ely Mines, Ely, Nev.

(*See also Example* 4.)

Irregular Lenses in Porphyry.—The leached capping at the Cumberland-Ely is more than 300 ft. thick. Not only the capping is weak, but it is sandy and runs in many places. Besides, the ore is not strong enough to stand without timbers. Owing to the depth of capping, under-

*Trans A. I. M. E., May, 1888, "Mining in Soft Orebodies at Low Moor."

ground mining had to be resorted to, while owing to the richness of the ore, which averages about 3.5 per cent. copper, and its weak nature the slicing system of mining was adopted.

The orebody is mined through the Veteran (Fig. 102) shaft, sunk in the limestone at some distance from the orebody. This shaft has four compartments placed side by side and is 475 ft. deep.

From the bottom of the shaft the orebody is blocked out into squares by a series of drifts and cross-drifts 50 ft. apart. These drifts are driven 7x8 ft. in size as mules are used for haulage. They cost from $5 to $6 per ft. From the drifts at each corner of the block raises are driven to the capping. These raises, owing to the ore being moist, are driven on a slope

FIG. 102.—Caved area, Veteran mine.

of 75 deg. so as to prevent the ore from packing in them. The capping is from 40 to 100 ft. above the main level, which is driven approximately at the bottom of the orebody, although in places ore extends below it.

These raises are timbered with 6x12-in. close cribbing placed flat ways. This cribbing is 5x7 ft. over all and is framed with a regulation half joint. The cribbing is divided by means of a 3-in. plank, fitting into a gain 1 in. deep in the wall plates, so as to leave a manway 20 in. wide in the clear. Round timbers have been tried, but they did not work as well as the framed timbers. These raises are connected by means of sub-level drifts so as to leave several ways of getting down to the level. They are generally driven on company account, but sometimes on a footage contract and cost about $5 a foot.

From these raises drifts are driven to the capping and then cross-drifts are turned along the boundary at right angles to the first drifts. These meet the cross-drifts from the next raise about the center of the block, and then another drift is driven parallel to the boundary and alongside the other. These drifts are timbered with drift sets having round posts 8 ft. long and 12x12-in. caps with a 2-in. plank 7 ft. long nailed to the under side for a spreader to keep the posts from coming in. As the drifts advance a floor of 2x12-in. planks is laid and also a track for the car to run on. These sets are lagged with 2x12-in planks. The posts rest on the ore. Whenever the ore extends above the top slice it is mined by double deckers, that is by drift sets placed on top of the lower drift sets. The sets are spragged one from the other.

After one drift has been mined along the boundary, the ore alongside is mined out by means of a series of parallel drift sets. Sometimes the post of one set is made to serve the set on the side, but generally this is not done. The ground is quite heavy and frequently sprags have to be put up to re-enforce the caps. After several parallel cuts have been mined and the timbers in the first cross-drift have begun to show signs of crushing, another layer of 2x12-in. planks placed crossways to the first is laid so as to form a mat to catch the finely crushed overburden. The posts are then bored almost through and loaded with half a stick of 40 per cent. dynamite. These holes are blasted by electricity, 25 to 40 at a time wired in series, this blasting being done retreating. It has been found that it is much better to blast these posts than to let the weight of the overburden crush them down, for when the posts are blasted, one is sure that all the overburden has been dropped. Besides, the mat of planks comes down more evenly and there is less danger of the overburden running while mining the next slice. Thus the whole slice is mined out. Sometimes the ore is mined in cuts parallel to the original drift, but generally that is not the case.

The ore is trammed in small cars holding about12 cu. ft. and dumped down the raise. Of course before blasting the timbers the tracks are pulled up. The ore is trammed by mule to the shaft pockets in trains of large cars holding 1.15 dry tons of ore each.

The ore is mined on contract at 65 cents per large car and all the men working from one raise are in on the contract. The men furnish their own powder, fuse, caps, and candles. The ore is soft and so the holes are drilled with a single jack or in the soft ground angered. Two short rounds are generally put in each day; one blasted at noon, the other going off. Each raise is provided with an air pipe from a suction blower.

For the second slice the men drop down 10 ft. and start from the raises next the boundary another drift to the capping. On this slice the drift sets do not reach to the floor above and so blocks and a bridge cap are necessary. The bridge caps are covered with 2x12-in. plank lagging. The second slice is mined in a similar manner to the first, and the floor is planked each time before blasting. The third slice is also 10 ft. thick, but the lower slices are 17 ft. thick. In mining these, double-deck sets (one drift set on top of another) must be used to support the floor above.

The overburden, owing to its being moist, is said to consolidate somewhat after it is dropped so that there is not much ore mixed with the capping when it becomes necessary to mine next to an area already caved. In the panel being caved there is always at least one open set between the unmined ore and the sets being blasted. Still sometimes in the first few slices, and especially where the capping is sandy, the overburden runs, and some ore is lost. Fortunately at the Veteran the ore and capping are separated by a very distinct line, the over burden being reddish.

The round posts are local timber that comes from Ward mountain or Duck creek and costs about 12 to 14 cents per running ft. for logs 9 to 12 in. in diameter. The square caps come from Oregon. Most of the development work is done on company account. The company employed about 400 men underground on the two shifts and mined about 1500 tons a day or 3 3/4 tons per man.

Example 45.—Oversight and Other Mines at Cananea, Mexico

(*See also Examples* 6, 18 *and* 34.)

Irregular Lenses in Porphyry.—Slicing System.—Formerly all the underground wide bodies of ore were mined by square sets, but this method of working an orebody is only used at the Kirk mine at present, where a narrow vein with weak walls and a soft ore requires this method. It is also used in mining the rich ore capping the upper part of the orebodies, such as is the characteristic condition at Cananea. This capping, which is generally somewhat strengthened by the precipitation of new material in it, is often tenacious enough so that the timbers can be robbed before the back comes in. The posts and caps are framed from 10x10-in. Oregon pine and the girts are simply 8x10-in. pieces sawed to the right length. The posts have a horn 5 in. long and 5 in. square on each end, while the caps are framed to have a 5x10-in. horn 1 1/2 in. long. The sets are 5-ft. centers, capways and girtways, while the posts are 7 ft. 4 in. over all in the stopes and framed to give 8 ft. in the clear on the sill floor. The total cost of mining with the square sets is about $3.25 a ton where the timbers are not recovered. The square-set method is also used in conjunction with the slicing system to mine tongues of ore that extend beyond the ore above, and in working under large masses of waste that occasionally occur in the stopes and are left unmined.

But the typical method of mining at Cananea is the slicing system. This is used in mining the Oversight, the Veta Grande (see Fig. 103), the America and the pillars at the Duluth mine, as well as the old gob-filled square-set stopes at the Capote. The ore in the rotten, kaolinized porphyry at the Veta Grande and the Oversight is so heavy that it is useless to try to mine with unfilled stopes, while if the square-set stopes are filled this costs almost as much as breaking the soft, rotten ore. Besides, owing to the fact that the ore consists of seams of chalcocite through a fractured rock mass, the fines contain most of the values, and with the square-set method much of the fines are lost through floors and go into the gob. In fact, it is because of these losses which attended square-set mining that the old filled square-set stopes in the Capote and the Elisa mine are being worked again by slicing. These old fillings as mined average about 3 per cent. copper, when about 30 per cent. is sorted out in the stope as waste.

In mining by the slicing method it is necessary to get the back broken and following down in a flat, regular mass over large areas. To accomplish this and to permit of getting all the rich ore that occurs on top of practically every orebody at Cananea, the ore above the first

Fig. 103.—Caved ground, Veta Grande mine.

level on which the ore is found is mined by means of square sets. In the rotten porphyry it is not difficult to get this back to caving, but in hard ground this is not so easy. Still this top weight is necessary in order to have enough pressure on the mat to hold in the posts when the ore must be blasted. The sill floors of these square-set stopes are tightly lagged so as to allow the waste mass that follows down to be caught up on posts when slicing begins.

The slicing commences at one of the bounding walls of the orebody, generally the hanging, and is carried back to the other wall in slices 30 to 75 ft. wide and 11 ft. deep. But before caving begins a series of two-compartment square-set raises are carried up through the orebody approximately 30 ft. apart. And as much as possible during the stoping connection is kept open with two raises at a time. These raises are lined with 3-in. planks.

As the extraction of the ore progresses, either by blasting or by picking, according to conditions, the mat of timbers that is following down after the ore is caught up by posts. These are round timbers from Texas, 6 to 9 in. in diameter and 10 ft. long. These are stood on 5x10-in. stringers for footboards and are placed under the stringers on the floor above. These stringers are 10 ft. long and are placed 5 ft. apart in the direction in which the work is progressing, for of course the stringers are placed parallel to the face that is being advanced. The supporting stulls are often 10 ft. apart in the other direction for only those needed are used.

As has been said the height of the slice is 11 ft. But as the sill pieces mash down into the soft ore frequently and as the posts must stand on a 5x10-in. sill on the floor below they are made a foot shorter than the slicing height so as to allow them to be put in easily. If required, blocking is put in between the post and stringer. As soon as the ore is mined out, a floor of 2x10-in. or 2x12-in. planks, 10 ft. long is laid so that the posts can be shot out as soon as the pressure becomes great. These planks are fitted around the posts and the posts are never stood on the floor as they might crush and raise the floors out of shape.

One miner can follow along with a second slice when the miner ahead has advanced far enough so that each is beyond the influence of the caving being done by the other. This varies according to the ground, but generally a distance of 50 ft. between the two is sufficient. If desired the following slice can be farther to one side or it can be on the slice to be taken next below. This is immaterial.

Whenever the pressure on the posts becomes great the roof is dropped. This is done by drilling with an air-driven auger an inch hole 3 or 4 in. into each post and loading these holes with about a third of a stick of dynamite. These posts are blasted with cap and fuse, and as soon as the roof has become quiet again the men come back to work.

The ore frequently is so hard that it has to be drilled by machine and in that case of course a piston machine has to be used. These are 2 1/2-in. drills, and two men work on one of them. Any waste that is found in the ore is thrown back on the floor in a part where the roof is about to be caved, while if a large mass of waste is met with it is worked around and then by means of square sets the ore below is worked out until finally again caving is resumed under the boulder.

In case the ore makes out farther into the walls than is the case on the floor above, the ore is followed out by means of square sets to the new boundary, a floor is laid, the posts blasted, and the slicing system resumed on the next slice. In case the pressure on the posts is not great enough and it becomes necessary to shoot a heavy blast, the posts are braced by means of 2-in. strips nailed to them. Before the slice is caved the strips are knocked off for use in another part of the slice.

Formerly the floors were lapped for about 12 in. instead of being laid on 5x10-in. stringers, but the use of stringers only entails a loss of a few inches of timber and makes a floor better and more easily caught up on the slice below. The round timber used for stulls comes from Texas and costs $.80 per 10-ft. stull, but this low price is obtained by not holding the sellers close to a fixed size. All pieces too small for posts are cut in two on a rip saw at the mill, and the half-poles used for lagging. In this way a stronger and cheaper lagging than split-lagging is obtained, while the cost of the posts is considerably less than formerly. About three tons of ore are mined per man underground at Cananea by the slicing system. The cost of labor and timber for placing ore in chutes is 60 to 70 cents a ton, making this system cost about half as much as square setting, calculated on the same basis.

The method is safe, the ventilation is fair, sorting can be done in the stope, all fines that go through the floor are gotten on mining the next slice and not lost, as in the square-set method. If the ore were saved, the grade would be lowered too much owing to the considerable amount of waste in the ore. About 30 per cent. is sorted out in the stope during slicing. If caved this waste would be crushed so fine that it could not be economically sorted on a picking belt. This sorting in the stope is one reason for the mining costs not being somewhat lower than they are, but they are much lower than formerly, even as it is. On the whole the slicing system is well adapted to the soft ore conditions at Cananea. Of course considerable timber is used in slicing, but not nearly as much as in the former square-set method, and the timber used is of cheap grade as it does not have to be especially strong.

CHAPTER XVII

SLICING UNDER ORE WITH BACK-CAVING IN ROOMS

Example 46.—Lake Superior Soft Ore Lenses on Gogebic, Mesabi and Menominee Iron Ranges

(*See also Examples* 2, 7, 8, *and* 38.)

Lenses under Glacial Drift or within Rock Walls. Timber Mats between Slices.—This system is applied everywhere on the Gogebic, with differences in detail according to the mine. Diagrams for East Norrie practice are shown in Fig. 104. Levels *A* and *D* are driven in the ore 60 ft. vertically apart, with sub-levels B and *C* at 20-ft. intervals. The main levels have two parallel haulageways (*M* and *M′*) for electric locomotives and are connected at intervals by curved cross-cuts D, D′, etc. Next, levels *A* and B are connected by raises *r* with two compartments (chute and manway), and vertical where practical. At each sub-level the manways are covered by a hinged door and the chutes by a horizontal grating of rails, set 6 in. apart, to prevent the entrance of boulders. At the chute bottom is placed a gate above the haulageways consisting of a vertically sliding steel plate raised by a lever. Finally the sub-level "*B*" is blocked out by drifts BB and *B′B′* and cross-cuts *aa′*, *bb′*, etc., though where the ground is difficult to keep open these openings need be put in only just before stoping.

All excavations must be supported, and round, unbarked timber is used. In haulageways the three-quarter sets are about 18 in. in diameter with 8-ft. caps and posts set 4 ft. to 6 ft. centers, while in the sub-levels there are 7-ft. caps and posts of 8 in. to 12 in. in diameter. The raises are closely cribbed with 4-in. to 6-in. poles and made 3 1/2 ft. to 7 ft. inside with a central cribbed partition.

After sufficient development, stoping can be begun, the ore being removed by a combination of drifting and retreating caving. Work can begin on sub-level B when the ore above level *A* has been extracted and the caved hangwall rests on the floor of *A*, which was previously covered with a floor of waste timbers. Starting at *a′*, a room with sub-level, three-quarter sets (7-ft. posts and caps) is driven sideways to wall at *s*, when the right-hand posts of the last set are blasted out to allow the 12-ft. height of overhanging ore to fall. When this ore has been removed by shoveling the next set behind is blasted out and the withdrawing continued until the cross-cut *aa′* is reached. At the Glenn mine (Mesabi) the side of every third room only is covered by 1-in. boards before starting to cave; at the Fayal (Mesabi) mine this boarding is put on each room; but here, at the East Norrie, no side-boarding at all is used.

The next operation is to similarly remove rooms *s'* and the process is then continued until the whole panel *a-a'-s-f* has been caved. When the first panel is suffic'ently advanced work can be started from cross-cut *bb'*, beginning at room *a'*. Afterward panels *cc'*, *dd'* and *ee'* can

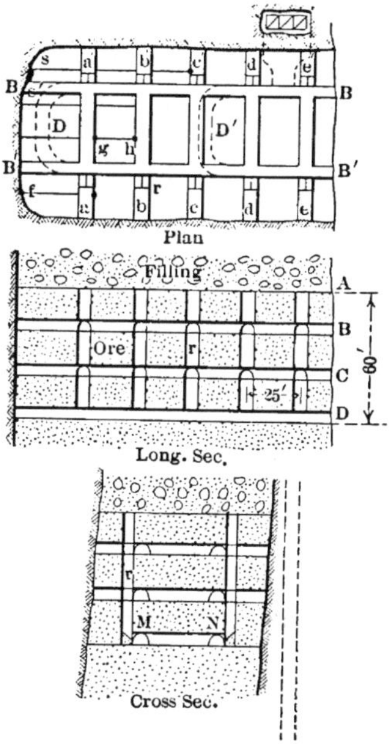

FIG. 104.—Stoping at East Norrie mine.

be started, so that the caved ground will finally advance with a bench face like *f-a-g-h-b'-c'*. As though overlying ground is caved, one chute top after another is covered over and abandoned. When the caving on sub-level *B* has retreated 25 ft. or so from the wall at *s*, the caving on sub-level *C* can be similarly attacked as soon as the caving above will allow the extreme cross-haulageway D to be abandoued in favor of D'.

In some cases it is easier to conduct this system of driving the rooms parallel to cross-cut *aa'* instead of to drift *BB*. In this case the rooms would be started on both sides of drifts *BB* and *B'B'* and would meet half way between them, while the caving face could be kept parallel to *aa'*, as the slicing receded from the end wall *sf*.

The distance of 20 ft. between sub-levels, at the East Norrie, is unusually large, and while it requires less development it makes it less easy to keep the ore free from filling, and in narrow veins or near horses the back may hang up. At the Newport Mine (Gogebic), *D* shaft (soft ore) the levels are 75 ft. apart with five sub-levels, making the latter only 12 1/2 ft. apart. At the Newport *K* shaft (hard ore) the haulageways and a few of the sub-level drifts are timbered, but the balance of the openings are unsupported and only a floor of timber débris is laid down before caving. Here the sub-levels are 12 ft. apart, and there are nine between the haulageways, which are 110 ft. vertically apart.

Mesabi Range, Minn.—Here underground mining is the choice for deposits too small to warrant stripping, or for those parts of great bodies with too deep a mantle to allow of its economical removal. It can be worked equally well in winter, except for the small extra cost of reloading the stock pile for shipment. In opening the mine the ore from development helps pay the expense, hence much less additional outlay is required than for a stripping system. Lastly, any intermediate bed of lean ore can be left underground. The chief extra operating expense over competing systems comes from the use of much timber and hand shoveling. It is also estimated that 10 per cent. of the ore is lost in the present "room-caving" now in vogue.

An underground system for this range must be adapted to a soft, friable bed 50 ft. to 200 ft. thick, with a hang wall of glacial drift. At first square-setting was tried, but it was found that if not closely filled a stope would collapse. As practically all the ore body is hoisted, such filling could only be obtained from the surface. To save this expense the slicing system was adopted, with square-setting as an auxiliary. This is now the only system, with a few exceptions, used in all the soft-ore mines of the iron ranges, and is universal on the Mesabi.

At Eveleth the main levels are 40 ft. vertically apart, with two sub-levels between, and as the drift sets have 7-ft. posts, a shell of about 5 ft. of ore is left to be caved in each slice. The best practice avoids heavy ore pillars by sinking the shafts in the wall rock; but the haulage-ways are in ore, and for these heavy, round timber-sets are necessary to withstand the roof pressure. Some older shafts are inclines, sunk in the foot wall under the sloping side of an ore trough, using common skips. But new shafts are often sunk vertically on account of cheaper maintenance and hoisting, since the introduction of the Kimberly skip, which dumps automatically in vertical ways. In the Fayal mine, with an output of 1500 tons daily, about 150 men are employed underground, thus giving 10 tons per man, which average is decreased to six or eight tons per man when the surface force is included. Owing to the great superficial extent of the lenses, care is taken that the sub-level drifts should not be vertically over each other on the haulageways, in order to prevent premature collapse. The Glenn mine (Oliver Co.) near Hibbing presents peculiar conditions, and has been most systematically laid out. As the ore body lays in a trough and is but 80 ft. thick at its center, only one electric-haulage level is used, following the bottom of the trough and having parallel haulageways, *A* and *A′* (Fig. 105) 50 ft. apart. Above this is a level *BB′* at a 44-ft. vertical interval, with two sub-levels between *AA′* and *BB′*.

On the first level above the haulageway the cross-cuts *BB′*, when they strike the wall rock, are continued in a vertical raise (as *Bc″*) to the next sub-level (or level). Then *c″c* is driven out horizontally till it in turn strikes the side of trough at *C* and is continued in another

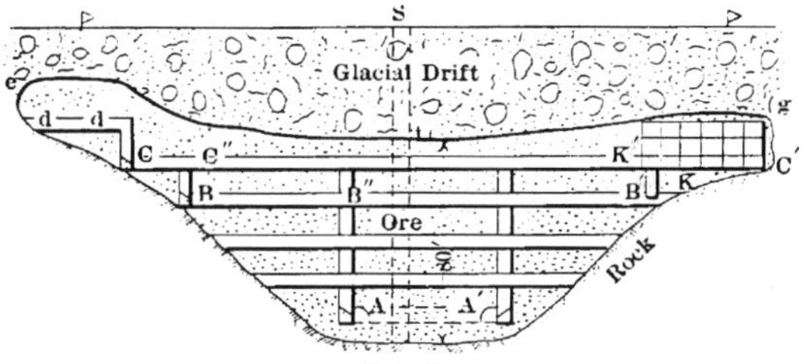

FIG. 105.—Cross-section of stope, Glenn mine.

raise (*cd*) up to sub-level *dd′*. Each of these raises is used as a chute into which ore, won from the sides of the trough, can be dumped. Thus the highest placed ore may have to pass through several raises in alternation with cross-cuts (as *d-d′-c-c″-B-B″-A*) before reaching the electric trains in *A* for transmission to the shaft.

Chapin Mine, Menominee Range, Mich.—Since its opening in 1880 the Chapin has been one of the largest and richest of the "Old Range" mines. The deposit consists of a series of lenses extending 6100 ft., canoe-shaped in horizontal section and conforming with the enclosing iron slates, which here strike N. 75° W. and dip 70° to 80° N. About 300 ft. to the south the Randville dolomite overlays the hangwall slate in an overturned fold. The lenses pitch 30° W., and in the past four have been worked while a fifth has recently been found to the south of the others, which lie end to end with only small offsets. The largest lense is 2500 ft. long and 130 ft. wide and was bottomed at about 1400 ft. depth. The westernmost lense was formerly divided between the Ludington and Hamilton companies, but is now worked with the Chapin under the management of the Oliver Co.

Former Stoping Methods.—The history of this mine well illustrates the evolution of underground mining at Lake Superior. The first work was done by stull stoping, in which the hangwall is supported by stulls and pillars of ore, the rest being stoped overhand. But this method proving impractical, because of the width of vein and softness of ore and walls, it was replaced by modified square-setting.

With the latter system the vein was laid out in transversal rooms 18 ft. apart and timbered with a square-setting, that involved the use of continuous sills and caps 16 in. by 18 in. and 18 ft. long. After three years' trial the growing scarcity of huge timbers and the collapse of some rooms compelled the abandonment of square-setting.

Next came rock-filling, which involved the removal of the ore by breast stoping (beginning at one level and raising upward to the next), each slice being filled with rock close behind the extraction. This stoping resembled that now in vogue at the Soudan mine (see Example 20), except that here, with a softer and wider vein, the breast-stoping and filling had to be carried on in 8-ft. drifts instead of the whole width of the vein at once. Rock for filling was also quarried an an adjoining sandstone quarry (when dead-work rock was insufficient), instead of letting it quarry itself from a caving hangwall, as at the Soudan. An additional expense at the Chapin was the great amount of shovelling, due to the necessity of packing the filling tight against the back of vein.

Present System.—All these disadvantages of rock-filling has caused the substitution for it at the Chapin of slicing, as the latter is not only more flexible but saves the expense of digging and handling filling and of blasting nearly half of the ore. The main haulageways here are 200 ft. vertically apart on account of the expense of cross-cutting from the hoisting shaft under the difficult drainage conditions. Between two haulageways there are three levels 50 ft. apart, and between these latter are the sub-levels (12 1/2 ft. apart), from which the rooms are turned off. Ore on the levels and sub-levels is handled in half-ton cars by the miners, who dump it into the chutes leading to the haulageways.

The raises are about 50 ft. apart and are the first thing to be started from the haulageways. The system of levels and sub-levels drift is only completed when needed for stoping. The raises are cribbed up in two compartments—a continuous chute and a manway, which is made non-continuous (by transposing it to the opposite side of chuteway at every sub-lateral) in order to lessen the danger of rock falling on ascending men.

Output.—When running at capacity the output is 3500 tons daily, with a total of 900 to 950 men, or nearly four tons per man. Most of the vein is first-class ore (58 per cent. iron), but bunches of second-class, which are made low grade by intercalcated bands of jaspilite, also occur especially as a transit between the good ore and the wall rock. In 1907 the shipment was 800,000 tons and the timber consumption of 1,750,000 bd. ft., while 15,250,000 tons was the total output to January, 1908.

Shafts.—At first inclined shafts were sunk in the ore, as many as 10 having been thus started in the main lense alone. Next, vertical shafts were sunk in the dolomite hangwall to cut the ore in depth, while the latest is the "Ludington C" or Cornish pump shaft put down vertically in the foot wall to 1525 ft. depth. The last is 10 ft. by 21 ft. inside of its steel sets, whose wall plates are 6-in. I-beams set 2 ft. to 5 ft. apart, vertically, according to ground. There are four compartments, the shaft being divided crosswise into three divisions; the first, 10 ft. by 6 ft. for two 5-ton Kimberley skipways; the second, 10 ft. by 5 ft., for a cageway, and the third 10 ft. by 9 ft., for the Cornish pump pipes.

Haulage.—In the main haulageways the ore is usually handled in electric trains of ten 3-ton steel cars, on a track in the East Norrie, or 24-in. gauge and 25-lb. rails. On the sub-levels the ore is trammed by the miners in cars, which usually hold one-half to one ton, but occasionally (Fayal mine) are of 2-ton size.

As advantages or slicing may be given (1) safety, (2) adjustability to varying output and irregular bodies, (3) small consumption of explosives, (4) large productive capacity, (5) good ventilation. As disadvantages may be mentioned the ore lost through mixing with the filling (which is around 10 per cent.) and the timber consumption. The latter is as large as with square-setting, but less costly, for no squaring or accurate framing is necessary. Though the rooms are excavated by the expensive system of breast stoping, the large fraction of the output that is caved requires no blasting.

This system is adapted to any friable ore body or sufficient width to permit free descent of the filling. It is especially applicable to thick, flat bodies (like those of the Mesabi) and, the caving being all done in narrow rooms, it is probably the most flexible and easiest controlled of all the caving systems.

Square-Setting.—This method is now only used, as an auxilary to room-caving, in places where the latter would be awkward to apply. On a lower level of the East Norrie a portion of the ore extends only 50 ft. above the sill floor, as it is cut off from the filling above by a thick dike. The insertions of square sets for this block obviated the difficulty of starting a new cave overhead by blasting.

The sets here are of round timber 8 ft. high by 7 1/4 ft. square (to centers), with posts about 18 in. and girts and ties of 12-in. diameter. No filling or planking (except for working platforms) is used. The sets are inserted in rooms, extending from foot to hanging of the ore body and from the level up to the dike. Rooms are only three sets long and a pillar of like length is left between adjoining rooms.

After several rooms are finished the robbing of the pillars is begun by starting their excavation from the hanging side and inserting square sets. In case the hangwall should cave before all the ore is extracted the men have time to escape and the balance of the pillar can be recovered by raising from the level below. When finished, the square-set excavation will be caved by blasting out a central post, and the ground below can then be attacked by the regular room-caving system.

In a comparatively thin body like the Glenn mine (see Fig. 107) square setting is also a convenient aid: It is most applicable near the top to extract the ore under the irregular roof *efg* of the lense, so that room-caving can be started below sub-level *cc'* from a horizontal back. Here the sets are also round timbers and 8 ft. square (to centers) and are set up first in rooms like *c'K*, which measure three sets back from *c'*, three sets wide on each side of cross-cut *cc'* and as high as top *g* of the lense. When room *c'K* is finished the floor and the sides showing solid ore are covered with 1-in. boards (culls) and the room caved by blasting out a central post. Another room, on the same cross-cut as *KK'*, can then be started behind the cave and the ore won without danger of contamination.

Drilling.—In soft ore, hand augers and jumpers are used. The augers are 3 ft. to 6 ft. long and made by welding a T handle of 5/8-in. round iron to an auger, twisted from 3/16-in. to 1 1/4-in. steel and having a double point 1 1/4 in. to 1 1/2 in. wide. The jumpers are the same length and of two kinds, one of 5/8-in. round steel, with 1 1/2-in. chisel point, and the other of 1-in. round steel with a moil point 4 in. long, set on an angle with the shank, in order to bore by picking into soft seams. In case a hard place is struck with a jumper the whole can be finished by double-jacking; and the latter method is extensively used by itself in the hard ore at Newport *K* shaft on the Gogebic. Air drills, usually with 3-in. pistons and 60 lb. to 70 lb. of air pressure, are extensively employed for drilling in hard ore or wall rock.

Example 47.—Mercur Mine, Mercur, Utah

Bedded Lenses in Sloping Limestone; Timber Mats between Slices.—The Consolidated Mercur Gold Mines Company operates the Mercur, Golden Gate and Brickyard mines, together with a cyanide plant of 800 to 1000 tons daily capacity, at Mercur, Utah. Mercur, a town of but a few hundred inhabitants, is near the southern end of the Oquirrh mountains, 62.5 miles by rail south of Salt Lake City at the terminus of the Salt Lake & Mercur railroad.

The ore deposits occur in the lower portion of a bed 5000 ft. thick, known as the Great Blue limestone. It has been intruded by several sheets of quartz porphyry, from 4 to 40 ft. in thickness, which in general correspond with the bedding of the limestone in both strike and dip.

The ore deposits underlie the sheets of porphyry and are made up of both altered porphyry and altered limestone. Their lines of greatest mineralization coincide closely in direction with a series of nearly vertical fissures which have a trend toward the northwast. The mineralizing agents are believed to have ascended in gaseous form through these fissures, depositing the ore minerals along the lower contact of the porphyry and impregnating both it and the limestone, forming a total thickness of workable ore varying from 4 to 70 ft. As a rule only a few feet of the porphyry is mineralized, the greater portion of the ore being in the limestone.

The unaltered porphyry is fine-grained, compact and nearly white in color, showing a lew inconspicuous phenocrysts of quartz, feldspar and biotite. This unaltered rock, however, is not seen near the ore deposits. There it is almost black, quite soft, and contains small crystals of gypsum, while upon exposure to the air it crumbles and has the semblance of a dried mud. The mineralized limestone is hard and cherty, but the evidence points to the fact that the silicification antedated ore deposition.

The Mercur Mine

The Mercur mine is at the southern end of the property. All of the ore is now taken out through one adit which has a single 18-in. track, 1200 ft. of which are equipped with electric haulage, while on the remaining 1900 ft. horses are used.

The principal vein is the Mercur, which also runs through the other mines, and which outcrops plainly on the hillsides on both sides of Lewiston canon. It is dark gray in color and is characterized by fine networks of quartz or calcite with more or less barite running through it. The vein as worked, including the so-called "lower" vein which in most places is mined with it, varies in width from 12 ft. to a maximum of 70 ft., with an average width of from 20 to 30 ft. The dip varies from 10 or 12 deg. up to 30 deg.

In the narrower portions of the ore-body, the levels are driven about 20 ft. apart vertically, and the ore is mined in open stopes which may or may not be timbered with stulls. From the drifts inclines are put up in the vein, and the ore is broken down on both sides of them. These stopes have no fixed dimensions, the width depending upon the extend of the shoot and also upon the character of the roof. About 4 ft. of ore are left next to the hanging-wall to support the roof, and stulls are placed wherever necessary. When the stope has attained its proper size, the ore next the hanging wall is extracted by the retreating system. The roof is then allowed to cave.

The above method is employed only where the ore has a thickness of less than 15 ft. Most of the orebodies in this mine have a thickness considerably in excess of this figure, and these are mined by a modification of the caving system of Example 46.

From the main levels raises are put up at intervals of 25 to 50 ft. (see Fig. 108) and, so far as possible, are kept as near the foot-wall as is permitted by the angle of slope at which the ore will run in the chutes. In most places the dip of the orebody is so low that a series of chutes, offset from one another, is necessary, the ore being drawn from one chute and trammed into another nearer the hanging-wall.

Where the orebody is wide, as many as three drifts approximately parallel may be run on each main level, chute-raises being carried up

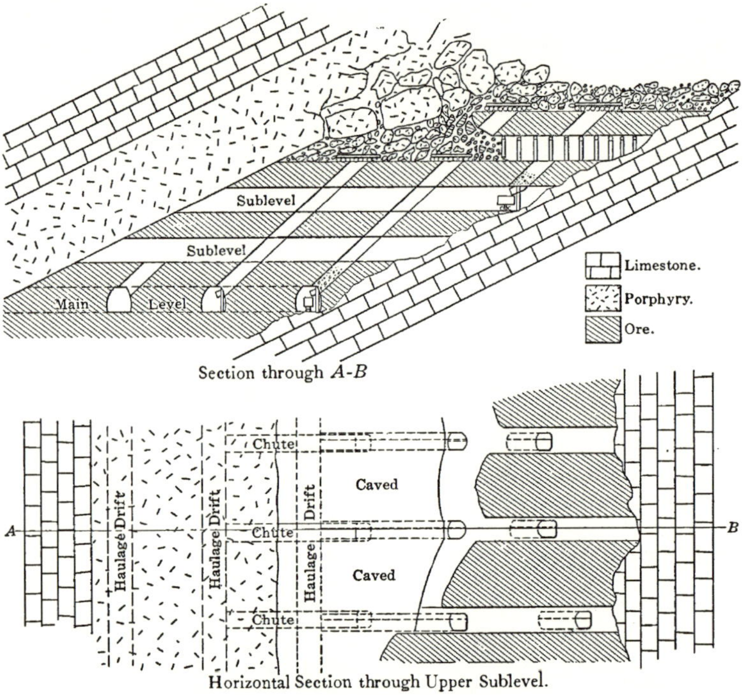

FIG. 106.—Stoping at Mercur mine.

from each drift in order to reduce to a minimum the labor of handling the ore in the stopes. As a rule, only the foot-wall raises are used for both manway and ore-chute.

From the raises, sublevels are started at vertical intervals of 14 ft., the development on each sublevel consisting only of cross-cuts driven to both the foot- and hanging walls. The work of caving is commenced on the upper sub-level and continued downward in successive steps. In starting this work the ends of the cross-cuts next the hanging wall are widened out until two or more of the cross-cuts are connected. A few holes are then drilled in the roof and blasted to start the caving. The broken ore is then shoveled into wheelbarrows and conveyed to the chutes. When waste appears in the broken ore the intervening pillars are successively sliced in rooms to the hight of the cross-cut toward the foot-wall, meanwhile holding the roof temporarily by stulls. Practically all the work is done by hand, so heavy timbers are not needed to withstand the shocks of blasting. As the working faces of the slices recede, the roof is allowed to cave, only maintaining a safe working place for the men along the new outline of the pillar. After the second slice has been taken out, the back over the working places must be supported by three-quarter sets as in Example 46.

When the ore from one sub-level or slice of ground has been extracted, caving is commenced on the sub-level immediately below, where the process is carried out in the same manner. It is not necessary that all of the ore be extracted from one sub-level before caving can be started on another, for caving is carried on at several levels in the mine, but not in the same block of ore.

Recovery and Costs in Caving System

Where the sub-levels are 14 ft. apart the thickness of the back which remains to be caved varies from 4 to 7 ft., the lesser thickness having proved the more satisfactory. It is questioned by the mine superintendent whether better results would not be obtained if the sub-level intervals were not reduced to 12 ft. By caving a small thickness of ore, a higher percentage of recovery is made, and the operation is kept under better control. It will be seen that in this mine from 50 to 70 per cent. of the ore is extracted by room-stoping and slicing, leaving but 30 to 50 per cent. to be obtained by caving proper.

Where the caving is begun next to the hanging-wall it is sometimes difficult to obtain all of the ore along the foot-wall, particularly where the dip of the orebody is low. This method, however, is well adapted to the conditions prevailing in this mine, where the ore is firm and stands well without timbering but the hanging wall is weak. In case the caving was commenced next the foot-wall the pressure upon the pillars would become almost insupportable by the time the caving had proceeded for two-thirds the width of the deposit.

The system requires little powder or timber, but as employed here requires a large amount of development work and a great deal of shovel and wheelbarrow work. The ore is rather soft and almost all of the drilling is done by hand, there being but seven machine drills in use in the three mines. Mining is done largely by contract, the price in the Mercur mine varying from 60 cents to $1.40 per car of 1 1/4 tons delivered at the electric-haulage station, while in the Brickyard mine the costs in some instances run as high as $4 per car. On the Mercur proper considerable ore was taken out on the surface through glory holes, at a cost of about 20 cents per car. The ore, however, was of low grade. The limit on ore mined is placed at $3 for oxidized and $4.65 for base (su'phide) ore, though much ore of less value is mined and milled. Six-dollar ore is considered first-class, while $10 per ton is high grade.

Development is done largely by contract, the price for a 5x7-ft. drift varying from $2 to $4 per ft., the average cost being $3.50. The contractors furnish all their supplies and deliver their ore or broken rock to the electric-haulage station. With hand drilling the drifts are run at a rate of up to 70 or 75 ft. per month. Miners receive $2.75, and machine men $3 for an eight-hour shift, while contractors average about $3.50 per day. The mine is operated two eight-hour shifts every day in the week. From 270 to 290 men are employed underground during the 24 hours and the daily output ranges from 700 to 800 tons of ore. The average figure is, therefore, about 2 1/2 tons of ore per day per man.

Example 48.—Kimberley Diamond Mines, South Africa

Sub-vertical Volcanic Pipe. No Mats between Slices.—The mineral deposit here is a great vertical volcanic neck or pipe 200 yards across. The excavations were carried to a depth of 200 to 300 ft., by open-cut, before the caving of sides of the huge pit made further work at the bottom so dangerous that underground mining had to be adopted. Vertical shafts were then sunk in the country rock or "reef" at a safe distance from the open pit and cross-cuts driven from these to enter the solid diamond-bearing or "blue ground" below the surface workings.

Instead of attempting to withstand, even for a time, the pressure of the superincumbent mass of broken reef, the first system introduced was a caving in the back and a filling of the excavations after precious blue ground had been extracted. When numerous small tunnels had been driven to the margin of the mine, that is, to the point where they reached the sides of the crater, the blue ground was stoped on both sides of, and above, each tunnel until a chamber was formed extending along the surface of the rock (wall) for 100 ft. or more, with an average width of 20 ft., and about 20 ft. high. The roof of the chamber or gallery was then blasted down or allowed to break down by the pressure of the overlying mass of broken diamond-bearing or barren débris.

In the early stages of underground mining there was an enormous amount of diamond-bearing ground which had been left behind when open mining was discontinued, and which had been crushed either by the moving sides of the immense opening or by the collapse of the underground pillars when mined by the old system (of pillar and room). It happened frequently, after breaking through to the loose ground above, that clean diamond-bearing ground would run down as fast as it was removed for weeks or months at a time. The galleries would at times become blocked with large pieces of blue ground, which had to be blasted, and then a further run of blue ground would follow. When the blue-ground was worked back toward the center of the crater, large boulders or fragments of basalt which had come down through the loose reef from the surface would be met with. This system of working would be continued until reef alone came down, the waste or reef removed being sent to the surface by itself and piled on the waste dump. It formed only an inconsiderable proportion (1 to 4 per cent.) of the total output. When the roof caved in, the gallery was nearly full of blue ground. Only a part of this ground was removed by the men working on that level, the miners prefering to take it out on the next level below. This process of mining was repeated from level to level until finally there was no more loose ground to be recovered. The cost of extracting blue ground while loose ground existed was very low.

When the underground work had reached a depth of 800 ft. or more, a new danger appeared. The huge open mines are filled with débris form the sides, caused by the removal of the diamond-bearing ground by open quarrying. This débris was composed of the surface red soil, decomposed basalt, and friable shale, which extended from the surface down to a depth of about 300 ft. In addition to the débris from the surrounding rocks, there were huge masses of "floating shale," resembling indurated blue clay more than shale. Large heaps of yellow ground and tailings, which the early diggers had deposited near the margin of the mines, and west-end yellow ground, contributed to the mud-making material. The black shale which surrounds the mines disintegrates rapidly when it falls into them. It contains a small percentage of carbonaceous matter, and a large amount of iron pyrites. When the huge masses of shale fell into the open mine, they frequently ignited, either by friction or, more probably, by spontaneous combustion, as they have been known to do on the dumps, and burned for months and even years at a time. These masses of burned shale become soft clay and form a part of the mixture which fills the open crater. This débris moves down as the blue ground is mined from beneath it, and becomes mixed with the water which flows into the open mine from the surrounding rock, and with storm water, and forms mud. This overlying mud becomes a menace to the men working in the levels below. Frequent "mud rushes" occurred suddenly, without the least warning, and filled up hundreds of feet of tunnel in a few minutes, the workmen being sometimes caught in the moving mass.

It became evident that the first system of working was dangerous, the men sometimes being, when a mud rush took place, either shut in or buried in the mud coming from the opposite end of the mine. It was decided, therefore, to work the mines from one side only, and to have the offsets to the rock connected one with the other at as few points as would be consistent with the ventilation of the working faces. Main tunnels are driven (about 100 ft. apart) across the crater upon its longer

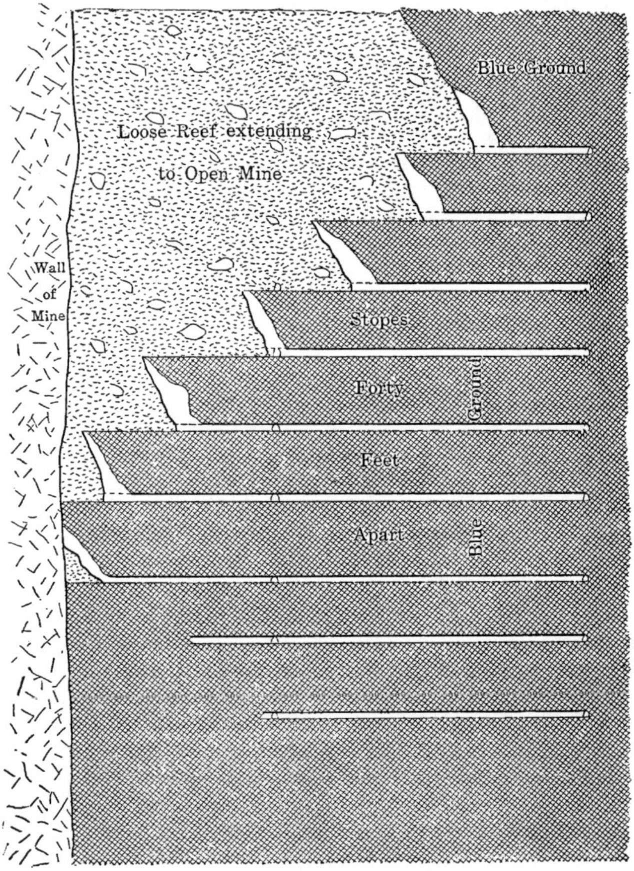

Fig. 107.—Vertical section of stopes, Kimberly mine.

axis, and at right angles to these small tunnels are driven out every 30 ft. until they reach the hard rock an the south side of the mine. These tunnels are widened, first along the rock, until they connect one with another, and at the same time the back are stopped up until they are within a few feet of the loose ground above, thus forming long rooms or rather galleries, filled more of less with the blue ground, upon which the men stand when drilling holes in the backs. The working levels were at first 30 ft. apart vertically, but for greater economy the distance was soon changed to 40 ft.

The broken blue ground in the galleries is taken out, as a rule, before there are any signs of the roof giving away. At times this is impossible, and the roofs cave upon the broken ground, and the blue ground is covered with (barren) reef. Fig. 107 shows the arrangement of cross-cuts and drifts and the manner of stoping under the conditions described.

As the roof caves or is blasted down, the blue ground is removed, and the loose reef lying above it comes down and fills the gallery. Tunnels are often driven through the loose reef, and the blue ground which has been cut off and buried by débris is taken out; but it is sometimes left for those working the next level below to extract.

After the first "cut" near the rock is worked out, another cut is made and in this manner the various levels are worked back, the upper level in advance of the one below, forming terraces as shown in Fig. 107. The galleries are not supported in any way with timbers, but all tunnels in soft blue ground are timbered with sets of two props and a cap of round timber, and are covered with inch and a half lagging. Soft blue ground is drilled with junper-drills sharpened at both ends. In hard blue ground drills and single-hand hammers are used. The native workers become very skilful in both methods of drilling, and do quite as much work as white men would do under similar conditions.

CHAPTER XVIII

PRINCIPLES OF MINING SEAMS

(*a*) Comparison of Longwall and Pillar Systems

In the mining of seams of coal or of other bedded deposits of similar regular thickness over large areas like iron ore, gypsum, salt, etc., the two systems largely used are "longwall" and "room and pillar" or simply "pillar." The longwall system extracts all the coal in the first operation along a long stretch of "wall" or face and allows the roof to settle gradually behind the miners upon a "gob" or partial waste filling. The only excavation kept open besides a narrow passage at the working face are a few roads, to the bottom of the shaft or other exit, for ventilation and for the tramming of coal and supplies. In the pillar system, on the contrary, the first operation consists of driving roadways to extract only part of the seam in rooms, while leaving the balance in the form of pillars to sustain the roof. Later the pillars are drawn or "robbed" so as to finally recover as much of the seam as possible.

Either system may be pursued "advancing" or " retreating." If advancing, the attack of the longwall or the robbing of the pillar system begins next the safety pillar, left to protect the shaft or other entrance, and advances outwardly toward the boundry; while if retreating, the roadways of the longwall or the roads and rooms of the pillar system are driven to the outer boundary of the mine before the "attack of face" or the "robbing of pillars," respectively, is begun.

The two broad divisions of the longwall system are "continuous-face," in which the face is kept in the form of a circle or similar closed figure, and "panel" where the face is handled in panels or blocks, along a sufficient stretch for free roof subsidence, without forming a closed figure. These divisions have each several varieties and often shade into each other.

The pillar system has three varieties: "room and pillar," where the rooms are wider than the pillars; "stall and pillar," where the stall or room is narrower than the pillar; and "panel," where the mine is divided into sections or panels, separated from each other by peripheral pillars, and each is divided into a number of rooms with corresponding pillars. In some mines it has been found advantageous to combine the longwall and pillar systems or even to operate them separately in different portions of the property.

The longwall system is adapted only to uniform seams with roofs of an elastic material like shale or sandstone rather than those with a blocky fracture like limestone. Hitherto, longwall has been most used for working seams of coal, but it is likely hereafter to be widely applied to other deposits like the Appalachian iron beds, the Michigan copper amygdaloids, or the Transvaal gold banket, in order to overcome the obstacles incident to great depth. In coal mining, longwall has a particular advantage in thin seams over the pillar system, because the robbing of pillars in such seams in usually unprofitable, and longwall reduces to a minimum the expense of driving and maintaining the roadways. Longwall also gains in desirability with increasing depth where the pillars of the rival system must continually widen and thus proportion ately be dearer to recover.

Longwall is adapted to beds containing considerable waste, for the waste can all be stored underground and if suitable for pack walls will obviate the use of timber cogs. In Europe, with plenty of waste for gobs and packs, seams as thick as 10 ft. have been worked in one slice by longwall. Less timber is consumed in the longwall than in the pillar system because in the latter the props can seldom be used again. The subsidence of the longwall roof is gradual so that it does not inflict such breaks in the formation, to let in water or to damage surface structure, as ensue from pillar-robbing. By longwall, a mine can be developed more quickly and more cheaply, and more lump coal and a higher percentage of the seam can be excavated in less time than by pillar work. In longwall, the ventilation system is cheaper to construct and to maintain, for the mine's resistance is less; few or no explosives are needed; and there is less dnger from falls of the roof. Longwall requires better trained miners than pillar work but a miner's output is greater. After a strike of nearly six months recently in a Western coal district, it cost the pillar mines nine times as much to clean up and to get started again as it did similar mines where the longwall system prevailed. This result was strictly opposite to the opinion previously held by many on the subject. Longwall, however, is unsuited to fluctuating outputs, for the roof, when being moved at all, should subside uniformly along the face. For coal mining, longwall is gradually superseding pillar work in Europe wherever conditions are suitable. America is bound to follow suit as soon as her mines become deeper.

(*b*) Comparison of the Retreating and Advancing Systems

The retreating system of mining seams is rapidly supplanting its rival, the advancing system, in European mines, but in America the author knows of no case of its use in longwall and of comparatively few cases in pillar working. On inquiring why the advancing system is still preferred, the only two reasons to be found for its use in opening a mine are that it requires less capital and less time. Everything else is against the advancing system which has a smaller percentage of mineral recovery, a worse control of roof, ventilation, and drainage, a greater liability to gas and dust explosions, a higher cost of maintenance of roadways, a larger timber consumption, and a smaller output from an equal developed area at the face.

The mistake of sacrificing safety, mineral and profit per acre in order to get an output quickly at minimum cost may be unavoidable for small weak enterprises, but no valid excuses can be made by strong companies which continue to persue such a penny-wise, pound-foolish policy. The driving of entries to the boundary to inaugurate the retreating system for either longwall or pillar working completely explores the traversed territory. These entries expose the seam's faults, rolls and irregularities, and thus indicate both the lowest points of the floor for the location of sumps and pumps, and the high points of the roof where may be placed churn-drill holes for the escape of gas or safety shafts with ladders to serve as natural ventilators when the fan is idle.

With the retreating system not only are there no old gobbed areas within the active workings to generate foul gases and fires, but before stoping begins and fills up the mine with men, the seam has been perforated everwhere by the entries, and most of its water-channels and pockets or feeders of gas have been discovered and placed under control. In retreating, when stoping begins, drainage, ventilation and tramming are covering the whole area of the property, and are at a maximum; and all, especially the two latter, tend to grow less as the working area is contracted, while the advancing system implies a continual extension of the area covered by each. The maintenance of entries is a serious expense in an advancing system, as they must not only be constantly rebrushed, but are liable to develop irregularities from squeeze which make uniform tramming grades difficult to maintain. The final capital cost of entries is the same in either system but the cost of maintenance for retreating is only a fraction of that for advancing. This gain alone will often more than offset the earlier outgo of capital requisite for the former system.

With the advancing system, coal is apt to be lost even by longwall, while the history of even recent pillar working in America indicates that an average of hardly 70 per cent. of the seam is recovered. To obviate the only two drawbacks to retreating, the need of much capital and time, the two systems can, in pillar working, be easily combined temporarily, by opening off enough rooms from the advancing entries to maintain a modest output until the boundary is reached, where pillar-drawing can then be started, and the regular retreating system inaugurated. In longwall working, the combination of advancing and retreating is less simple because of the complications it is liable to cause in the control of roof, especially with the continuous face method; but with the panel layout, it can be effected in those cases where the roof is flexible enough to permit the longwall operation of isolated panels of moderate size.

The longwall practice cited in the next chapter is all on the advancing system owing to the lack of retreating examples in America, but the advance layouts described can readily be transformed into those for retreat by merely starting the initial longwall face at the boundary of the property instead of at its entrance.

(c) Mining by Roof-Pressure

Blasting must ever be a danger in a colliery; and all practical substitutes not involving the creation of flame or high-temperature gases are to be welcomed. The different forms of wedges, hydraulic cartridges, lime cartridges, and other appliances of like purpose have all received full attention, but little has been written on Nature's own solution of the difficulty; viz., roof pressure, and its systematic and scientific utilization.

Any bed or seam is subjected to a certain compressive force owing to the weight of the superincumbent strata: if a portion of the bed is removed and no artificial means of supporting the excavation attempted, a "center of relief" is established, the roof and floor of the cavity move together, and the coal (or other material) round and about the cavity is cracked and crushed by the roof weight, and eventually some of it forced out into the open space. If the coal surrounding the excavation had been undercut, it would have fallen under the action of the roof weight sooner and in better condition; but, had the undercut been too deep, the coal would have fallen en masse, and have necessitated manual labor in breaking to a size suitable for removal.

The cleavage of the coal must be studied in order to determine its behavior under roof pressure. The terms bord (or face) and end (or butt) are pretty universally employed in application to a coal face advancing with its length parallel to the planes of main cleavage or cleat, and perpendicular to those planes respectively. It is well known to every collier that bordways is the easiest direction of advance; but coal so hewn is most likely to result in a high proportion of slack. On the other hand, the coal is strongest end-on; is hardest to hew, but is most likely to result in a large percentage of lump when so obtained.

The mode of fracture of the roof also needs attention. The forces which induced the cleat into the coal had, in the generality of cases, a similar effect on the strata above, causing an incipient cleavage in it coincident in direction with the cleat of the coal. For this reason, the maintenance of a long straight face absolutely bord is almost an impossibility; at such a face the roof would be beyond control, would break off "short" against the face (Fig. 108), and would not only be a constant source of danger, but would take most of the useful weight from the face.

This last fact was recognized very early by coal miners and wishing to combine the easy bord direction of advance with a better control of roof, they instituted the stepped face (Fig. 119). Since the mean line of advance in the case shown in the figure runs some 30 deg. from bord, it follows that the roof will break parallel to this line. Stepped long-wall has many disadvantages; first among which must be placed the fact that the stepped face is unsuited to machine holing. Secondly, out-standing points of coal, such as *K*, Fig. 119, receive an undue roof weight and become crushed. While at the other extreme we have points such

FIG. 108.—Effect of pressure on roof.

as *m*, Fig. 119, too far back and too well protected for the roof weight to act usefully there, where also the coal is bound on two sides (along the face and down the step), and correspondingly hard to hew. Again, since the packs have to be built close against the side of the step to support the coal (the space between the two is seldom more than 2 ft., often less) the ventilating current suffers from such restrictions—an effect which is further augmented by a frictional loss brought in by the air being forced to travel a zig-zag path. Stepped longwall is giving way in places to the straight "half-on" face, which allows of machine holing and a well-controlled roof.

Further factors which must receive consideration in a discussion of the effects of roof pressure, beyond those outlined above, are:

1. The nature of the seam.
2. The nature of the floor and roof.
3. The rate of advance of the face.
4. The amount of dip of the strata and the direction of dip as compared with the direction of the cleat.

To utilize the roof weight to the best advantage, the coal must be undercut to a certain uniform depth, such that when the sprags are withdrawn the coal falls with a vertical fracture from the back of the undercut, without any extraneous aid by blasting, or even wedging.

To achieve this, the undercut must generally be deeper than what is considered advisable by hand; hence, we must depend on the machine to make this desideratum an actuality. At the Altofts Colliery, Normanton, Yorkshire, they have succeeded in almost dispensing with blasting by holing 5 ft. 6 in. under in a flat 3 ft. 3 in. seam, 1500 ft. below the surface. With a hard seam a much deeper undercut than 5 1/2 ft.—perhaps 7 1/2 ft. or 8 ft. in some cases—would be found necessary if blasting were to be abolished; such a depth would if it became anything like general, cause the abandonment of disk machines in favor of those of either the bar or "puncher" type.

A most important advantage of the coal-cutting machine lies in the straightness and length of face necessary for successful application: the straightness of the face enables the timbering to be absolutely systematic; and this factor together with the great length of the face allows of the roof pressure to be controlled and utilized with precision. Where faults are absent, the longer the longwall face, the more effectively may the roof pressure be employed as the means of breaking down the coal.

The weight of the roof is not the only force in action on the coal; before the coal is worked the pressure of the floor is exactly equal and opposite to that of the roof on the seam; when an excavation is made in the seam, the floor, expanding on being relieved of much of its compression, exerts an upward force which at first is of the same intensity as the roof pressure but which becomes dissipated sooner than the latter; nevertheless the floor pressure is often of service to the miner and is taken advantage of at many collieries where, owing to there being a suitable band of dirt at or near the top of the seam, overcutting is resorted to in lieu of undercutting. Coal so obtained often is in better condition than coal obtained by undercutting.

A treacherous roof, which breaks and falls immediately the weight comes on it, rendering timber of little avail, is an undoubted evil. Much may be done in the way of palliation, however, by quickening the rate of advance of the face, and proportionally shortening it to maintain a uniform output (the same is also advisable in the case of a soft seam). It has often been found effective in keeping up a bad roof to leave a thin strip of coal against the roof. The device seems to act something like a plaster on a wound: it has an effect out of all proportion to the slight increment of strength it supplies: its action is to prevent the slacking and slipping of the roof; to maintain it in its entirety.

Just previous to installing machine cutting into a colliery, experiments must be made to ascertain the depth of holing, such that when the sprags are withdrawn the roof pressure, aided by the weight of the coal undercut, is sufficient to break off the coal at the back of the holing. The result arrived at will be somewhat (6 in. or a foot) short of the correct figure, inasmuch as the coal face, when undercut by machine, will advance two or three times as speedily as when the holing is done by hand, and hence the roof, not having the time to weigh so heavily will require a larger surface on which to act.

The coal when the sprags or wedges are removed must fall not in a solid block (Fig. 109), but well cleaved (Fig. 110) and ready for immediate

Fig. 109.—Undercut coal, fallen en masse.

filling. Should the coal fall as exemplified by Fig. 109, the defect can generally be remedied by turning the face more toward bord, or by lessening the rate of advance (the former method in preference), and experiment in that direction should be made at once.

Fig. 110.—Undercut coal, fallen in blocks.

Judging from present-day experience, little need be feared on the count of coal cutting when the depths of our mines become excessive; indeed, under the heavy roof pressures then in action the use of explosives at the coal face is likely to be abolished, and the depth of holing necessary will, if anything, be less than that at present in vogue. In very deep mines, however, it has been found that, although the character of the seam remains the same, the percentage of slack increases with the depth. Before the Royal Commission on Coal Supplies, Mr. Martin opined that an increase of depth from 1200 to 2400 ft. would result in an increase in the percentage of slack from the same seam of 5 per cent.; and at Pendleton Colliery, while the coal was worked at depths of less than 2 500 ft., the percentage of slack was 21.5, but when the workings had reached the depth of between 3000 and 3500 ft., the proportion had increased to 39 per cent. This is merely another way of stating that the

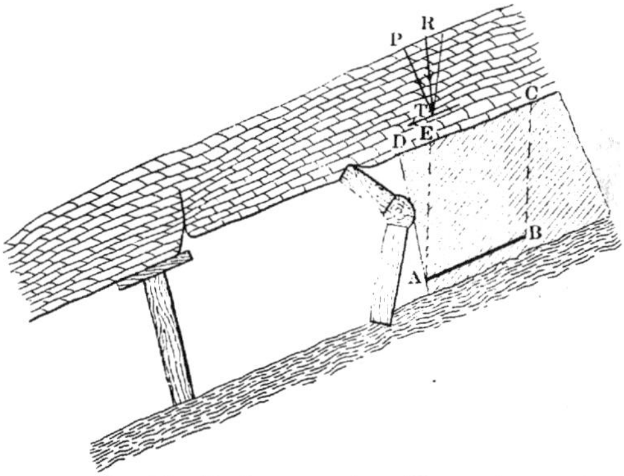

FIG. 111.—Roof pressure when mining to rise.

roof pressure has been too severe for the coal, and to mitigate such an effect the coal should be got, as far as possible, by machine, working end-on, and with a rapid rate of advance.

The effects of dip on the action of the roof pressure is important. In a working proceeding full rise, experience tells us that, other things being equal:

1. Hewing is easier.
2. Work is more dangerous (from falls of roof and face).
3. More slack is produced than in a similar working in a flat seam.

Hewing is easier for the reason that both the roof pressure and the weight of the coal have more total useful effect in a rise working than in any other. In Fig. 111, by means of the undercut *A B* a wedge-shaped block of coal *A B C* D is undermined, if sprags or wedges be placed under the mouth of the undercut, the triangular block *A* D *E* is still unsupported, giving us at once the reason for the liability to falls of face in such a working, and also demonstrating the need for the cocker sprag (shown) or equivalent means of supporting the face. The action of the roof is two fold. There is a pressure P, acting normally to the plane of the seam; there is a thrust T, acting in the direction of dip, tending to make the roof slide over the face toward the empty space behind it.

The force T is evidenced in the fact that a fracture in the roof of a rise working "gapes," owing to the lower side having moved slightly down under the influence of T. Thus it is that falls of roof are more prevalent in rise workings than in any other; the side thrust T, not only quickly breaking up the roof, but also widening the joints the better to allow severed slabs to fall.

It is largely to this side thrust that we owe the production of slack which is one of the disadvantages of rise working; grinding is introduced, a far more effective slack producer than mere normal pressure.

The resultant action of the forces P and T on the coal may be best represented by the single force R. The direction of R cannot be accurately assigned, but it lies somewhere between the normal (P) and the perpendicular (shown dotted), and its position is probably somewhere as shown; its magnitude, by the parallelogram of forces is simply $\sqrt{P^2 + T^2}$. Considering the coal acted on by R aided by the weight of the coal itself, no further demonstration is needed of the reason of the ease experienced in working coal to the rise.

FIG. 112.—Roof pressure when mining to dip.

Rise working would be rendered safer, and less slack would be produced if a rapid rate of advance were maintained, and to compensate for the lessening of roof pressure which would result, a deeper undercut would be necessitated. Carefully built pack walls are also highly advisable in a seam liable to produce slack, and thus especially in rise workings.

In the dip working (Fig. 112), the action of T, the side thrust, is much less important; the tendency is there, but the action, so far as the grinding of the coal is concerned, is nil. Also, any line of break appearing in the roof is closed, instead of opened, by the slight lateral movement of the roof over the gob or goaf: hence, working to the dip of the seam is, generally speaking, the safest of all directions of working the coal. As before the resultant roof pressure acts slightly down-hill (shown at R), causing the coal in this case to be difficult to hew.

The coincidence or noncoincidence of the direction of cleavage and the direction of dip is a factor of importance, influencing the behavior of the coal under the roof pressures. If the directions bord and dip coincide a rise working is doubly easy, but the face will need stepping or the roof will be beyond control; also, under these conditions, dip working will be facilitated, and generally the face in such a working may be maintained straight, owing to the side thrust closing the jointings in the roof. On the other hand, should the directions end and dip coincide, the easiest mode of advance will not be full rise but in some direction between that line and the strike of the seam, while the difficulty of working directly to the dip will be intensified. Intermediate between these extremes there is an infinite number of angles at which the directions of dip and cleat may lie, every case needing special consideration and experiment.

CHAPTER XIX

ADVANCING LONGWALL SYSTEMS FOR SEAMS

EXAMPLE 49.—SPRING VALLEY BITUMINOUS COLLIERIES, BUREAU COUNTY, ILLINOIS

(*See also Example* 5.)

Thin Flat Seam at 500-*ft. Depth. Advancing with Continuous Face by Scotch System. Loading into Cars.*—The Illinois coal reports show that over 5,000,000 tons are produced in the longwall field or about 12 per cent. of the States' total output. Most of the longwall mining is done in the prairie-like counties of Bureau, Grundy, and La Salle. Here the coal seams are remarkable not only for their variety and quality, but in their freedom from horse-back, faults, and other irregularities, which are encountered elsewhere. The mines are developing a 3 1/2-ft. seam, called commercially "third-vein coal," which is 350 to 500 ft. beneath the surface. Overlying the seam is a flexible shale 4 to 9 in. thick and underlying it is 6 to 24 in. of fireclay.

The Scotch system extracts all the coal in the first operation, commencing at the periphery of the shaft pillar and mining out the whole seam toward the property limits. In Fig. 113, the ideal plan of the layout of a Spring Valley mine 350 ft. deep, *h* is the hoisting, *a* the air-shaft, and *km* the shaft pillar, 600 ft. square, over which are located the shaft house, shops, and other necessary surface structures. To open out the seam for longwalling, the pillar is first cross-cut by the two headings *gn* and *pq*, the former being double-tracked to permit the handling of cars to and from shaft *h*. Roads are uniformly 9 ft. wide, excepting at the partings where they are 14 ft. wide to hold two tracks. These peripheral headings are driven from the points *g*, *n*, *p*, and *q*, and by widening them inbye, the first longwall face is begun. At first the face is not continuous, but in arcs like those dotted around points *g*, *n*, *p* and *q*, but as the arcs move inbye they finally meet and form one continuous circle. Soon the face has advanced sufficiently to allow the first break to be made in the roof around the shaft pillar, so that thereafter advantage can be taken of the pressure from the descending roof to mine the coal as in Fig. 114.

Fig. 113, shows the longwall face after it has advanced some distance from the shaft-pillars and appears as the circle 1-4-7-10. The face is reached through the main roadways 1, 2, 3, etc., by which it is divided into approximately equal spaces. The essential feature of the Scotch system by which it differs from the other continuous-face longwall system, "the Rectangular," is the turning off of cross-roads at 45-deg., angles from the main directions for roads, *g-n* and *p-q*. The road *e*-3 was formerly the extension of *gk*, but its present position makes possible a gentler curve for haulage into *gh* and permits the use of the 40-deg. frog of the other 45-deg. turnouts. The layout of roads is based on a room about 42 ft. wide at the face (as at *f*) which corresponds to 60 ft. along the road

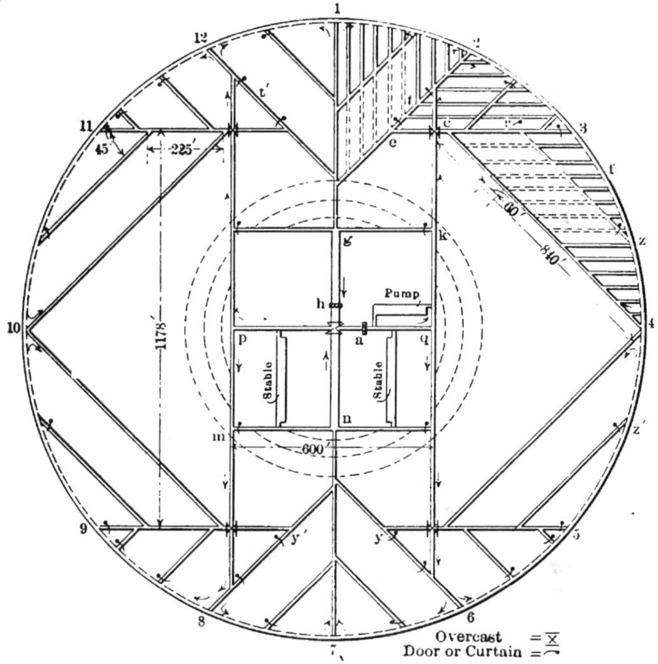

FIG. 113.—Plan of Scotch Longwall system, Illinois.

c-4. The angle-roads are turned off at 225-ft. intervals from a cross-road as *e*-3. Experience has shown that a room can not exceed a length of 225 ft. with out having its track rebrushed before completion. Hence, the rooms of an angle-road like *cd* are abandoned as soon as they are cut off by the next road *r*-4. Crossroads *e*-3 and *y*-5 are 1178 ft. apart, so that angle-roads *c-d* and *d-g* can each be 840 ft. long to contain exactly 14 rooms.

The direction of the air current is shown by arrows. The shaft *a* is the downcast, the shaft *h* is the upcast; roads 12, 2, 6, and 8 are intake, and the other roads are return airways along their inbye portions. In the intake roads, double doors are placed at *t*, *t'*, *y*, and *y'* while single doors are placed at points with strong draft like *z*, *z'*, etc., to hold the air along the working face. Fire-proof burlap curtains are used in rooms where the draft is not strong enough to force them up. The ventilation is excellent and easily regulated. Trap boys are stationed at the double doors and also at cross-roads where collisions are liable to occur from trains approaching in opposite directions.

FIG. 114.—Roof sinking behind Longwall face.

For haulage the grade is nearly level; big mules are used on the main roads and small ones for gathering. At No. 4 mine, 7 gathering mules with two cars apiece haul 250 cars daily from the face to the second parting, thence 4 mules haul trains of 30 cars to the first parting whence they are hauled in similar 4-mule trains to the shaft bottom. The total output of 1000 tons is hauled in the day shift by about 40 mules, the face being about a mile distant from the shaft. The cars are of wood, weigh 1100 lb., hold 2700 lb., of coal and run on a track of 42-in. gauge, laid with 16-lb. rails except at the bottom of the shaft.

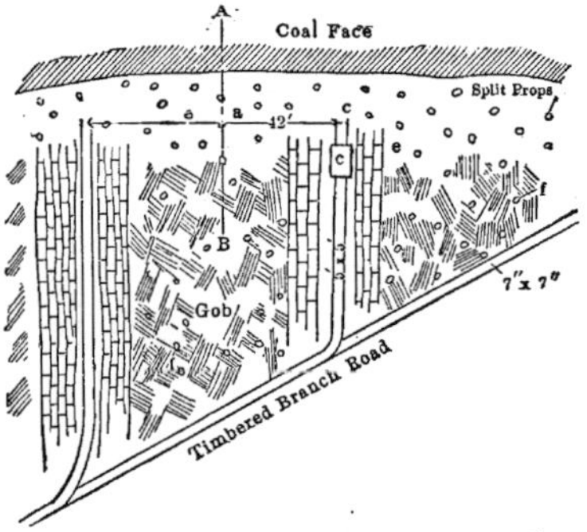

FIG. 115.—Plan of packs and tracks at Longwall face.

Fig. 115 is a plan at the face showing two room "gateways," or room roads with tracks turned off at 45 deg. from the "Timbered Branch Road." Halfway between the gateways is the "mark" *a* which separates a room into two 21-ft. halves, each assigned to one miner who both mines and loads his coal into a car on the nearest track at *c*. The undercut is made by hand pick, from a crouching position, in the floor; when the latter is of sandstone, the coal itself must be grooved. If the roof is working properly, which is ascertained by sounding the face with a hammer, the undercut need only be 20 in. deep. Otherwise a depth of 5 ft. is sometimes necessary. The clay cuttings are thrown back into the gob.

The packwalls along the gateways are 6 ft. wide and do not approach nearer than 2 ft. to the face in order not to obstruct the air-current. They are built of slate brushed from the roof (see Fig. 109), by a hand pick to a height of about 6 1/2 ft. above the tracks. The roof of the haulage roads sinks as the face advances and must be continually rebrushed for slate, which is partly stored in an abandoned room and partly hoisted in cars to the extent of 10 per cent. of the coal-car hoist. The rebrushing of the roads, the repair of the track and the retimbering, is done by company men on day's pay, but initially the miners brush the gateways, lay the tracks, build up the packwalls, and set the props in their own rooms as part of their contract price per ton of coal loaded.

The undercut is held up by sprags (see Fig. 116) until the whole 42-ft. face of a room has been completed. When the sprags are knocked out

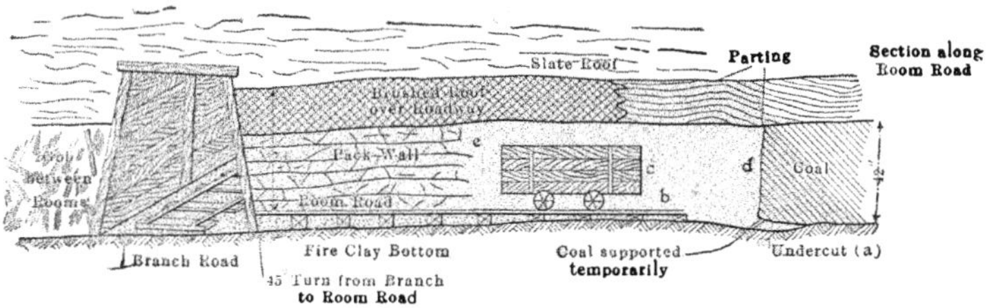

FIG. 116.—Section of Longwall face, gateway and branch road.

and the undercut coal does not fall, it is wedged down in two layers by steel wedges starting from a shear made at the breast center or "mark." To reach the car from the mark, the coal must be reshovelled twice in the narrow alley between props and face. Lines of 8-in. by 4 1/2-ft. props, about 3 ft. apart, are set along the face (see Fig. 114) at necessary intervals and some of these are not recovered. The haulageways are timbered with three-quarter sets which in certain places are seen to be cribbed 10 ft. high above the caps to catch up roof-caves that broke down the original timbering. Steel I beams are used for caps over some of the double track partings and the shaft bottom road is walled with masonry. Wooden cogs are placed at acute road corners as c (Fig. 113) and also to replace props along the coal face where the roof is unusually weak.

Work can cease on part of the coal face without injury to the balance if care be taken to keep the whole face regular and convexly curved, for trouble ensues if corners or concavities are allowed to develop. A fall or creep of roof at the face, which closes up the space between gob and coal, may occur from uneven advances, from failure to build up the pack

16

walls, from unusually weak spots in the roof, or from long shutdowns. The face is reopened as "yardage" work by driving the gateway into the coal and turning off a heading along the face with a 2-ft. rib between it and the old gob.

The haulageways are simply gateways selected because they occur at the proper intervals of the layout of Fig. 113. Near the face, rooms are always advancing in two directions which intersect each other at a 45-deg. angle. Thus room *zf*, as it advances, is cutting off all the rooms between *r*-4 and *zf* which are then abandoned and their occupants assigned new working places. It is always arranged to give an experienced miner a 21-ft. face to work for himself, but occasionally he takes a green hand as helper and pupil. There is little gas, though two fire-bosses are employed. The mining and hoisting is all done on day shift. The output of coal is about 2 1/2 tons per man employed above and below ground.

EXAMPLE 50.—MONTOUR IRON MINES, DANVILLE, PA.

Thin, Sloping, Shallow Beds. Loading into Cars.—The long-wall method of minin g when introduced into these mines was but little used in this country, and seldom in beds as thick as these, with breasts frequently 4 to 5 ft. high. Figs. 117 and 118 show the general method.

Levels are driven 90 ft. apart, and the face of each gangway should be kept in advance of all higher gangways, so that of the gangways C, *E*, and *H*, for instance, the face of *C* should be the farthest from the slope or the mouth of the drift; but in fact this is seldom done here, as it necessitates the outlay of large capital before any return is realized.

The gangways are driven 7 to 10 ft. wide and 5 1/2 to 7 ft. high, so a man or mule can go erect. The lowest gangway *C*, Fig. 117, is known as a "fast-end" gangway, as it is driven entirely in the solid, while *E*, *H*, etc., are "loose-end" gangways, with but one side in the solid and the other side formed by the stowing. The face of the gangway *C* should be kept far enough ahead so that blasting there will not interfere with the workmen in the breast D, and the same consideration should determine the distance of the breast *F* from the face of *E*.

To facilitate the loading of the ore into cars from chutes, the gangways are so driven that the roof of the bed will lie at the top of the upper rib, Fig. 118, and to secure the proper gangway height the bottom rock must be taken up. In doing this along the lower gangway *C* a drainage-ditch is left upon the lower side. When first driven, the gangway *C* is timbered upon the upper side, but, as settling takes place, the props are usually broken, and it is necessary generally to renew them and to blow down the roof of the gangway, which frequently settles sufficiently to obstruct the haulage-way. Often, however, the stowing becomes so tightly packed in settling that retimbering is unnecessary. In the soft ore, by reason of the creeping of the bottom, the gangway-props must sometimes be renewed several times. Breasts or rooms D and F are turned off at an angle of 35 to 45 deg. with the direction of the gangway, depending on the dip of the bed. The breasts are 24 to 30 ft. long, and usually there are five breasts in a tier between two gangways. The height of the breast varies, with the nature of the ore, from 2 to 5 ft. In the hard limestone-ore are three streaks of ore which are taken out if sufficiently rich; but if the ore is lean the central streak alone is taken out, with just enough rock to allow the mine to work his breast. The hard limestone-ore and the block-ore have to be blasted, but the soft-ore is scraped out in the form of mud.

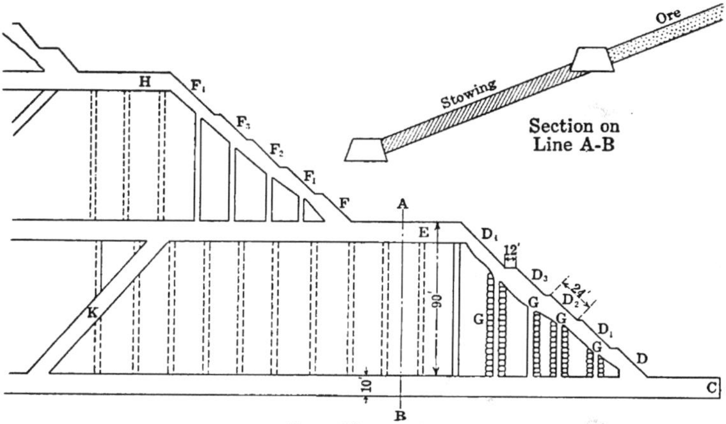

FIGS. 117 AND 118.

FIG. 117.—Long section of stope, Montour mine.
FIG. 118.—Cross-section of stope, Montour mine.

Each breast is worked by a miner and one laborer; or two miners will combine and work two breasts; and, sometimes, one miner and two or three laborers will work two breasts. The miner working the top breast of a tier, such as D_4, Fig. 117, also drives the gangway E, takes up the bottom rock to give sufficient height for haulage, piles the stowing carefully on the lower side of the gangway, and prepares the road-bed for the track-layers, for which additional work he is paid extra.

A ditch is not left along the loose-end gangway, as the water should drain through the stowing to the fast-end gangway C. The gob is thrown loosely between the breast and gangway below, excepting along the chutes G, where it is carefully piled to support the roof. A chute or gateway G, 2 1/3 ft. wide, is left for each breast, down which the ore is thrown to the gangway below; it is sometimes lined with boards, but generally a carefully-built dry wall of gob suffices. The ore from each breast is carried to the chute by hand, and drawn out from its bottom into cars as desired.

There is usually a platform at the chute bottom to facilitate the loading; in the soft ore it is placed directly above the car, but it is nearer the bottom in the hard ore, and sometimes the ore is simply allowed to pile up along the gangway. When necessary to prevent the air-current drawing up a chute, a canvas curtain is hung loosely over its mouth; but ordinarily only the last five inside chutes are kept open, the others being boarded up and filled with gob when the gangway *E* has advanced enough to receive the ore from the next higher tier of breasts. The gangways *E*, *H*, etc., are connected with the fast-end gangway *C* by "pitching gangways" *K* driven through the gob, and back of these last the gangways *E* and *H* are abandoned, so that the fast-end gangway *C* is the only one kept open through its entire length. The gangways in the soft-ore are timbered and lagged on sides and top, but in the hard limestone-ore and in the block-ore it is generally necessary to timber the sides only, as the roof is of good slate or sandstone.

The props are placed 2 to 6 ft. apart, depending upon the nature of the roof. In the hard fossil-ore and in the block-ore the breasts are not timbered, excepting when necessary to protect the chutes, as the gob fills up the space and supports the top. In the soft fossil-ore small props, 3 to 5-in. dia., are used to keep up the top, as the gob does not fill more than one-third of the vacant space. Heavy timbers are usually placed along all chutes. The timber is furnished by the company, but the miners set it, both in the breasts and along the gangway, and as it is cut on company property it is cheap.

All general work, such as track-laying, the clearing away of "falls," etc., is done by miners, detailed for each separate piece of work, instead of by laborers, and for such work the miners are paid *per diem*. All drilling is done by hand, and the ventilation is secured by natural draft through chimneys. In the drifts, the cars are either pushed by hand or hauled by mules to the mouth, while in the slopes the mine-cars are hoisted to the top by second-motion engines.

The mine-cars are 2 ft. deep, 4 ft. long, and 3 1/2 ft. wide, and hold about one ton of ore. The gauge is 30 in. and the wheels are 14-in. dia. and loose.

Cost of Mining.—For breast-work, miners are paid by the ton and for gangway-work they are paid tonnage and yardage.

Tonnage payments depend upon:

(1) The nature of the ore.

(2) The height of the breast.

Yardage payments depend upon:

(1) The nature of the ore.

(2) The kind of gangway.

Since the nature of the ore in the fossil-beds and the height of breast vary so irregularly, it is almost impossible to give exact figures so that chiefly ratios will be given. Upon a basis of $1 per ton for mining block ore, the following are the prices paid during the past 15 years:

Block ore per ton	$1.00
Block ore per yard, fast-end gangway	4.00
Block ore per yard, loose-end gangway	1.70
Hard fossil ore per ton	0.95
Hard fossil ore per yard, fast-end gangway	6.25
Hard fossil ore per yard, loose-end gangway	2.40
Unskilled labor per day	0.73

Soft ore costs to mine one-third to one-half the above hard ore prices. One ton per day for each man working in a breast is considered an average output for a shift of 10 hours. The miner pays his laborer, or laborers, *per diem*, at the above rate. In gangway-work the average rate of advance was 15 ft. per month for loose-end gangways and 7 ft. per month for fast-end gangways. Owing to the many conditions affecting the rate of advance along the gangways, it was necessary to employ a system of "allowances" in payment of gangway-yardage so as to equalize as nearly as possible the pay of gangway-miners.

Example 51.—Bull's Head Anthracite Colliery, Providence, Eastern Pa.

(*See also Examples* 5 *and* 59.)

Thin, Sloping, Shallow Seam; Panel System; Loading into Buggies on Endless Rope.—The coal property is about 1200 ft. square, and a section of the measures just above and below the seam being mined longwall is approximately as follows:

20 ft., slate and soil;
2 1/2 ft., fireclay;
1 ft., bone;
5 *ft., coal seam;*
40 ft., slate;
2 ft., sandstone;
6 in., slate;
"30 - *in." coal seam;*
18 in., hard slate;
9 ft., soft shale;
18 in., hard slate;
3 1/2 *ft., coal, 4-ft. seam;*
90 ft., sandstone and slate;
8 *ft., coal,* Diamond *seam.*

Below the Diamond seam occur the Rock seam, the Fourteen-foot, and the Clark, all of which and also including the Four-foot and Diamond seams had been worked out by room and pillar prior to begining to mine the Thirty-inch seam by longwall.

Consequently, the rock above and below the Thirty-inch seam was cracked and in many cases out of place, the cracks often extending to the surface. In consequence the footing for props was most insecure, and although the cover above the Thirty-inch seam was only about 75 ft., it was impossible to hold it by timbering and it would have been probably impossible to take out the coal by room and pillar. The longwall method of working as developed by Supt. Vipond is shown in Fig. 119. A rock slope was driven up from the Four-foot seam at a slight pitch so

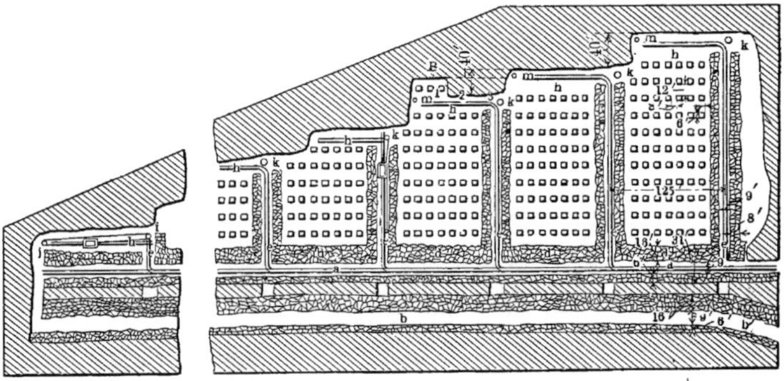

FIG. 119.—Plan of Longwall system, Bull's Head colliery.

that empty cars can be hauled up the pitch by mules. From the head of this pitch the gangway *a* was driven 31 ft. wide and 5 to 6 ft. high, bottom rock being taken up to give sufficient height. At the same time the parallel airway *b* was driven and ventilation secured by means of the headings shown through the gangway pillars. The airway *b* connects by a passageway *b′* with a ventilation shaft from the underlying Four-foot seam. The rock obtained in driving the airway is piled in walls along both sides of the airway. The rock resulting from taking up the bottom during the driving of the gangway is built into a continuous wall *c* along the lower side of the gangway and into walls *d* 16 ft. wide along the upper side of the gangway. Through the upper walls *d* are passageways *e* which are 9 ft. wide and are spaced 125 ft. between centers. These passageways, called gateways, have loose walls *b* 8 ft. wide on each side, thus making the total width of the gateway and the walled space 25 ft. The gateways are driven the same height as the gangway for a short distance in from the gangway so as to provide a place for the mine car to stand while it is being loaded and out of the way of traffic along the gangway. This distance depends upon conditions, but is usually not over 40 ft. Above this point the gateway is made only the height of the coal and the overlying slate, that is, about 36 in. The coal is overlaid by about 6 in. of slate which is always taken down, and it is this which furnishes the greater part of the material needed for building the pack walls along the greater length of the gateways and along the face as will be described later.

The method of opening out a face is shown at *A*. Strips are taken off the face parallel to the gangway and the gangway walls *d*, and as soon as sufficient width is secured between the gangway wall *d* and the face of the coal, a track *h* is laid as near to the face as possible so that it will not interfere with the work of the miners. This track has a gauge

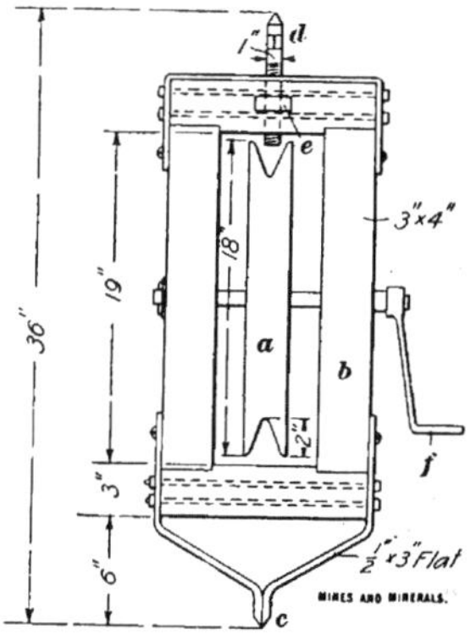

FIG. 120.—Single winch, Bull's Head colliery.

of 2 ft. 3 in., is laid with 25-lb. rails, which are 10 ft. long. The rail sections are joined by two fish-plates, one placed on each side of the flange. The rails are held together by iron bridles which are laid directly on the bottom. On this track is a small buggy into which the coal is shoveled. This buggy is moved by an endless wire-rope operated from a point *i* on the gateway as follows: A cast-iron wheel *a*, Fig. 120, 18 in. in diameter and having a groove 2 in. deep, is held in a wooden frame *b*. At the bottom is a pointed iron *c* fixed on the frame. This rests upon the bottom rock. At the top is an adjustable pointed round iron *d* the lower 9 in. of which is threaded so that by means of the nut *e* countersunk as shown in the frame, when a wrench is applied to the squared portion above the thread the point can be forced up against the roof and the frame thus held securely in place. A 3/8-in. wire rope is wound two or three times around the wheel *a* so as to give it sufficient grip on the wheel. At the other end of the track along the face at *j* this rope passes through an ordinary 6-in. iron block and tackle which is hooked to a chain placed around the prop. One end of the rope is attached to the front end of the buggy and the other to the back end of the buggy. By turning the handle *f* of the wheel, the buggy can be moved forward and backward along the face. This buggy is made of timber, holds about 20 cu. ft., and one side is 20 in. high and the other 18 in., this difference being made to allow room for loading over the side. The wheels are 6 in. in diameter placed on 1 1/4-in. axles. In the bottom there is an iron plate which slides in and out sideways, being moved by a handle. The track *h* along the face *A* which is just being started extends out over the track *e* and by pulling out the slide in the bottom of the buggy the coal is dumped from the buggy into the mine car standing on the track *e*.

The coal is not undercut, and in general does not need to be drilled or blasted, the weight of the cover being generally sufficient to loosen the coal with an occasional shot when the roof pressure is not sufficient. Owing to the broken conditions of the measures due to the mining out of the underlying seams the coal in many places is loose and simply needs to be picked out. Six men work along each face, three miners and three laborers who load the coal into the buggies. The face is worked in several sections as shown at B, each section being taken out for a certain distance, about 12 yd. depending upon the ease with which the coal can be broken down; but no section of one face is allowed to get far ahead of any other no matter how easily the coal can be mined. By the time section 3 has been mined out the coal in section 1 will have again loosened by the weight of the cover and can then be taken out after the cogs have been built. The track *h* is moved near the face after each section is mined and a row of cogs is kept close up to the track. Each face of coal is kept about 40 ft. in advance of the next following face. The cogs are 6 ft. square and the rows are 8 ft. apart parallel to the face and 12 ft. apart perpendicular to the face. These cogs are built from the slate overlying the coal, and as it comes down in large slabs a very firm cog is formed. The space between the large rocks is filled in with dirt and a perfectly solid cog thus formed.

As already noted, after the gateways *e* have been driven in full height, a distance sufficient to allow the mine car to be placed in the gateway out of the way of the haulage on the main gangway, the height of the gateway is decreased to the thickness of the seam and overlying slate, that is, about 36 in. The coal is moved from the face to the car at the mouth of the gateway by means of a buggy similar to that used along the face and already described, but instead of the winch shown in Fig. 120, it is moved by means of a double winch placed at the point *h*, Fig. 119, on the gateway where the height is decreased to the height of the seam. This winch, Fig. 121, has two drums *a* and *a′* which run loosely on the axle *b*, but by means of the clutches *c* and *c′* by means of a lever not shown, either drum may be made to turn when the handle *d* is moved. If the load is too great to be moved by turning the handle *d* the winch may be operated on second motion by means of a pinion *e* attached to a movable axle *f*. One box of this axle at *g* is loosely bolted to the framework allowing a little play of the axle, while in box *h* is an elliptical instead of a circular hole through which axle *f* passes so that it can be pushed over to the dotted position *f'* throwing pinion *e* out of gear. The winch is set on a framework of timbers one end of which rests directly on the bottom, while the other end is let into a groove in a prop. There are two ropes

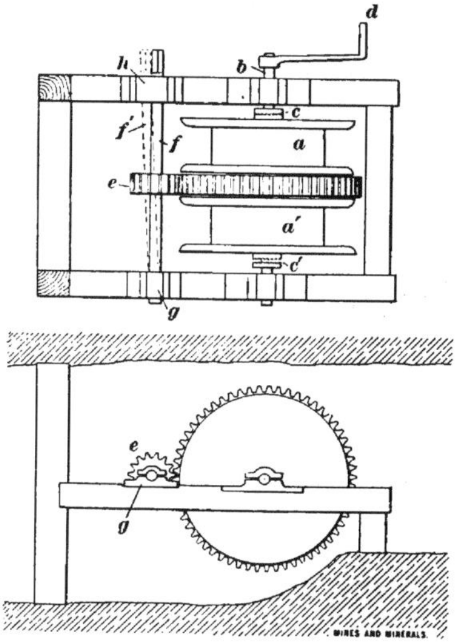

Fig. 121.—Double winch, Bull's Head colliery.

which wind upon the drums *a* and *a'*. One of these is attached to one end of the buggy, while the other passes to the upper end of the gateway, thence through a small 6-in. iron block and tackle *k* fastened to a prop by a chain and back to the other end of the buggy, or by means of guide pulleys or rollers at the inby end of the gateway the rope may be carried along the face to a return pulley *m* at the extreme end of the face and the gateway buggy taken along the face and the coal brought directly from the face to the gangway.

The conditions for operating the longwall system of mining are particularly unfavorable, for the bottom is badly broken and a stable footing for props is often unobtainable. At the face of one of the gateways at the time of our visit the bottom had dropped away entirely from beneath the coal leaving the coal supported only by contact with the overlying slate. The top is also badly broken, allowing the surface water to enter the mine and giving a roof that cannot be controlled. This roof settles down over the gangway packs about 2 ft. so that while the gangway is driven about 6 ft. high it is only about 4 ft. high after the workings have settled. Over the gateways and cogs the roof settles about 2 ft. Thus far an output of 1800 tons per foot-acre has been obtained, a much better yield than is usually obtained in anthracite mining.

EXAMPLE 52.—VINTON BITUMINOUS COLLIERY, VINTONDALE, PENN.

Thin Stoping Seam, 800 ft. Deep; Panel System; Loading into Pan Conveyors.—Transporting coal from the working face to main haulage roads by means of mechanical conveyors is a comparatively recent departure from ordinary mining methods. This system, which was first introduced in England, was early recognized by leading operators there as possessing superior advantages over the usual manner of working, especially in thin coal seams where the roof has to be brushed to allow all but the tiniest cars to reach the longwall face.

The coal worked at Vintondale is bound tight to roof and floor and is the "B" or Lower Kittanning seam, 42 in. in thickness, which lies on a pitch of 8 per cent., with an average of 200 feet of cover. The coal is of a soft and friable nature, free from slate bands and bony coal, but interspersed with sulphur pyrites, which, at times, cause considerable annoyance in cutting and drilling. The bottom is a mixture of coal and fireclay, while the roof is composed of from 8 to 12 ft. of black slate, overlaid with sandstone. The slips in the slate are well marked, and lie at an angle of 25 deg. with the line of greatest dip; the longwall face is kept normal to these slips. The present panel modification of longwall mining was first started in No. 3 mine in 1900. At the outset cars were run around the working face and loaded. This method brought only fair results, owing to the necessity of using small cars, steep grades, and difficulty in keeping roadways open.

Arrangements were then made for the placing of a conveyor along the face, allowing the cars to be run under the head-end to be loaded. The first conveyor, which was made entirely of wood, was a cumbersome affair, and much time was consumed in moving it laterally along the face after the cut had been loaded out; but, after a year's trial, the results obtained were so gratifying that metal conveyors were designed and ordered, and preparations were made to employ this system on a much larger scale (Fig. 122).

The metal conveyor consists of a trough or pan, made of sheet steel 1/8 in. thick, 12 in. wide at the bottom, 18 in. wide at the top, and 6 in. high, set on strap-iron standards as shown in detail in Fig. 123. A conveyor is made up in sections of 6-, 12-, 15-, and 18-ft. lengths, connected together by means of 1/2-in. flatheaded bolts, countersunk. The front is inclined for a distance of 45 ft. to allow clearance for mine cars to pass under (see Fig. 114). The rear end is inclined for 15 ft. to compensate for the size of sprocket wheel. A return runway for the chain is afforded below the pans by angle irons.

A cast-iron driving sprocket, 18 in. in diameter and 13-in. face, is

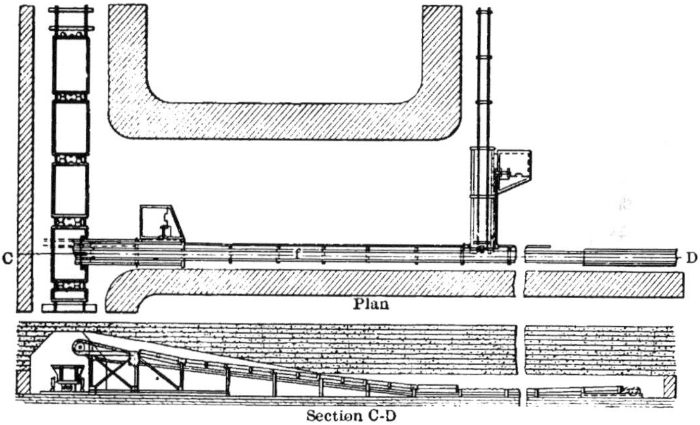

FIG. 122.—Plan and section of Vinton conveyor system showing head of main conveyor.

attached to the front end. On the shaft of this sprocket, which is extended 12 in. beyond one of the bearings, is keyed a 12-tooth, 16-in. diameter sprocket, which connects with the driving mechanism. The rear-end section (*c*) consists of a framework made up of two I-beams, 6 ft. long and strongly braced, on which rest the take-up boxes for keeping the chain in adjustment, and the rear sprocket wheel over which the

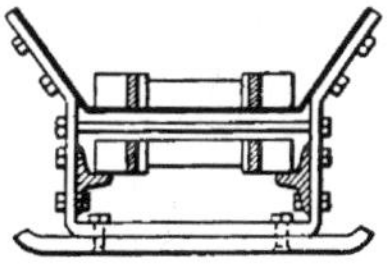

FIG. 123.—Cross section of conveyor, Vinton colliery.

chain returns. There are two conveyor chains, held apart, the width of the trough, by crossbolts which act as scrapers to replace the usual plates. The chains are of steel and are designed for quick repairing.

The triple conveyor system (Fig. 124), was finally designed and installed as an improvement over the single type. In laying out a mine for this system, the main entry and airway are driven up or down the pitch, and cross-headings are driven off them at intervals of 400 ft. at such an angle as will give a 2-per-cent. grade; 75-ft. barrier pillars are left on each side of the main entries. The cross-heading is driven 20 ft. wide and gobbed on the lower side. The air-course, which afterward is used as the panel or block face, is driven 20 ft. wide, but no bottom is lifted; a 40-ft. pillar is maintained. Block headings are run perpendicular to cross-headings at 518-ft. centers; they are driven 18 ft. wide, with bottom lifted in the center 5 ft. wide, and deep enough for a 5-ft. clearance.

When the block is ready for operation, a conveyor 350 ft. long is placed in the block heading, and along the face of the air-course on each

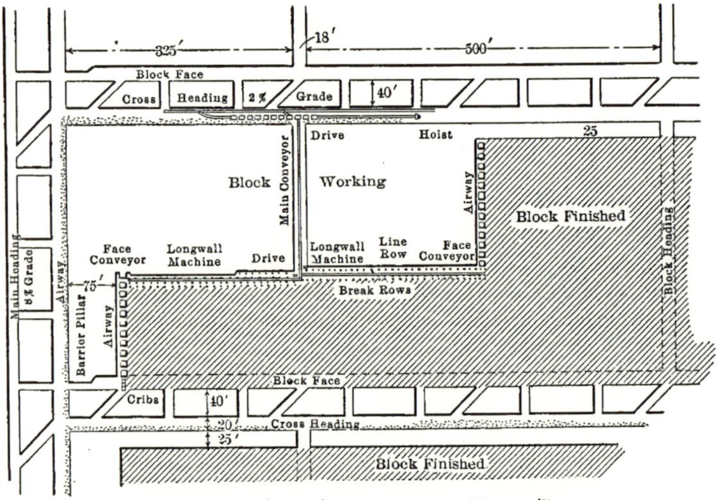

FIG. 124.—Plan triple conveyor system, Vinton colliery.

side is placed a conveyor 250 ft. long, with delivery ends directly over the main conveyor, one being 5 ft. in advance of the other. Each conveyor is driven by a 20-horsepower, 250-volt, series-wound motor, encased in a sheet-iron frame mounted on steel shoes, so as to be easily moved.

Airways are maintained on the blocks by driving two places slightly in advance of the block face, 6 and 4 ft. wide, respectively, with a 10-ft. pillar between. The first place acts as a stable for the machine, and is driven by the machine. The airway is pick-mined, and one man manages to keep these places going on the rear end of both blocks. By this arrangement no cribbing is necessary.

The blocks are worked to within 25 ft. of the cross-heading, when the conveyors are removed to another block. The remaining pillar is brought back along with the heading stumps.

The power is carried to the top of the block heading by a 00 wire. Here are attached two insulated twin cables, one to furnish power to the machines, the other for the drives and hoist.

The cables are carried down the block heading, one on each side of the main conveyor, being attached to it by means of malleable-iron brackets. At the junction of the conveyors connections are made with the drives, also with a cable that is attached to each of the face conveyors.

Stations are established 50 ft. apart on the face conveyor cables, to which connections are made with the short cable attached to the longwall machines and electric drills. Switches are placed at the head-end of the main conveyor, by which the power is controlled.

The method of handling the cars to the conveyor is simple. A side track is laid 300 ft. long, of which the block heading is the center. Connection is made with the main track at the lower end, and a cross-over switch is placed directly under the conveyor. At the upper end of the siding is placed an electric hoist. A trip of 14 cars is shoved into the empty track, and the rope is attached and the trip pulled up to the conveyor. Signal wires are hung between the conveyor and the hoist, and and as each car is loaded the trip is pulled forward. When loaded, trip is dropped on the loaded siding, the rope disengaged and attached to the empties.

The crew operating a double block consists of 17 men, *i. e.*, block boss, machine runner and helper, driller, shooter, two conveyor men, hoist boy, five loaders, and four timbermen. Two longwall machines of Jeffrey or Sullivan make are used, one for each side, although one machine can keep up the work in case of emergency. The machine men finish cutting one block, in five hours, and then put the machine in position to start back on the cut and move over to the next block and begin cutting. They are followed by the shooter and loaders.

When a block is cleaned up, the timbermen move up the conveyor.

This consists of setting a line of props, called the line row, about 8 ft. apart, and a distance from the conveyor equal to the depth of the undercut. As these are placed the old line row, which is now against the conveyor, is withdrawn. The pulling jacks for moving the conveyor are distributed along the block 40 ft. apart and placed in position.

The shot firer keeps closely after the machine, and is through shooting shortly after the undercut is finished. The driller then starts from the far end of the block to drill holes in the new face. It usually takes him about two hours to drill the entire width of the block.

Each loader is supplied with a pick and shovel and a piece of sheet iron 9 in. wide and 6 ft. long, which he attaches to the conveyor to act as a sideboard. As each loader cleans up his place he moves forward to the head of the line. This continues until the coal is loaded out, which usually requires about six and a half hours.

When cleaned up, the drive is reversed and the timber which has arrived on the last trip is run through on the conveyor to such points on the block where it is required. When this is accomplished the power is shut off, and the conveyor is moved up to the line row. This lateral move of the conveyor requires very little time, very seldom exceeding five minutes. A break row, consisting of two rows of props set 2 ft. apart, is now placed along the lower side of the conveyor. These props are set on a cap piece, placed on a small pile of slack, and wedged at the top. Two break rows are all that is necessary to protect the block. In the meantime, a portion of the crew are engaged in pulling out the extra break row. This is the most hazardous work on the block, and is given personal attention by the block boss. Axes are used in this operation, and about 75 per cent. of the props recovered are practically uninjured.

While part of the crew are employed timbering, the rest make the necessary connections, and go along the conveyor with a pump jack and level it up. They also build a crib at the head end, which is placed to prevent the roof from breaking over into the block heading. All the dead work is taken care of by the four timbermen; thus not hindering the steady flow of coal, which averages 150 tons daily, from a 5-ft. undercut.

For the purpose of keeping the machinery in as good shape as possible, a skilled mechanic is attached to each mine. He assumes charge in case of an accident and makes necessary repairs, although most of the breakdowns are easily taken care of by the block boss and machine man.

In the starting of a block is where the best results are obtained, as the roof requires little attention until about 100 ft. have been extracted. It then begins to weigh heavy on the posts, and it is found necessary to carry three or four double break-rows with cogs in anticipation of what is called the "big break." This usually occurs when the block is advanced from 100 to 150 ft., although in several instances a 500-ft. face has been carried up 200 ft. before the overhanging strata broke. After the sand rock is down, only two break rows are carried, and the roof keeps breaking behind the last row as the face is extended.

The men are paid day wages and, as they become accustomed to the work and machinery, are advanced accordingly. The "block" boss, as an incentive to secure the best results, is paid a small bonus per ton besides his regular day rate. The cost averages for the last two years show that block coal is loaded on the mine cars 35 per cent. cheaper than the district mining rate for pick work with loading into cars.

The above conditions prevailed in Nov., 1907, but on the author's visit in Sept., 1910, he found the longwall system superseded by the former room and pillar system for the following assigned reasons. 1. The frequent breakage of the conveyors caused a very irregular output.

2. The miners preferred the contract payments of room and pillar to the time wages of the longwall system. 3. The timber consumption was excessive, because many of the props could not be recovered.

None of these disadvantages are irremediable, and the cost of timber can always be obviated wherever enough slate can be cheaply got from the floor, parting, or roof to build pack walls to replace part of the props and cogs. The conveyor-longwall system has proved profitable for thin seams in Europe, and the Vintondale method should prove commercially successful in other American fields where the natural conditions are suitable.

EXAMPLE 53.—DRUMMOND BITUMINOUS COLLIERY, WESTVILLE, N. S.

Thick Stoping Seam at 2000-ft. Depth. Loading into Cars handled on "Jigs."—When coal workings extend beyond a vertical depth of 1500 ft., it generally becomes unprofitable, if not impossible, to work by one of the "pillar" methods, for the enormous weight of the overlying strata will not only break and crush the timber, but will also either crush the pillars or force them into the strata immediately above or below the seam, resulting in a "creep" and the closing up of roads.

The size of the pillars must increase with the depth, until at about the depth noted above, the pillars become so large and the amount of coal that can be safely worked so small, especially if it is of a friable nature, that the operations become unprofitable, and another method must be adopted or the mine closed up.

Such was the situation in 1896 at this mine in working by the room and pillar method a 17-ft. seam of friable, gaseous coal, with a very weak roof of black carboniferous shale, the seam dipping from 18 to 27 deg.

The mine had been developed by two parallel slopes on the dip, and from these double-entry "lifts" were turned off every 400 ft. These entries or levels were 9 ft. wide by 7 ft. high, and as depth was gained it was found difficult to support their roofs. At a depth of 1200 ft., the first "chocks" or cogs had to be built on each side of the level. In the next lift, 400 ft. below, it became necessary to change the working system if the coal were to be mined at a profit. The change was made without great expense, any interruption of the regular output or any considerable variation in the ventilation, etc.

The advancing panel system of longwall was adopted and it has proved quite successful considering the depth reached, which is 7,870 ft. on the slopes or over 2000 ft. vertically. The new system has been particularly free from fatal accidents at the face, those occurring happening in the roadways, etc. The slopes are sunk as formerly, diverging slightly to increase the pillar of solid coal between them. They are supported on either side by pillars also increasing in width with depth so that they are now about 350 ft. wide.

Either one or both of these slopes is used as the intake airway, Fig. 125, while return airways are maintained, one on each side along the slope pillars. Two levels are driven as formerly which form a lift with about 400 ft. of solid coal between pairs of levels. The upper level of each pair is used as a haulage road, and the lower level forms the intake airway for each lift. This intake carries fresh air from the slope, where it is split, to the inner workings first; from there, returning and ascending, it passes through each of the working places to the lift above, and thence to the return airway; it is also used for drainage, and generally there is a dam built on it near the slope which catches all the water from the lift.

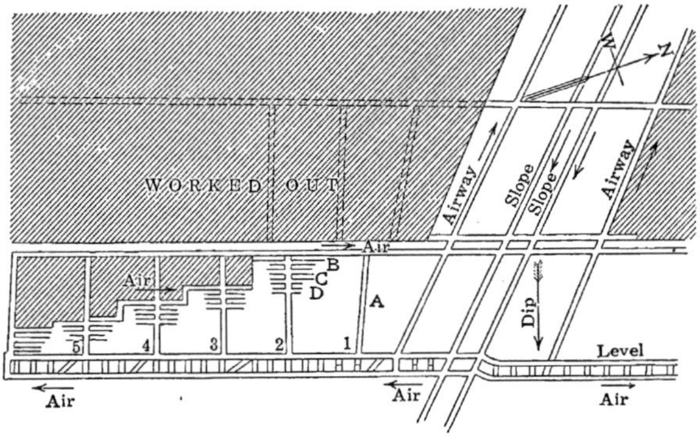

FIG. 125.—Plan of layout, Drummond colliery.

The levels are driven as nearly parallel as possible, rising about 1 ft. in 130 ft. with from 15 to 20 ft. of solid coal between the chocks. This pillar is often removed and the space filled in with stone from the roof, the result of "brushing" which must be done very shortly after the levels are driven. These levels are driven 8 ft. wide and 8 ft. high, they are first made about 18 ft. wide and 7 ft. high. This leaves a "bench" on the bottom which is only cut in the case of roadways. On this bench chocks are built quite close together on each side, and about 8 ft. apart across the road, Figs. 126 and 127, with sided timber over them across the road about 3 ft. apart, and slabs over the timber to support the roof. The chocks are built of blocks of wood over 5 in. in thickness and 5 ft. in length, making them 5 ft. square. After these are built (similar to logs in a wharf) the bench is cut along the chocks and the bottom lifted to give the 8-ft. height.

Off these levels, "jigs" are driven up on the full pitch of the coal, Figs. 127 and 128, not more than 400 ft. apart; they are chocked as well as all other roadways. An airway 5 ft. wide is carried up on the side of this jig farthest from the slope, and the chocks on this side must be made air-tight. This is done by filling them with stone and fine coal, etc. Owing to the very heavy pressure required in maintaining ventilation at this depth, canvas doors can only be used as a temporary arrangement. A wooden door is placed in an air-tight frame across the level to direct the air up this airway between the coal and the air-tight chocks, passing

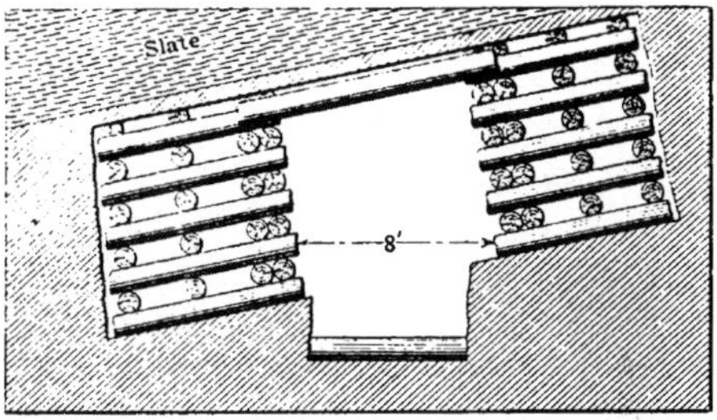

FIG. 126.—Cross section of road, Drummond colliery.

around the face and returning down the jig which is 8x8 ft., Fig. 127. This practice has been proved many times to be the only practical way, as the air will not pass up the large and down the smaller airway in sufficient quantity to keep the face clear of gas. This method is continued until the jig is driven through to the lower level of the lift above,

FIG. 127.—Plan of gateway and face, Drummond colliery.

when the door is removed and the air passes up the jig and out to the airway, and the airway along the chocks is allowed to cave.

Working the Rooms.—Beginning at the lower entry of the lift above on one of these jigs, rooms are broken off with about 41 ft. between the centers. The "gateway" or road in the rooms is much the same as the levels already described. They are timbered in the same way, Fig. 128, except that the chocks are built about 2 ft. apart on each side, and only about 6 ft. apart across the road. When this gateway is driven in about 25 ft., work is started on the breast. From it the coal is all taken out to a thickness of 7 ft. and up to the room or level above. The breast is then timbered with upright timber props with cap-pieces between them and the roof. Sometimes sided timbers are placed with one end on the high-side chock of the roadway and props under the middle and upper end. The gateway is kept 15 ft. to 20 ft. ahead of the breast, the roof of which is allowed to fall in as the face advances; generally when about 40 ft. from the jigs the roof falls, often causing a great smashing of timber on the road below, the bottom rising up as well. The face of the

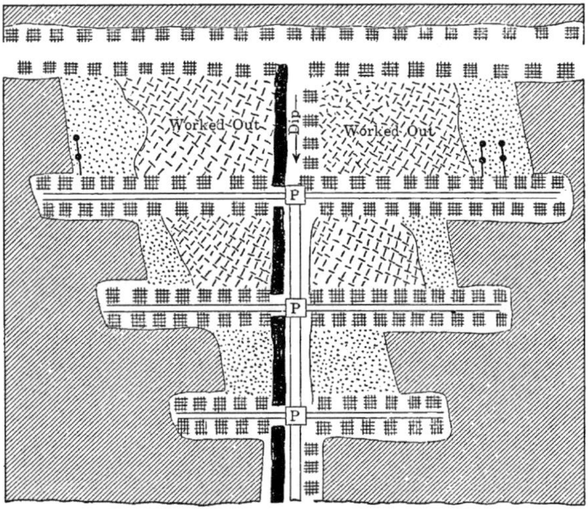

FIG. 128.—Plan of jig road, Drummond colliery.

gateway is kept a short distance ahead, for if the face of road were in line with the face of the breast, it would be very apt to fall solid across the face of the road as well, and take a week or more to get into working shape again. Through carelessness of the miners this sometimes happens. No explosives are used, for if these places are properly timbered and the weight thrown on the face the coal is easily worked with hand picks, but wedges are required in lifting the bench in the roadways.

Quite often the roof falls in solid to the face, then it is necessary to drive a heading up in the solid coal at the face and start the breast over again. This perhaps is the greatest difficulty met with in the whole operation, for when a fall like this takes place the ventilation is cut off and generally some gas accumulates and when the heading is started up, the gas, also rising, follows the miner and causes trouble before it can be driven 20 ft. or 35 ft. to the place above.

Generally three of these rooms are worked on each side of the jig simultaneously, the upper ones leading and the others following in step-like order from 20 ft. to 40 ft. behind the preceding one. In this way the upper 7 ft. of the 17-ft. thickness of coal in the seam is taken out in one operation and in future years the balance may be mined similarly.

Three miners and a laborer work in each room, as the success of this method requires that the breasts be kept steadily, if slowly, advancing. With depth it is necessary to shorten the gateways in order that the coal may be all taken out before they become entirely closed up, for on every side may be seen examples of both squeeze and creep. The roof pressure is so great that thin clay partings in the coal squeeze out like clay from a brick-machine, while the lateral pressure on the coal walls reduces the space of the openings 30 per cent. in a few months. The combined pressures so break ordinary booming in about a month that it becomes necessary to brush and retimber again. Here places may be seen so closely timbered that neither rock nor coal are to be seen for long distances except at the working faces. Studies are constantly made of the faults, cleats and clay partings encountered, (so as to make the pressure mine the coal with the least labor), of the proper setting of timbers, of the breaking away of the ribs and how to avoid it; and of how to distinguish the actual sounds of danger as there is always some cracking of timbers heard in the working places. So expert do those extracting the coal from the breast become that they work on to the last minute before a fall takes place amid appalling conditions.

Haulage.—There is no mechanical haulage on the levels, the work being done by horses. At the bottom of the jigs and at the mouths of the rooms—which are opposite each other, three on each side of the jig—large metal plates are laid on timbers placed horizontally and made solid in that position. When laid they form a smooth surface from 6 to 8 ft. square. On these the cars can easily be turned in any direction.

The road on the jig is either a double track or three rails, with a passing turnout halfway (see Fig. 128). Over the two lower plates on the jig, lifting rails about 8 ft. long are fitted into clips. When running coal from the lower rooms, these rails are removed and a tail-rope the necessary length is attached with a safety hook. A drum, controlled by a boy, is placed at the top and the weight of the full car running down takes the empty car up. The cars run on their own wheels from the surface to the face, their gauge is 2 1/3 ft., and their capacity is 1600 lb. of coal.

Output and Timber Used.—No timber is drawn, the great difficulty being to get enough timber in to keep sufficient room for the proper ventilation of the mine. The mine produces about 1200 tons of coal a day and consumes two thousand 5-ft. sticks, but of a small diameter.

CHAPTER XX

PILLAR SYSTEMS FOR SEAMS

Example 54.—Advancing-system Layouts for Room and Pillar, Pillar and Stall, and Panel Methods

Room and Pillar System.—The pillar system is also known as "room and pillar," "pillar and chamber," "bord and pillar," etc. It is applicable to all classes and conditions of mining where the roof pressure is not such as to destroy pillars of reasonable sizes, subject, however, to such modifications as serve to adapt it to the varying conditions of weak

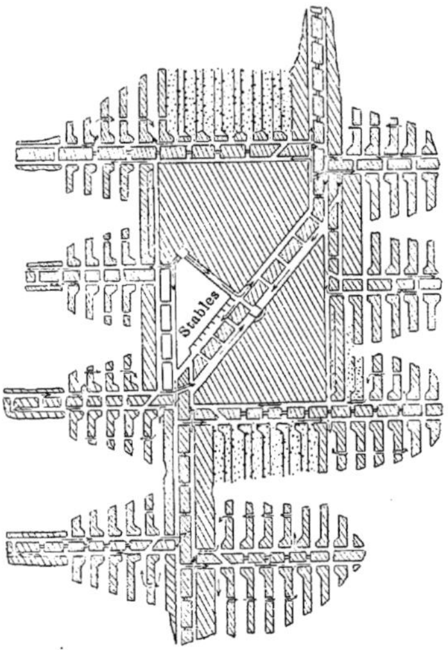

Fig. 129.—Layout for Room and Pillars system, flat seam

or strong roof or floor; tough, friable, or gaseous coal; predominance of face or end cleats; inclination of the seam, etc. The features of the system are openings driven square from or at an angle to the haulway. Such opening may be driven wide or narrow, and may be a roadway, incline, or chute, as best adapted to the existing conditions.

Room and Pillar System (proper).—This layout as applied to flat seams, or where the inclination does not exceed 3 deg., is illustrated in Fig. 129. The shaft bottoms, including the stable, are here shown crossing the shaft pillar at an angle conforming to the surface tracks. thereby giving a straight dump and tipple in line with the shaft. The stables are located close to the shaft bottom, where the mules can be rescued in case of accident, and where the daily feed and refuse can be conveniently handled. Free access is had to the stables from the main haulage roads without passing through a door; while immediate access is had close to the shaft through a curtain or canvas. Good ventilation is secured by a small separate split of fresh air, while the return air from the stables at once enters the return from the mine and passes up the shaft without contaminating the mine air. Another feature of the arrangement shown in Fig. 129, is the small number of doors. The coal coming from any room upon the main road of a pair of entries has no

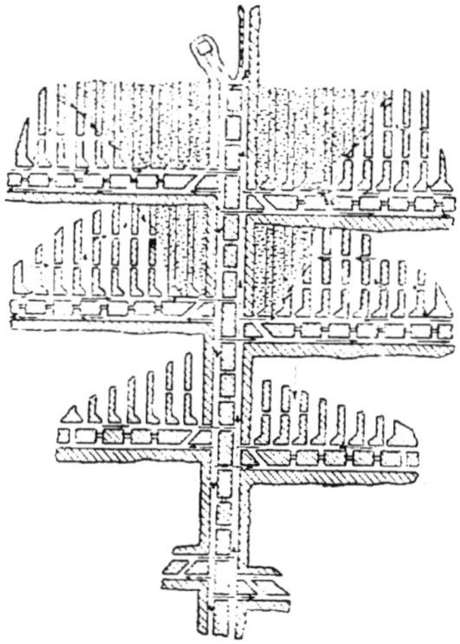

FIG. 130.—Layout for Room and Pillar system, sloping seam.

doors to pass through; while that coming from the back entry of each pair has but one door to pass through on its way to the shaft. This is a great saving of expense and trouble and may often avert possible disaster arising from the derangement of the ventilation by doors being left open. The chambers or rooms are here turned square with the entry narrow for a distance of 4 or 5 yards and then widened out inbye, the road in each room following the straight rib. The waste from the seam is stored in the room. The rooms are spaced, under normal conditions of roof and floor, from 40 to 45 feet apart, center to center. The breast is usually 8 yards wide and is driven up from 60 to 100 yards. When the breast is abandoned the miner starts to draw back his pillar unless for special reasons this is delayed for a while.

In Fig. 130 is shown the application of the room and pillar system to seams pitching from 3 to 5 deg. It differs from the method shown in Fig. 129 by turning rooms to the rise only. When the pitch of the seam is from 5 to 10 deg., the car may still be taken to the face and loaded by driving the rooms across the pitch, or at an angle with the level or gangway. This reduces the grade of the track in the rooms. When the inclination of the seam is still greater, buggies are sometimes used, the track being built upon the refuse of the seam and raised at its lower end where a tip is arranged by which the coal is dumped from the buggy into the mine car ready to receive it. Where the coal is soft, this method cannot be used. It is employed on pitches not exceeding 15 or 18 deg. in thick seams.

Seams pitching more than 15 deg. are usually worked by chutes or self-acting inclines. When the pitch is less than 30 deg. sheet iron is

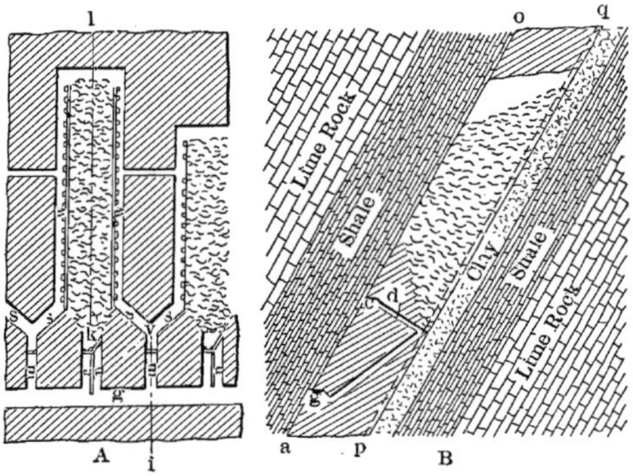

FIG. 131.—Room and Pillar system for steep seam.

usually laid in the chute as a floor, to enable the coal to slide more easily; but on inclinations of less than 20 deg. it is usually necessary to push the coal down the chute by hand or by mechanical means, as it does not slide readily. On pitches steeper than 30 deg. sheet iron is not necessary, as the coal will slide without. Fig. 131 shows, in plan *A* and section *B*, the arrangement of the breasts, chutes, and manways, and the position of the gangway and air-course at the roof of the seam, in thick, steep-pitching anthracite beds. This position of the gangway and air-course secures a better inclination of the loading chute and manways, and presents less danger from squeeze. In the figure, *g* is the gangway and *m* the manway leading to the dividing at the floor of the seam into two branches *s*, *s*, which lead to either breast. At this point, or slightly below it, a small cross-cut *d* is driven up to the airway *c*. This is bratticed off and used only in case of need, as the air is regularly conducted up one breast manway and down the other side to the highest cross-cut and thence to the next breast. Brattices with small doors are also placed in the manways to keep the air from taking a short circuit through the manways. Small manways are bratticed off the side of each loading chute for the use of the loaders.

Self-acting inclines are used, sometimes, upon steep pitches in preference to chutes. In this case, butt headings are usually driven to the full rise and rooms set off on the strike from these rise headings, buggies being used in the rooms to convey the coal from the face to the incline. It is hardly necessary to state that dip inclines are rarely ever introduced as a permanent feature, it being better to sink the main slope far enough to permit another level from which the coal can be worked to the rise.

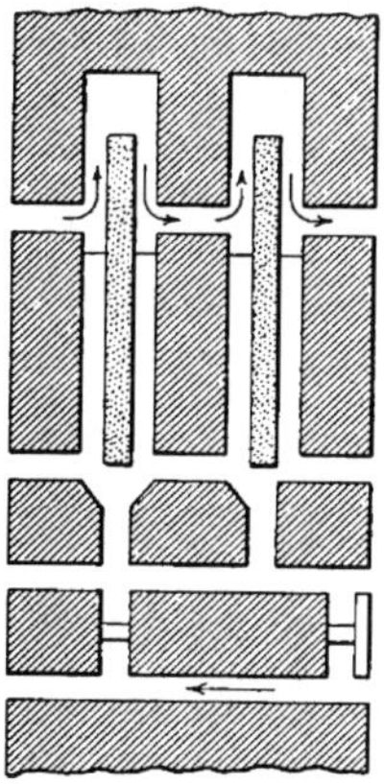

FIG. 132.—Single Stall system.

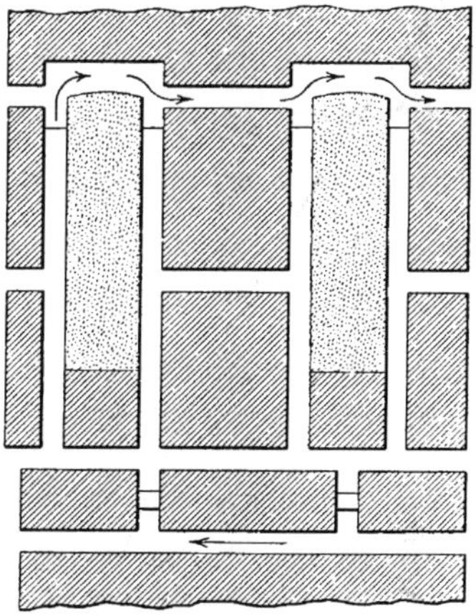

FIG. 133.—Double Stall system.

Stall and Pillar System.—This is similar to the system just described, except in the relative size of pillars and breasts. It is adapted to weak roof and floor, or strong roof and soft bottom, to a fragile coal, or to other similar conditions requiring ample support. The stall system is particularly useful in deep seams where the roof pressure is great. The stalls are usually opened narrow and widened inside to furnish a breast which varies, according to conditions of roof, floor, coal, depth, etc., from 4 to 6 yd. wide in the "single-stall" method. The pillars between the stalls are usually about the width of the breasts.

Fig. 132 shows the method by single stall and Fig. 133 that by double stall. The former is more applicable to flat seams or seams of small inclination; while the latter is used on steep pitches. The single-stall method affords but one road to a breast; and, hence, does not permit of the concentration of men possible in double stalls where there are two roads to each breast. In the double stalls the breasts are wider, ranging from 12 to 15 yd.; while the pillars sometimes reach a width of 30 yd.

Panel System.—It is advisable to mine in panels: 1, When the seam contains much gas, making it essential that the ventilation of the entire mine be under absolute control; 2, when the coal is readily affected by the air, and disintegrates with long standing; 3, When the roof pressure or the conditions of the roof are such as to require extreme caution to prevent squeeze or creep. The panels are formed by driving entries and cross entries so as to intersect each other at regular intervals of, usually, about 100 yd. The entire field is thus ultimately divided into separate squares or panels, each of which has practically its own system of ventilation. Each alternate haulway may be made an intake to supply air to one tier of panels, while the next succeeding passageway may be used as the return to conduct the air from each panel to the foot of the upcast. If more air than usual is needed in any one panel, it can be obtained at once by enlarging the opening in the regulator which controls the air for that section. In case of an explosion in any one panel, it is not usually communicated to the other panels. Extractions can be commenced as soon as a panel is formed; and usually consist in driving a heading across the panel and opening the coal by single or double rooms or stalls. Next, the room pillars are carefully drawn and the roof inside the peripheral pillar of the panel is allowed to fall. A high extraction of coal can thus be safely secured with a small loss of timber.

Example 55.—Nelms' Retreating System

Room and Pillar Layout for Flat Coal Seams.—This method insures the operator a greater amount of coal than when the seam is worked advancing on the room and pillar system. Since mining men in the United States now recognize that our supply of fuel is exhaustible, it certainly behooves all operators to mine every ton of coal possible.

In this retreating system, the main entries are driven 50-ft. centers with cross-cuts every 100 ft. The middle entry, in the three-entry system, is used for the haulage road, being also a main intake airway. After turning a pair of butt entries off the main, the second crosscut, 200 ft. from the last butt entry, should be a 45-deg. chute for motor haulage. The dotted lines on the main entry at the bottom of the butts show the position of the "parting." The motor, hauling 25 1 1/2-ton cars comes in the middle main entry, swinging its trip of empties in the chute, the motor running up the straight where the drivers have stocked their loaded coal. The motor can then pull its loaded trip outside and the drivers proceed to distribute their cars, two drivers going in each butt entry. The drivers make two trips, while the motor makes one.

The butt entries are driven on a 90-deg. angle from the main entries, and at a distance of 1400 ft., they intersect a set of three-face entries running parallel to the main entries. The butts are driven 50-ft. centers, with crosscuts every 100 ft. This system of turning butts off the mains is an ideal one for haulage and ventilation. Instead of driving rooms off the butts beginning near the main entry, the rooms are started from the face-entry side and all coal is worked toward the main entries.

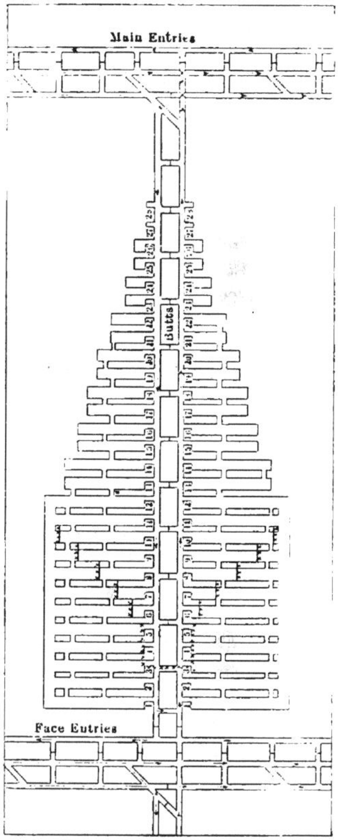

FIG 134.—Plan of Nelms' Retreating system.

Usually 60 ft. of solid coal is left to protect the face entries, and 60 ft. to also protect the mains. The rooms are started four at a time, and as soon as the first four have been driven 50 ft., the next four are started on both butts. The rooms are all driven on sights 90 deg. off the butt entry and driven 25 ft. wide for a distance of 240 ft., there being a 15-ft. pillar left in each room. The crosscuts in the rooms are from 80 to 100 ft. apart, and should be "staggered" across the different rooms so as not to make a weak place in the roof by having the breaks all opposite.

Four Pairs of Butt Entries Will Produce 1200 Tons of Coal Daily

After driving the rooms the full distance, they should be cut over to the next room by the mining machine, the cut being 20 ft. wide. The great advantage to be gained in this system is the method of not having work scattered all over a mining territory. Four pairs of butt entries, thus mined, will produce 1200 cars of coal each working day.

In Fig. 134, there are 32 "machine" rooms (Nos. 13 to 28 on each side) working on the pair of butts, and requiring 32 men (loaders), two men having two rooms and working them together. It is the general practice to clean up one room at a time and so always have coal to load in one room or the other. Each machine loader receives six cars, thereby producing 192 cars per day.

There are 10 pillars being robbed (Nos. 6 to 10 on each side), and these produce 30 cars of coal, as the pillars are worked by one man. In some places two men work the pillars. The "turn" in coal mines is such that a machine loader receives two cars to the pick miner's one, thereby equalling each other's wages, as pick costs about twice as much as machine coal. The chain pillar and stump will produce 12 cars, with four men working, and the two butts yield 234 cars per day.

The engineer can advance the work in such a standard way that his machine coal will always total to the proper amount. A mine foreman should find this an easy way to keep his men standardized, the machine loaders always having machine places and the pick men pick places, thereby increasing the safety factor of his mine, as his machine men would never have to do pick work.

The ventilation shown by the arrow heads is the most practical to use; the splits are shown and also the overcast at the bottom of the butt entry, there being a regulator in this overcast. The motor road is clear of doors on the main entries. The arrangement of chutes on the left side would be slightly different.

Example 56.—Nelms' Advancing-retreating System

Room and Pillar Layout for Flat Coal Seams.—The layout for the advancing-retreating system is as follows in Fig. 135: Three face entries, on 50-ft. centers, are driven parallel to the main entries at 1400-ft. intervals. The sectional area of the face entries is kept as nearly as possible to a 60-ft. standard in a 5-ft. seam.

It is advisable where possible to use all three entries for intake airways; then the middle entry can be used for a haulage road and should be confined to itself and not enter into the ventilation at all. No. 1 room on the butt entry can be driven 16 ft. wide and used as a return airway. When the No. 1 room is maintained for an airway, it should be widened toward the face or main entry and be driven on 70-ft. centers with the face entry. A 30-ft. pillar of solid coal should be left between No. 1 and No. 2 rooms; No. 1 rib can then be easily extracted when robbing is commenced on this butt entry.

The gob in No. I room should be kept as low as possible and if easily handled, it should all be loaded and dumped outside; the result is a return airway, 16x6 ft. =96 sq. ft., which is usually large enough.

Butt entries should be turned every 450 ft. A chute is driven on a 60-deg. angle, from the middle main entry to the outside main for haulage. The butts are turned on a 90-deg. angle, and are driven on 50-ft. centers for a distance of 1400 ft. A 60-deg. chute should connect the butt entries at 25 ft. from the center of the outside main, for haulage from the butt entry. It is bad practice cutting corners off break-throughs.

Each butt entry should maintain a sectional area of about 50 sq. ft. and be driven perfectly straight so as to overcome the troubles of track laying, cars jumping track, etc. Entry sights should never be more than 180 ft. apart, so as to allow the mine foreman a good chance to keep his sights well up. In providing ventilation, two pairs of butts on one split are suitable. The rooms should be turned 90 deg. off butts and driven as shown in plan.

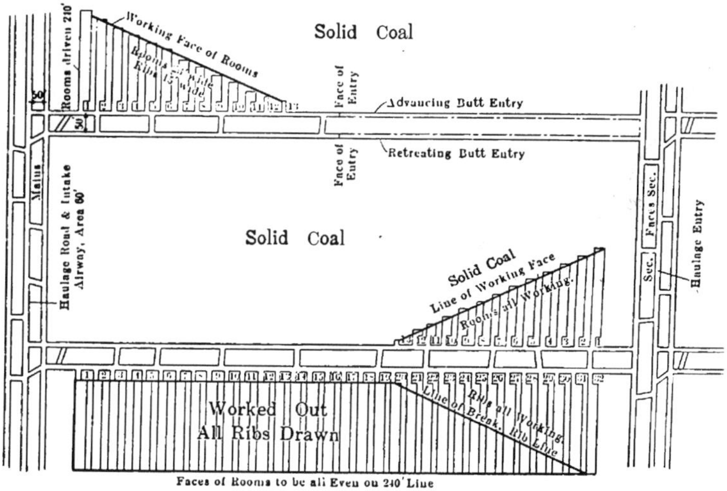

FIG. 135.—Plan of Nelms' Advancing-retreating system.

General Layout

No. 1 room should be started as soon as possible, then No. 2 and so on from the advancing butt entry. When No. 2 room is finished working on the face, No. 14 should just be started; when No. 2 rib is out, No. 14 room should be just finished and No. 27 room just starting. The ribs must be extracted as soon as each room is finished, no matter whether the next room on the advance side is finished or not; the rib is started while the next room is still 30 ft. from being finished. When No. 32 is finished working on the face, No. 19 rib is just finished, also No. 1 room on the retreating butt is about finished and No. 13 room just started. As soon as No. 2 rib on the retreating entry is finished, it is advisable to start extracting immediately the butt-entry stumps and chain pillar, bringing everything along with the retreating butt and closing entry in tight, knocking out the brattice in each succeeding break-through for ventilation.

This method allows the rib men and the machine loaders to be always separate, the workings are confined to the smallest space possible for a large tonnage, and ventilation is easy. The mine will not be dotted with old abandoned workings if the method is consistently maintained.

When a set of butts are thus worked out they are off the operator's hands. There is never any danger of a squeeze as every movement of the rock runs up against solid coal, and for this reason it is impossible to have a squeeze swing across a set of butts to another set, as it generally does where both entries are worked advancing. Five pairs of butts developed on this plan can produce from 1000 to 1500 cars of coal a day.

Small mule "partings" should be at the bottom of each pair of butt entries and the distance for mule haulage will then be at a minimum. The coal from these partings can be gathered by a 6- or 8-ton locomotive, and delivered to a longer parting whence a larger locomotive can take it outside.

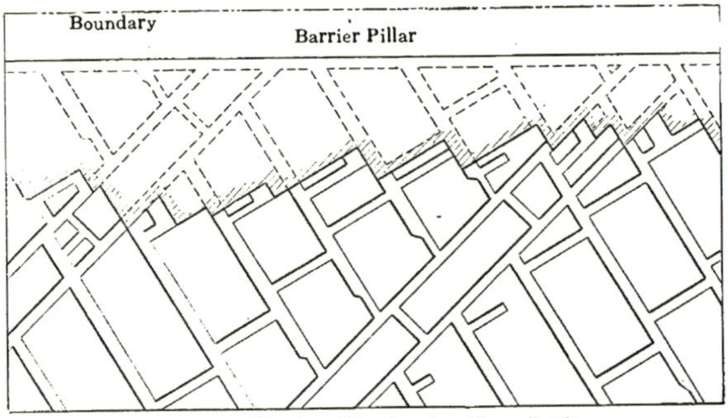

FIG. 136.—Plan for pillar-drawing, Connellsville.

EXAMPLE 57.—CONNELLSVILLE COKE DISTRICT, WESTERN PA.

(See also Example 58.)

Retreating System on Thick, Flat Pittsburg Seam.—The plan of mining in the Connellsville region has grown from the primitive methods, suitable to the favorable mining conditions when operations were first started, to the scientific methods which became necessary as the cover increased, and when most of the difficulties that are likely to be met in deep mining were encountered and overcome. The conditions here are that all the coal is coked so that fine coal is an advantage; no machines are used, but the coal is dug with pick; the seam averages 7 1/2 ft. of clean coal mined; the roof is friable and some coal is left in the top to support it; the overlying stratum is generally 6 to 10 ft. of slaty coal with sandstone above.

Where the acreage owned or assigned to each mine is too large to admit of going to the extreme boundary before starting to draw the ribs, it is customary to divide the field into panels 1000 or 1500 ft. square, by face headings driven off the main butt headings as shown in Fig. 136. The coal is first removed from the extreme side of each panel away from the main butt headings, a diagonal break line is established, as shown, and the coal withdrawn, retreating toward the near corner, keeping the break line straight, and the coal between where the drawing is being carried on and the main butt headings as nearly solid as possible, the butt headings and rooms being drawn only fast enough to open up the coal for the drawings. The break line of roof is kept as nearly perpendicular as practical to direction of rooms. The recent tendency in the large mines is not to start pillar-drawing till the boundary is reached.

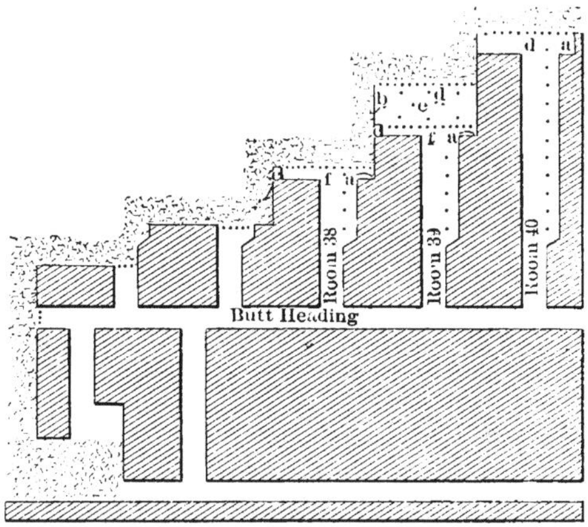

FIG. 137.—Pillar-drawing with thin cover, Connellsville.

The face of the coal in this region is well defined on a line running N 17° E. Where the grades are not too great all headings are driven square on the face or on the butt, and the rooms always on the face and only to the rise. The rooms are driven 10 or 12 ft. wide and a line of posts set as the room advances, as shown in Fig. 137, the posts being set about 4 or 5 ft. apart. A track of wooden or steel rails is laid by the miner close to the upper rib. The width of the pillar, which varies from 10 to 70 ft., is governed by the softness of the bottom and the thickness of the overlying strata.

There is some variation in the method of drawing the individual ribs but the principle is the same. On account of the nature of the roof, short falls are necessary, two or three being made before the overlying rock is broken. When the rock breaks it will crush posts so that it is necessary to break the roof against the end of the ribs and not over posts. If care is taken the digger knows when to expect a fall and very few posts need be lost.

Fig. 137 shows the method in use when the overlying strata are under 200 ft. thick and where a curved track and track along the face are not put in. A slab *a* is taken off the right-hand rib, the whole of the left-hand rib is taken out excepting enough only left in to keep the gob from mixing with the coal as shown at *b*. A small shoulder *c* is left in the far corner of the rib to take the brunt of the weight. A row of posts *f* is set from this shoulder across the room to preserve a working face. The posts *e* in the back of this row and between it and the previous fall or break are drawn and any which cannot be drawn are cut so as to get a

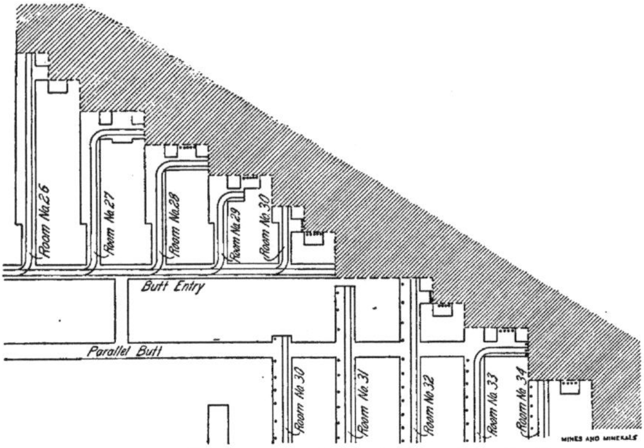

FIG. 138.—Pillar-drawing under thick cover, Connellsville.

good fall. Two men work in each room. Room 40 shows the situation in the last room of a tier of rooms along a butt heading. Slabs are being taken off each side pillar while the men are protected by the row of posts *d*.

Room 39 shows the condition of a room just before producing a fall by withdrawing the row of posts *d* and separate posts *e* between the rows *d* and *f*. The row of posts *f* and the stump of coal *c* protect the face during the fall. The stump *c* is next removed excepting for the small slab *b* and the work proceeds as shown at the face of room 40.

Fig. 138 shows a method of drawing pillars where there is over 200 ft. of surface above the coal and where a curve is used to run the track along the face. The rooms are driven 12 ft. wide while the pillars between the rooms vary from 34 ft. up to 60 ft. All of the rooms are driven on sights so that the pillars may be of uniform thickness. After the last room on the heading is driven the required length, which is about 300 ft., the pillar is cut across at the face of the room and 20 or 30 ft. removed before drawing the posts and getting a fall of the roof. The usual method is then for two men to work on each pillar, while one man cuts back in the center of the pillar on the face of the coal as far as he can conveniently shovel as shown at *a*, room 27, the mine car being on track *b*, the other man is drawing stump *c* and shoveling into the same car.

When *c* is all removed a fall is made and the situation is similar to that shown in rooms 30 and 26, the curved track having been removed to one side and a straight one substituted. Now one man cuts into the side of the pillar 8 ft. from the end at *d* as shown while the other is removing the stump *e*. When this is accomplished a fall is made and the curve put in, the conditions being then as shown in room 29; the two men then continue to cut over toward the gob in the next room as shown in room 29, the curved track having meanwhile been put along the face. Room 28 shows the situation when the two men have driven this cut through to the gob in the adjoining room. Room 27 illustrates the next

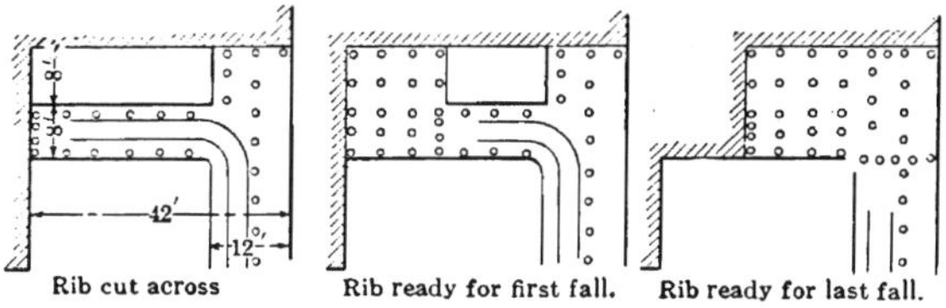

FIG. 139.—Second method—pillar-drawing under thick cover, Connellsville.

step. While one man works on the pillar *c* in the far corner of the room the other starts the cut *a* back into the face as shown. The curved track *b* is then lifted and a straight track put in as shown in room 26 in order to get out the pillar *e*. While one man is removing this pillar the other one starts to cut into the rib as shown at *d*, room 26.

When a fall is to be made posts are set 18 in. apart, as shown in room 26 across the end of the room and along the end of the hole into the pillar. All of the other posts beyond the break rows are drawn. The curved track is laid into the new cut in the side of the pillar at *d* and by the next morning the roof has fallen. In places where the rib is wider than shown on the plan a couple of falls can be made in the width of the pillars by placing break rows of props similar to those already described.

A third pillar-robbing method, for a heavy cover, is shown in Fig. 139. It cross-cuts the pillar for a track and uses the same curve as the previous method, but instead of cutting up the resulting pillar-slab into blocks, it begins at the gob and withdraws the slab gradually roomward, meanwhile recovering most of the props in the figure that were put in to protect the excavation of coal. A ratchet-chain puller for props is used where necessary. The use of this method at the Continental No. 1 mine of the Frick Coal Company gives a recovery of about 90 per cent. of the coal. The losses arise from a 6-in. coal layer, impure with sulphur, left on the floor; a coal layer, of 4 in. in the rooms and 9 in. in the entries, left on the roof; and some occasional stumps lost in robbing the pillars.

The foremen of the district are guided by the following 10 rules in extracting pillars. (1) Pillar robbing must not be stopped or diverted from the line of fracture without the consent of the chief engineer. (2) Robbing must proceed from the new toward the older gob to prevent uncalculable pressure on the working face. (3) Ribs must be robbed within one month of driving rooms. (4) Room centers must be at the prescribed distance apart. (5) In robbing entry-pillars, a length of only 2 room-widths must be attacked at once along the line of fracture. (6) Water ditches must be made for entry-drainage and especial care must be taken on soft bottoms. (7) A 200-ft. pillar must be left on each side of the main or flat entry during its life. (8) A wide barrier pillar must be left and care must be taken in approaching a neighbor's boundary. (9) Before permitting a fall of the roof, all timber must be drawn and a passage left for the escape of the miners. (10) A miner should keep the pillar he is drawing between himself and the gob instead of working between gob and pillar.

Example 58.—Pittsburg Bituminous District, Western Pa.

(*See also Example* 57.)

Advancing-retreating or Retreating System in Panels on Thick Flat Pittsburg Seam.—In those mines of western Pennsylvania, extracting the thick Pittsburg coal seam for shipment to market, the mining layout is different from that of the coking district of Example 57. Since the policy of the market-coal mines is to obtain as much lump coal as possible, the bulk of the coal is obtained from the rooms, for pillar coal is bound to be more or less crushed. This policy requires wide rooms and narrow pillars and results in a lesser total recovery of coal, but as an offset, more coal can be won by machine cutters which work advantageously in this thick seam. By the deep and fast undercutting possible with machines, blasting, with its ensuing slack, is at a minimum; and progress is rapid enough to preserve the coal faces from long exposure to the atmosphere and to allow of systematic timbering and an even subsidence of the roof.

The Monongahela River Cons. Coal and Coke Company is a very large producer, and a composite drawing of its method of working is shown in Fig. 140. Above the upper "butts" or butt entries, the rooms have not advanced far from the "Face Entries." On the middle butts, the rooms have reached the end of the upper panel and pillar-drawing has advanced halfway. On the lower butts, the lower panel is being worked by retreating from the panel-end, the rooms are nearly completed, and the line of roof-fracture, across both upper and lower panels, is following the pillar-drawing and is not far behind the finished rooms.

In the advancing system, the panels off the upper butt entry would be attacked first, and the rooms of this advancing panel would end in the old gob, while the rooms of the lower or retreating panel would end against the solid. As shown in the figure, the room-centers are 39 ft.

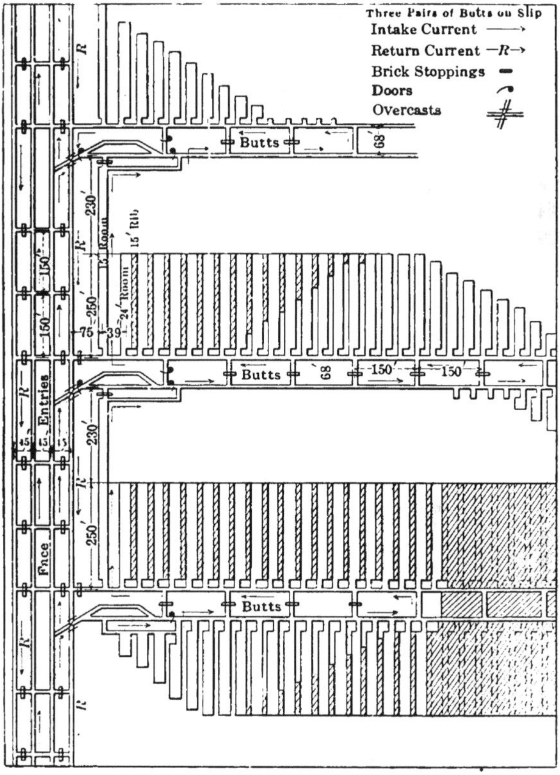

FIG. 140.—First layout (Monongahela collieries) at Pittsburg, Pa.

apart, of which space the pillars occupy 15 ft. It will be noticed that the room-stumps of each upper panel are left undisturbed on the advance so as to protect the return airway, but when the pillars of the lower panel are being drawn, the upper stumps are also pulled, as the receding line of roof-fracture passes them, along with the butt-entry pillars. The last pillars, however, must be left undisturbed in the advancing system along their whole length until all the adjoining coal has been exhausted up to the boundary. The use of four main entries, by this method, allows the two outside gangways to be return-airways and the two intake-airways to be on the inside and thus gives an ample main-airway area and a minimum interference with transport. The room-work is in the fresh air and pillar-drawing is on the return-air side of it. The room-track is always laid along the straight rib, and in many mines the refuse between the track and the other rib fills the room nearly roof high.

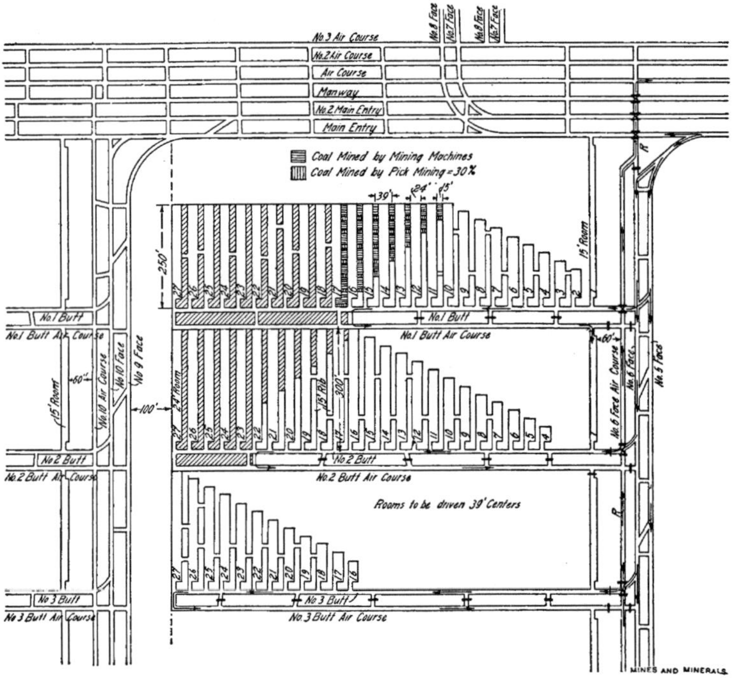

FIG. 141.—Second layout for large output, Pittsburg.

Fig. 141 shows a second layout for large output, used in the Pittsburg seam, with six main entries. There are three face entries, nominally, but four actually, as the nearest room on the butt is advanced along with them so as to give an additional airway. As shown, the rooms are only worked on the outbye side of the butts, and the first room is started from the far end of a panel and followed, at the proper distance on the retreat, by pillar-drawing. By starting work from No. 2 and the following butts at the proper time, it is possible to keep the line of roof-fracture of a panel continuous, for entry-pillars and room-stumps are removed as shown from the panel-end back to the butts.

The method of Fig. 141 has permitted the extraction of 70 per cent. of the pillars by machine cutters under an average cover of 200-ft. thickness. For this purpose machine cross-cuts, 21 ft. wide, are made in the pillars so as to leave for each a stump only 9 ft. wide to be removed by hand-pick. This cross-cutting is shown by the different cross-hatching of the figure which also illustrates the overcasts and brattices for ventilation, and the chutes, etc., for transport.

A third method of attack by which one company mines over 2,500,000 tons yearly is shown in Fig. 142. Here the room pillars, after the room

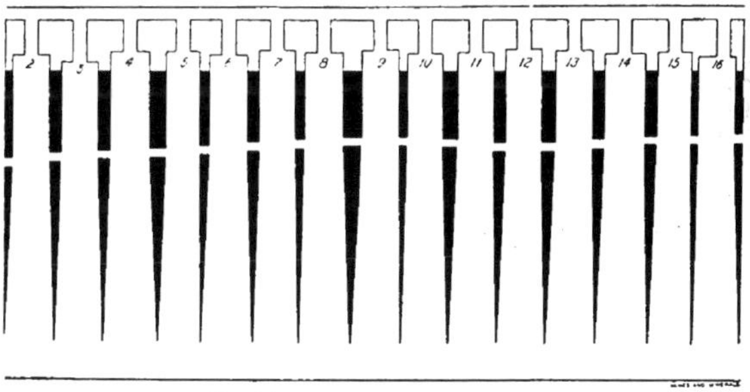

FIG. 142.—Third layout, with tapering pillars, Pittsburg.

has advanced 100 ft., or to the first break-through, are gradually tapered off to a point at the room-end. This causes the roof to fall along the tapered parts of the pillars and the latter are lost, but much of the thicker pillar near the room-neck can be recovered by subsequent careful pick-work. This method gets nearly all the coal by room-work, and a total recovery of 90 per cent. is claimed by its advocates. It is more dangerous, however, than the two previous systems, requires more timber, and squeezes are more liable to occur.

Where this high Pittsburg seam is dirty, so that much gob must be stowed along the rib on the advance, it is customary on drawing the pillars to leave a vertical shell of from 12 to 18 in. of coal next to the gob to prevent any pollution of the broken coal.

CHAPTER XXI

FLUSHING SYSTEM FOR FILLING SEAMS AND RECOVERING PILLARS

Example 59.—Anthracite District, Eastern Pa.

(*See also Examples* 5, 51 *and* 59.)

Parallel Seams of Various Thickness and Dip Filled with Refuse from Breakers and Dumps.—The flushing system was first developed in 1891 at the Dodson mine near Wilkesbarre, Pa., and was later copied and extensively used in many German collieries. Three conditions made flushing a valuable innovation in the Pennsylvania anthracite region, namely, the numerous large dumps of waste available for filling, the parallel and superincumbent seams to be extracted, and the overlay of much workable coal by townsites. The gravity of the urban situation is evidenced by the report of April, 1911, made by the Scranton Commission. This report states that a large part of Scranton is already undermined and that for the stability of the present dangerous area of 15 per cent. of the city and for the recovery of the coal pillars from the balance the flushing system is the only remedy.

The following description in based on the author's visits to mines of the following coal companies: Philadelphia and Reading; Delaware and Hudson; Delaware, Lackawanna and Western; Plymouth; and Lehigh Valley.

In mining the flat seams to the north of Wilkesbarre by the pillar system of Fig. 130 much of the waste broken with the coal can be left in the rooms; but in the seams of the southern districts where mining is done by "overhand stoping with shrinkage and chutes," as in Figs. 132 and 133, all the waste has to be hoisted. The crude coal reaching the surface is a mixture of pure coal, "slate," "slate-coal", and "bone." The "slate" corresponds to the shale and clay of the partings and beds of the bituminous regions, the "slate-coal" consists of lumps, part pure coal and part slate, and the "bone" is a coal containing too little carbon (present limit 60 per cent.) to be marketable. All crude coal is put through a dressing mill or "breaker" in which impure pieces are broken sufficiently to detach the slate and bone from the pure coal, so that all the latter may be screened for separation into commercial sizes and the former, along with the "culm" or coal dust, be sent to the waste dump or mine stopes.

FIG. 143. Breaker washery and culm-digging, Dodson colliery.

The limit of size between fine coal and unmarketable "culm" has so decreased in recent years that now all the fine sizes of the breakers, as well as many old waste dumps, are being washed over shaking screens in special mills called "washeries," for their content of fine coal of commercial value. The present upper limit for "culm" is a diameter varying from 5/64 to 3/16 in., but some independent operators use also some larger sizes for flushing. This rejected fine coal, as mixed with the slate and bone tailing from the breaker and the ashes from the boiler plant, forms "slush," the chief material now used in filling the mine stopes by the flushing system. At some mines, the larger pieces of bone are saved on a special dump as of possible future value. Fig. 143 shows the Dodson colliery at Plymouth, Pa., with the waste dump at *A*, the breaker at B, and the washery at *C*.

In digging an old waste dump for passage through a washery, in order to separate the marketable coal before flushing, a system of chain conveyors as at D and *H*, Fig. 143, is used. The usual conveyor has a single chain and drags its steel plate scrapers, 18 in. long, 12 in. high, and set 3 ft. apart, in a trapezoidal trough made by lapping the ends of 3-ft. lengths of steel or cast-iron plate. The maximum length of a single conveyor trough is about 500 ft. It is supported within square wooden frames, *E*, set 8 ft. apart, and built of 4x6-in. pieces. Near the top of the frames *E*, run two 25-lb. steel rails to support the scrapers on their return trip.

Each conveyor is run by an independent steam engine, as at *F*, connected by gearing to its head end, and its capacity of 100 to 200 tons of dry material per hour is fed in, anywhere along the trough, by hand shovels or by hydraulicing with hose. Obstacles between the dump and washery are passed by using several conveyors, set at an angle, of which only the conveyor at the feed end need be shifted as the dump dwindles. The driving engine is set on a timber frame so that it can be easily pushed into line, by screw jacks, when the conveyor is moved over by levers; both engine and conveyor are elevated on rollers before shifting. This is done by the regular attendants who consist of two men for feeding and one man at each driving engine.

So much fine marketable coal can now be saved from the present breaker-tailing and from the old "culm" dumps that the final rejected waste can fill only a fraction of the space left above the gob in the underground rooms. The huge dumps which formed such a prominent feature of the landscape, as late as the early nineties, are rapidly disappearing and filling is now being won even from river beds.

Thus the Plymouth Coal Co. has a plant to bring sand and fine mine waste, now settled at the bottom of the Susquehanna river, to its Dodson mine No. 12. A suction pump on a barge located across the river from the Dodson breaker delivers into a pipe which crosses the river on barges and discharges into an elevator which lifts the material to the flushing flume for the mine.

As no pieces larger than 1-in. dia. and few over 1/4-in. dia. are used in flushing, the coarser pieces of bone and slate, from breaker or dump, are passed through a pulverizer usually of the Williams' or Jeffrey's type, before reaching the flume *G*, Fig. 143, where they mix with the fine waste from washery *C*. Enough water is put in the flume to transport the slush along the flat pipes above the stopes, so the liquid slush carries only about 20 per cent. of solids by weight. The descent of the slush is through wrought-iron pipes, 4 to 6 in. dia., following either a shaft or a special bore-hole into the workings, perhaps 1000 ft. beneath. Over the top of the descending pipe is placed a funnel and a screen with 1-in. holes, and at each flushing station are three gate valves, to regulate the flow into the stopes, connected by electric signals with the surface. One of these station valves regulates the flow horizontally, another cuts off the vertical column below, and by a third the column can be drained up to the nearest flowing point above, in case of a stoppage.

The pipes used for transport along the levels are 4 to 6 in. dia., and of either wrought-iron or wood. In upgrade levels or where the pipe is under much pressure, new iron pipes with screw or flange couplings must be used, but when these are somewhat worn they are transferred to the downgrade levels. In the latter, the iron pipe has standard couplings on tangents, but on curves it has 7-in. unthreaded nipples for couplings slipped over the pipe ends and made tight by wooden wedges. These wedged couplings enable the pipe to be rotated, when its bottom gets thin, so that it can be worn to a mere shell all around before rejection.

The wooden pipe is made in Elmira, N. Y., of tenoned maple staves about 2 1/2 in. thick which are bound with spiral steel hoops and coated with tar. It comes in 2- to 8-ft. lengths, with male and female ends for slip-jointing with cement. It can be joined to cast-iron fillings by inserting special cast-iron nipples in its ends, and its own joints can readily be made to follow easy curves. It is lighter and cheaper than iron pipe, is found to last well on downgrade levels, and is preferable for use with acid water. In one mine, the iron pipe is used on a 2-mile permanent transportation line and the wooden pipe in the neighborhood of the actual filling.

Plugged cast-iron tees are placed at 100-ft. intervals along all lines in the levels, so any obstruction can be easily located and removed. When a pipe is upgrade, a special precaution is taken against clogging by passing fresh water alone through it, for 15 min., before stopping the flow. Care must be taken to provide air escapes at high points of the lines in order to avoid water hammer.

The openings filled by flushing are old rooms opened on the pillar system of the last chapter. A room on a dip is easiest filled, as it requires only one dam or barrier at its lower end. One disadvantage of increasing steepness is the greater strength of dam necessary to resist the correspondingly higher water head. In the Dorrance mine the old rooms had been opened on the rise from double flat entries as in Fig. 130. Every ten rooms along the entry were separated by panel-pillars following the dip. For filling, the flushing pipe was laid along the airway above the rooms and its discharge placed at the head of the central room of a panel of empty rooms. The latter had been prepared for filling by erecting dams across the necks at the room-bottoms and behind the break-through brattices of the ninth room's pillar, for the last room of the panel was to be left open as an air and manway. The brattices of the break-throughs of the intermediate rooms had been removed to permit of a free flow of filling along the panel.

Room dams are made of either stone or wood. The former are thick walls of roof slate laid up with mortar of slush and straw in a similar

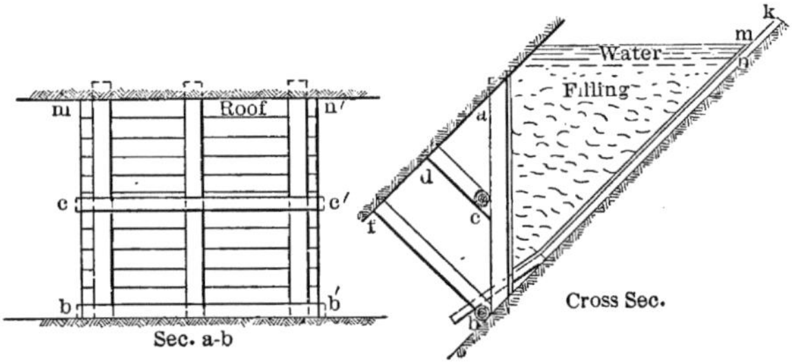

FIG. 144.—Dam for holding slush, Eastern Pa.

form to the wall of Fig. 145 described in the next Example. The favorite dams are of wood and a typical one is shown in Fig. 144. Round props *ab*, of sufficient size for the expected strain, are covered on their upper side with 2-in. plank and backed, as an extreme case, with stringers *bb'* and *cc'* with corresponding angle braces *bf* and *cd*. If the seam-walls are strong, the hitches alone will hold the props, so that the pieces *bb'* and *bf* can be omitted, and in thin seams even *cc'* and *cd* are left out.

When wetted, the seams between the planks soon close up sufficiently, but the irregular spaces around the periphery *mn'b'b* are caulked with straw in one mine, and in another, with a weak floor, the props are set in a low concrete wall, 12 in. wide. In one seam of the Dorrance mine on an 18-deg. slope with rooms 300 ft. long, the wooden dam of Fig. 144 is strengthened by a dry wall of roof slate, 3 to 5 ft. thick, laid above the plank *ab*. By slow flushing at first, this dry wall gets packed solid and keeps the plank from bulging under the heavy final pressure due to a vertical water head of 90 ft.

A room in the Dodson mine in the 22-ft. Red Ash seam was worked in two slices, the first taking only 8 ft. of coal from the floor. When preparing for flushing, the upper 14-ft. slice of coal was not taken down over the neck for 24 ft. from the room's lower end, so that the subsequent wooden dam needed to be only 8 ft. high. Holes are bored into the plank of the dams near the top, if necessary, to let the overflow water escape, but a better arrangement for steep dips is a wooden drain-launder *bk* laid on the floor up through the dam into the room. The top *m* of the cover of launder *bk* is kept a short distance above the top of the settled slush at *n* by adding new cover-boards as the filling rises. The overflow water then runs over into the launder at *m* and descends into the gangway ditch at *g* to flow to the sump; whence it is pumped to the surface, where, being acid, it is not reused unless fresh water is scarce. At the Dorrance mine where the rooms were being filled on the advance by extending the flushing pipe from one panel of ten rooms to the next, it was the practice to give each panel another dose of slush, while withdrawing the pipe, in order to close up the many spaces between the settled slush and the roof that had developed since the advance. For nearly flat seams, dams are built in the openings all around a panel of rooms, and the end of the flushing pipe shifted around inside the panel, close to the roof, so as to fill all portions equally. More or less methane is given off if the slush is exposed to air currents, but these are feebler, the smaller the spaces left between slush and roof. As another safeguard against gas, the filled panels are connected with the return airways of the active mine.

In the considerable areas where a subsidence of the surface is immaterial, the anthracite seams are best worked to the boundary, by that pillar system of the last chapter most appropriate to the given conditions; and the pillars then recovered on the retreat, allowing the roof to fall. Under the river flats where roof-falls might cause a crack up to the surface and flood the workings, one company's mines are laid out with permanent pillars of a size just sufficient to sustain the roof indefinitely, which means 16-ft. pillars and 24-ft. rooms for depths of less than 400 ft.

Flushing as a preliminary to pillar-drawing is beneficial in the anthracite region under two conditions. First, where the workings are overlaid by virgin parallel seams, and second, where they are overlaid by townsites. Former market conditions made the thin seams unpayable, so that the proper method of exhausting overlying coal seams from the top downward was not applied. Now the pillars can only be recovered from the lower seams, without wrecking those above, by a preliminary filling of the adjoining rooms. Formerly, it was not thought that the pillars left under townsites would ever be worth recovering, but higher coal prices have made them valuable and filling must precede their recovery.

The aforementioned Scranton commission recommends that as slush alone has insufficient crushing resistance for thick covers, sand should be used for filling under Scranton at depths beyond 500 ft. Also that filling should begin in the lowest seam of the series and continue upward until all are filled, care being taken to have the flushed areas over one another. After all the openings in all the seams have been filled, the pillars in the top seam may be removed and replaced at once by filling. The next seam below may not be attacked and handled in like manner until the pillars above, within a large panel, are removed and the overburden has come to rest on the new filling. In this manner several parallel seams could be robbed of pillars simultaneously, by panels retreating in vertical echelon, the robbing in the highest seam being farthest from, and that in the lowest seam nearest to the boundary.

In some mines with irregular layouts and small pillars, the formation had moved considerable before flushing was inaugurated. Thus in the 22-ft. Red Ash vein of the Dodson mine at Plymouth, the overlying formation moved so freely that gangways could only be kept open by using heavy timbers and brushing the floor. While in the Black Diamond mine at Luzerne, the walls of the 6-ft. Cooper seam were distorted with frequent roof-falls, and in the 8-ft. Bennett seam the roof had bent enough to badly squeeze many of the pillars.

The seams of the latter mine, which were excavated on the system of Fig. 130, dip about 10 deg. and the pillars of the flushed portion are now being robbed and replaced by slush. Where pillars are 20 ft. wide, or more, an 8-ft. heading is driven on one side of the pillar on the rise, often leaving a thin shell of coal next to the filling. Then, when the airway above is reached, the balance of the pillar is drawn on the retreat. The advance heading must be well propped, but the timber is mostly recovered on the retreat and, owing to the moving formation, the pillar coal is so squeezed that but little blasting is necessary. Too much roof pressure sometimes so crushes the coal that it falls to powder when extracted.

The Dorrance mine is under a suburb of Wilkesbarre and the policy of the owner, the Lehigh Coal Company, is to refrain from taking all the pillar coal, when robbing flushed areas under cities, because an unsupported cover will settle down at least 10 per cent. of the coal's thickness; and with flat seams, where filling close to the roof is impractical, the subsidence may be 20 per cent. In fact, the surface in some cases has subsided less from robbing pillars in open than in filled seams; for in the former case local breaks of roof may fill up the rooms with boulders and support the cover, while robbing pillars *completely* in the latter case starts the whole cover to subsiding as in the longwall system.

The flushed workings observed in the Dorrance mine were on a dip of 18 deg. and on a layout like Fig. 130 with rooms 20 ft. and pillars 40 ft. wide. A heading was first driven up in the pillar, to slab off 18 ft. of coal alongside the filling, and on the retreat from the room's upper end a 24-ft. cross-cut was put through the pillar, halfway between the original 12-ft. break-throughs, 100 ft. apart. Thus after flushing the new pillar openings, the roof was left supported by a line of coal pillars 22 feet wide by 32 ft. along the dip.

Under Mahonoy City, the 22-ft. Mammoth seam, dipping 55 to 60 deg., is being worked in two slices by a system like that of Fig. 131. The lower slice of 15 ft. is taken out in the room on the advance and the upper 7-ft. slice allowed to fall into the chute, by pulling the props, on the retreat. After flushing, the pillar is taken out, likewise, in two slices, by driving a heading through its center, leaving only a thin shell of coal on each side to keep out the room-filling. The entry pillars are drawn on the retreat, and all the new openings are flushed. In spite of this extraction of practically all the pillars, the surface here is stable, for with seams of steep dip, the subsidence upon the filling is not so serious as it is in the case of the flatter seams under Wilkesbarre.

As already mentioned, the 22-ft. seam in the Dodson mine is also worked in two slices but with the thin slice below. The pillars here are 26 ft. and the room is 24 ft. wide. The lower slice of both room and pillar is mined on the advance and the upper slice is recovered on the retreat as described in the last paragraph, except that the layout follows Fig. 130 to suit the 12-deg. dip.

Example 60.—Robinson Gold Mine, Rand District. Transvaal

Parallel Sloping Beds Filled with Mill Tailing

In spite of the extensive areas excavated since 1885 in the conglomerate of the Rand, but little filling has yet been done. At a few rich outcrop mines, it is true, the rooms were packed with rock to enable the pillars to be recovered. But packing is too costly a method for most of the area. As the mines reach depths exceeding 4000 ft., the former sized pillars are proving too small, and several unexpected collapses have occurred. Recently the flushing system has been tried with success at the Robinson mine, to permit of the removal of some rich pillars just under the stamp mill, in the following manner.

The tailing is washed from the dump by a 1-in. water pipe into a launder, 6 in. sq., which runs to the top of a winze. Here the pulp enters a similar launder which descends along the 40- to 50-deg. dip of the seam floor to the ninth level of the mine. The stope to be filled has been dammed at the lower end by a dry wall *W*, see Fig. 145, strengthened by poles P, and similar partition walls are built at right angles to cut it up into longitudinal panels. The fine waste B is piled above *W*, and covered with old matting *M* from the cyanide tanks. When flushing begins, the sand settles quickly, the water filters through the matting and dams, whence it runs to the sump to be pumped to the surface.

This water is used again after a little lime has been added to neutralize its acidity and render any entrained colloids harmless to hinder a quick settling. To save water, the launders are kept on a minimum gradient of 10 deg. The water used is 6 to 10 per cent. of the tailing by weight. The cost of filling is given at 2.1 d per ton, but as only 100 tons of tailing are sent underground daily, this probably does not include wear of the launders. The filling sets hard in 2 or 3 days. When a stope is completely filled, it only settles 10 per cent. of its height when crushed by the formation after the pillars have been removed.

The residual cyanide of the tailing leaving the mill has been destroyed by exposure on the old dumps, so that no poisonous results have so far ensued from using tailing as mine filling. In order to utilize fresh tailing, the cyanide must first be rendered innocuous. This is not

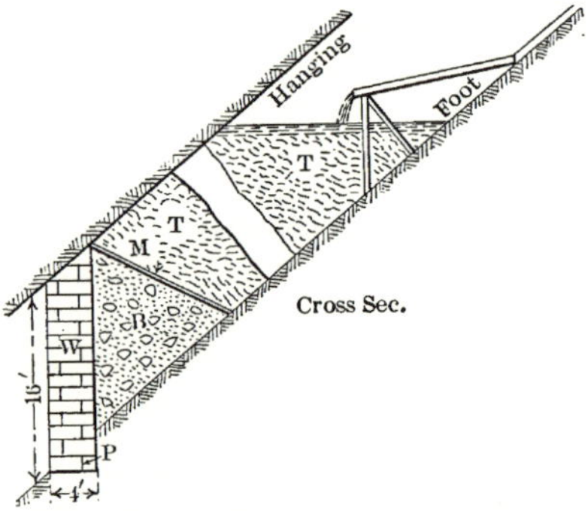

FIG. 145.—Dam for holding slush, Transvaal.

urgent at present, because the old tailing dumps are immense. Flushing the leading vats direct into the mine, however, would save the expense of conveying the tailing to the top of the very high dumps and of redigging it before flushing. Hence some mines are now getting ready for direct flushing.

The flushing system is now being freely used in the Rand to fill stopes *not* under buildings, in order to prevent the damage to the workings and the shaft pillars which is liable to ensue from pillar-drawing, especially as the mines get deeper. The rock tailing available for filling is much more resistant to crushing than the anthracite refuse of Example 59, and is strong enough for a filling at any workable depth. On the central Rand, there are two contiguous parallel beds, the Main Reef below, and the Main Reef Reader above, separated by a thin rock parting. As the Main Reef is the leaner, it has hitherto been neglected in many mines, but it is expected that the flushing system will now greatly facilitate its extraction under the worked-out stopes of the Main Reef Leader.

CHAPTER XXII

COMPARISON OF VARIOUS MINING SYSTEMS

The grouping of the practical Examples of this book has been based on the system described in each case in the general title of the chapter, while each example has also an individual title suggesting its own distinctive features. Mining being an applied rather than a pure science, the problem of choosing an appropriate system for any mine is a commercial question. Nevertheless, the problem should be viewed in the broad instead of the narrow sense. Greater temporary profits may mean lesser ultimate profits for the whole property owing to loss of ore. A given system may be cheap to operate but dangerous, and the cost of resulting accidents may greatly over-balance any temporary gains.

Besides the question of ore recovery and safety for life and property, the choice of mining systems involves such mechanical engineering problems as those of hoisting, haulage, pumping, ventilation, and lighting, and such miners' problems as those of breaking ground and controlling excavations under given conditions. Other things being equal, that system is preferable which minimizes the dead work for shafts, cross-cuts, and raises; which breaks the ore with the least drilling and explosives, and which keeps the mine open with the least artificial support. A system must adapt itself to the district's labor, whether skilled or unskilled, cheap or dear, scarce or plentiful, and must fit the mineral market conditions, whether steady or fluctuating. If shut-downs are likely to be periodic from strikes or glutted markets, the mining system chosen must allow the mine to lie idle with a minimum damage to its workings.

As the advantages and disadvantages of each Example have already been given with its description, only a general comparison need be made here between the different Examples.

The drag-line excavators of Cuba described in Chapter VI are only superior to steam shovels for deposits having a rough bottom on which the cost of track-laying for the latter system would not be justified by the depth of the ore to be shovelled from a given trackage. The shallower deposits of the Mesabi Range and of Ely, Nev., are well adapted to steam-shovel work, as neither the top nor the bottom of the ore lenses are sufficiently irregular to offer serious difficulties in reaching all ore or in laying tracks, and dumping ground for stripping is nearby. At Bingham, Utah, however, natural conditions do not favor steam shoveling because the top of the ore deposit is nearly parallel to the surface of a precipitous mountain and not only is it difficult to shovel stripping and ore separately, but much expensive grading is required for track connections to the numerous shovel benches. It is not likely that steam shovels will again be installed in such rough topography as that of Bingham, for surface-milling for thin, and underground-caving for thick, capping is far preferable. For properly laying out open cuts, preliminary prospecting is especially important; therefore its various details have been carefully described in Chapter VI. The opencut steam-shovel work of Example 5 avoids all the dangers and most of the expense of underground coal mining and is eminently suitable for all flat seams with thin covers; it has been used for the excavation of the Clinton iron ore bed of New York.

The open quarry of Puertocitos of Chapter VII is well suited to a sidehill location, but for other sites the milling system avoids the shoveling of the broken rock and should be adopted where the ore body is large enough to warrant the cost of the preliminary development necessary. The hour-glass chutes of Example 9 are especially suitable to work on a large scale and in winter, for the smaller boulders can be blasted at leisure in the warm chute with no danger to the pit men from flying pieces.

In Chapter VIII, the difference in the systems of Examples 10 and 11 are slight and are due to the fact that flat-holes are better adapted to the Joplin formation than the usual down-holes of quarrying. It is evident that deposits are only adapted to underhand stoping when they have considerable vertical height. Flat deposits must have a strong roof and sub-vertical veins must be of ore sufficiently strong to form a self-sustaining back over a large stope, unless a saddle-back support of timber can be rigged up as suggested for Example 13. Unless spots of lean ore occur so as to be utilizable for pillars, the ordinary underhand systems may involve considerable loss of good ore in pillars, excepting where auxiliary back-caving can be used as described in Example 13. The Mitchell system is an attempt to adopt underhand stoping to soft ground, by combining it with square setting. This system has promise of considerable usefulness in suitable formations.

Among the advantages of underhand stoping are the cheap drilling and breaking that go with underhand benches and the fact that no broken ore is locked up in the stopes as in all shrinkage systems, or lost or contaminated by mixing with waste as in the filling and caving systems. Timbering and developing work is at a minimum in underground quarrying but, as an offset, all broken ore has to be loaded by shoveling, which disadvantage may be overcome by the milling system of Example 13. In stoping veins only one stope can be worked at a time on one level on each side of the shaft, but several levels can be stoped at once by keeping the top ones farthest advanced and leaving a longitudinal pillar, or at least a shelf, of ore under each level to support the track over the stopes beneath. The underhand systems give no space for stowing waste in the stope where it is broken and are therefore best adapted to clean ore.

The first four examples of the simpler overhand stoping systems of Chapter IX are suited to strong ores and wall rocks. In both the Wolverine and the Homestake systems, all the broken ore has to be loaded by shoveling, but this has been preferred to the greater initial expense of building chutes for whose smooth working the Wolverine footwall's dip is scarcely sufficient. The use of triangular ore pillars or "rills" above the drifts, is an ingenious device to save timber. They can be as easily recovered when the level is abandoned as pillars in other places. By leaving enough broken ore or "shrinkage" close to the face, money is locked up it is true, but the expense of timber scaffolds is saved for the miners in steep veins. The broken ore also acts as a temporary support to prevent caving in case the stope walls are weak or the roof pressure is great. By using overhead chutes for filling the cars, a line of stopes can be worked simultaneously along one level, but this system gives no space for storing waste in the stope where broken. The stoping system of Example 18 has panel-cores inside of the square set frames like Example 12, but its method of attack is quite different and the system of the former is less well adapted to soft ground.

Where the walls are so weak as to cave on emptying the stope or where no ore can be spared for pillars and where much broken vein matter is worthless, the systems of Chapters X and XI, which use permanent waste filling instead of the shrinkage of Chapter IX, are applicable. In the American examples of Chapter X, the dry-walled drifts, the descending hangwall, the rill chutes and the auxiliary milling and square setting are noteworthy for Examples 19, 20, 21 and 22 respectively.

For the soft ore bodies of Bisbee with their irregular shape and their uncertain boundaries, square-setting in panels seems to be especially suitable for the reasons given in the description of Example 23. The soft walls and the fact that a large portion of the broken material has to be left in the stope as waste precludes the use of any caving system except perhaps one which, like Example 43, is well timbered and allows of stowage near the face.

Considering the foreign systems of Chapter XI, a similarity can be seen between the Australian Example 25 and the American Example 21, and between the Mexican Example 23, the Australian Example 26, and the American Example 22. In Example 27, the fragmentary nature of the vein filling has necessitated a system where no opening need be made wider than a cross-cut.

Not only does the adoption of rills save the cost of timbering the back of the drift where the ore is strong, but it saves much labor as the broken ore slides down to the discharge chute and the filling from the level above slides from the delivery winze to its place with little or no handling. For the filling systems, the rills should have a dip of 37 deg. to permit the waste to slide continually and freely, while for the shrinkage systems, where they are used as slides but once, the rills need dip only 23 deg. Filling with "flat-back" chutes, where the stope is carried horizontally and the drift roofed with timber for its whole length, has the advantage over rill chutes in heavy ground of a filling whose top surface is horizontal. This feature offers a better base for timber cribs or sets to support a weak back.

In comparing shrinkage with waste-filling by rills, the former needs only about one-fourth the number of winzes from the level above as the latter, and none of the expensive ore-passes of filling are required for shrinkage whose ore is always drawn off at the bottom of the stope. Also air drills do not have to be continually shifted in shrinkage to allow the filling to be placed. Nor is any ore lost by shifting down into waste. Nevertheless, shrinkage ore is often more or less contaminated by scaling off of the walls when emptying the stope.

Ore bodies with walls hard enough not to cave when the shrinkage is withdrawn, and containing ore clean enough to be hoisted without sorting, are best suited to the systems of Chapter XII which combine shrinkage and waste filling. Example 28 uses, where the ore is weak or irregular, auxiliary cribbing and waste filling. Both Examples 29 and 30 use some back-caving to cheapen the breaking of the ore. The latter finds some underhand stoping advantageous in breaking down weak and dangerous stope-backs. Example 30 is the only one which is now extracting its pillars but there is no reason why Examples 28, 29 and 30 should not also remove any pillars of ore by slicing, after the adjoining rooms have been filled with waste.

The systems of Chapter XIII, where stoping and shrinkage in the rooms is combined with caving of the pillars, may be considered hybrids between shrinkage and the true caving systems of Chapters XIV to XVII. By making the rooms of the systems of Chapter XIII quite narrow, we could have a block-caving system with the rooms serving as mere cut-off stopes. The Miami system resembles that of Example 30 except that the latter waits to fill a room before slicing an adjoining pillar, while the former withdraws pillar and shrinkage simultaneously and avoids any filling whatever. Mr. Lawton's idea of keeping the weak ore of a 60-ft. room supported by a sharply arched back whose haunches rest on broken ore blown under them by specially placed holes, is a great saving over the cribs under the back of Example 26.

The Boston-Con. system resembles the Miami in the stoping rooms, but its failure to similarly support the haunches of the weak back with broken ore caused its working to be dangerous for the miners. The Duluth system resembles the Boston-Con. but, the ore being strong enough to stand over the rooms, there is not the same danger in its operation from falls of the back. Conversely, to get the pillars to cave sufficiently for passing a chute, they have to be kept thin. The pillars when block-caved, are evidently less under control than when sliced as at Miami, and the saving of the timber and breaking costs of slicing, may be offset by greater loss of ore, more pollution by the descending capping and more hung-up chutes.

As already noted, back-caving is an auxiliary in the overhand stoping of Examples 29 and 30, but the chute-caving of Chapter XIV is an attempt to break all the ore by back-caving except that taken out by necessary development openings. In the Hartford system the only development needed in stoping is that of the spiral raises around the caving cores of the back. This system, however, can only be successful with well defined and strong walls and a regular deposit of clean ore. By the use of sub-levels in Examples 36 and 37, chute-caving becomes safer, more easily controlled and better adapted to irregular deposits with weak walls, though even with them there is no place to store waste in the stops. The subsequent uncovering by the steam-shovel work of a caved portion of the Utah-Copper mine showed a heavy loss of ore from mixtures with capping. Much of this loss, however, was due to unsystematic work and is not inherent in chute-caving itself.

The block-caving of Chapter XV allows the cheapest mining of any underground system, yet it is liable to the heaviest loss of ore with unsuitable deposits. The bigger the block, the less liable is the detaching drop to crush it completely. Success at the Pewabic is due to the fact that the ore is sandy and friable. At other Lake Superior iron mines with hard and tough ore, block-caving proved a failure because only the lower part of the mass was crushed and the solid core would not descend to be loaded. Although the Pewabic spstem has no chutes, which are expensive to construct and likely to clog, it has, as an offset, to cross-cut the broken ore of the block and load it all by hand-shoveling.

The Mowry ore being friable and located in a small vertical pipe offers ideal conditions for block-caving, and the plan of placing a square-settled floor at the base of each block, saves the hand-loading expense of the Pewabic. In Examples 40 and 41 the blocks are much smaller than at the Pewabic. Consequently they can be crushed sufficiently by one drop to be drawn through chutes. Conversely, as much breaking may be required as with chute-caving at the Utah-Copper where it is reckoned that 30 per cent. of the ore is extracted by development work.

On drawing caved ore from chutes, it is essential to provide means for reaching the top of the chutes in order to break up any boulders that cause stoppage. This has been arranged for in the various examples cited so to save delay in tramming and an expensive interference with the mine's routine. At the Inspiration mine, the chutes are expanded

19

into drawing-off stopes and have been made the central feature of the system. These chute-stopes are ingeniously arranged to aid the self-crushing of the descending block and thus enable larger blocks and tougher ore to be handled. The proposed use of a mat of ore is also an innovation and should greatly decrease the loss of ore which occurs when the main block breaks partly into boulders which may hang up in descending and permit the sliding beneath of quantities of broken capping. Timber mats are worthless for block-caving because they are ground up by settling and there is no chance to repair them as in the case of slicing. A mat of ore, on the contrary, breaks into boulders before the main block beneath does and prevents any waste descending into the interstices of the latter during its slow disintegration. The higher the caved block, the less proportional loss from pollution of the top layer with the descending capping, so that at the Inspiration, with a block over 200 ft. high, it is hoped to reduce the loss of ore to under 15 per cent.

In the slicing systems of Chapter XVI and XVII, the ore is broken entirely by breast stoping instead of by the underhand or overhand stoping of previous chapters. The "slicing under mats" of Chapter XVI takes almost as much timber as square-setting, but an inferior quality may be used. Slicing has also the disadvantages of continually breaking the ore from the bottom under the heavy pressure of the superincumbent filling, and of poor ventilation. But it allows the use of a larger proportion of muckers to miners and a larger output per man than square-setting, and is not liable to the collapses to which the latter is subject in heavy ground unless closely filled. Considerable ore may be lost by a little carelessness in square-setting by its sifting into the filling beneath, but in slicing any such siftings fall onto solid ore and are recovered later.

The "slicing under ore in rooms" of Chapter XVIII resembles the "slicing under mats in panels" of Chapter XVII, but there is a large saving in the consumption of powder and of timber at the expense of a slightly greater loss of ore which should, however, never exceed a total of 5 per cent. The use of rooms, one set wide, rather than broad panels allows the use of rough unframed timber, and the recovery of a half-set width of ore on each side and outside of the room-set without any timber whatever. "Slicing under ore" takes less skill and is quicker than "slicing under mats." Moreover, the timber floor mats of the former system do not have to be nearly as heavy or as well interlaced as those of the latter. As can be seen in the given examples, either system is flexible and applicable to various classes of deposits, but in ore bodies of any size or irregularity, "slicing under ore" is far preferable excepting where the ore is too valuable to permit of any loss at all, or where it contains much waste, for only "slicing under mats" permits stowage of waste in the stope where broken. At the Kimberley mine, the ore is sliced and back-caved in galleries without the use of timber. This system is applicable to any large deposit where the ore is homogeneous and of sufficient firmness.

The nature of both capping and ore affect the choice of a caving system. A capping which breaks large and chunky is favorable to block or pillar-caving, for fine capping sifts down into the large spaces in the disintegrating ore blocks and pollutes the ore. On the contrary, friable capping favors chute-caving and slicing because the roof settles more evenly in these systems. In slicing, any chunks above would tend to break through the timber mat every time one floor was dropped 10 ft. to the next below.

In dropping a block of ore for caving, it is crushed both by the superincumbent weight and by attrition on the side walls. In "slicing under mats," all the ore is broken by explosives, but in the other systems, greater or less proportion is crushed by dropping. It is evident that the larger blocks and the shorter and fewer drops, the less crushing action is exerted on the ore. The order of crushing efficiency for the different systems would generally be "slicing under ore," chute-caving, pillar-caving and block-caving. The last named system would require the most friable ore of all in order to be successful. If the capping looks different from the ore, it is separated more easily when drawing down a chute. Fortunately, this is usually the case in the examples quoted for if not of a different color, the capping is more silicified and breaks in larger pieces than the ore and differently shaped.

In adopting a caving system for a mine, after due consideration has been given the size, shape, grade, location, etc., of the ore body, a person should be able to narrow the final choice to two possible systems. One of these systems will mean a higher mining cost along with a higher percentage of ore saved than the other. If the net commercial result is practically the same in both cases, it is usually better to choose the higher cost system not only for reasons of conservation of resources, but because any future increase in the market value of the ore will be of most advantage to the system making the best saving. Of the caving systems, block-caving should usually cost the least. Then follow pillar-caving, chute-caving, slicing-under-ore and slicing-under-mats in the order named. The percentage of saving is in inverse order, but individual conditions may change the order of cost and saving, respectively, to a considerable degree.

The continuous-face longwall system of Example 49 is adapted to flat seams with little gas. The oblique angled roads which are the distinctive feature of the Scotch system are here at 45 deg., but have other angles (up to 60 deg.) elsewhere. In Grundy County, mines have been worked by the Rectangular system, with all roads at right angles, but, although the latter makes it simpler to maintain a regular face, it requires longer haulageways and is therefore less in favor. Where the seam is sloping or where much gas may make it advisable to specially regulate or shut off the ventilation of a dangerous locality, a panel system is advantageous. The limit of dip for panel-longwall is about 40 deg., for seams with yet steeper dips the systems in vogue follow overhand stoping, either "with shrinkage" or "on waste," as described in Chapters IX to XI.

Example 50 gives the advancing, panel-longwall method for stoping iron beds. Although the method was in general use decades ago, it is still applicable to ore beds that present the above-mentioned conditions favorable for longwall. It is especially suited to dips steep enough to allow the broken ore to descend by gravity down the footwall into the chutes. Example 51 illustrates how the panel system can be applied to a seam with an irregular floor and a cracked roof, features which are often stated to be fatal to successful longwall. The handling of the coal along the face by a buggy on an endless rope allows the gateways to be 125 ft. apart instead of the 42 ft. of Example 49 or the 20 to 30 ft. of Example 50, and thus saves a proportional expense in the building and maintenance of roadways.

Unlike the three previous examples, Examples 52 and 53 consume much timber, but this defect is not inherent in the systems but is due to the fact that neither from the partings in the seams nor from the strata of the roof was sufficient material secured for stone pack-walls and cogs. The Vintondale conveyors are certainly labor savers for thin veins, and where their liability to frequent breakdown is obviated the present dislike of the miners for the system can easily be overcome. The "jig" roads at Westville are well adapted to the average 22 1/2-deg. dip, and the 41-ft. width of room is nearly the same as for the similar loading in Example 49. Westville illustrates that for deep mines with weak coal, longwall is the only practical system and there should be no difficulty, when the upper 7-ft. layer has been extracted and the roof has come to rest on its floor, in similarly removing the lower 10-ft. of the seam.

The various figures of Example 54 illustrate merely the simpler outlines of the pillar system. To suit all the conditions of different mineral districts, an endless variety of layouts has been devised for modern mines. The entries vary from 2 to 8 headings abreast, and the angles at which the butt and cross entries meet, and at which the latter meet the main entries, is often made acute, to suit the dip, instead of 90 deg. The ratio of room-width to pillar-width is varied to suit the formation, the depth of cover, the seam, the method of mining and the desired size of mineral. The selection of mechanical equipment and the coupling of entries and shafts for effective and safe ventilation and for rapid movement of a large output of mineral, especially at the shaft bottom, are problems requiring special treatment for each case.

The Nelms' layouts of Examples 55 and 56 are well arranged to produce a large product from a small area. Such intensive mining, when thus systematically carried on, conduces not only to safety but also to the minimum cost of equipment and operation.

Examples 57 and 58 illustrate methods far in advance of much of the bituminous mining practise of America. The thickness, fine quality and accessibility of the Pittsburg seam has given it a value ranging up to $3,000 an acre, and this fact has tended to promote scientific working and a high percentage of recovery. In Example 57, illustrating mining for coking coal, the first method of recovering pillars without the use of curves is suited for narrow pillars, while the two last curve-methods are suitable to the wider pillars necessary under heavy covers. The three methods described in Example 58 are those for the market-coal mines and are designed to produce a maximum of lump coal with the use of machine cutters. They are laid out for a large output and with careful work make possible a total recovery of 90 per cent. of the seam.

The flushing system of Chapter XXI has been a boon to coal mining in thickly populated districts. It is really nothing but the old European filling system, using fine surface waste for filling and putting it underground without any of the manual labor that makes the old system so expensive. The cleaning up of the great culm heaps in the anthracite region, by washing and flushing, has rendered usable large areas of valuable surface land as well as made possible the recovery of much pillar coal, otherwise irremovable, under townsites. Less liability to accidents from roof falls is another advantage of flushing. Example 60 illustrates the first application of flushing to metal mining, and it is already having many imitators in the Rand. The flushing system is applicable to all metal mines where the ore lies in beds of large area, and where plenty of tailing, or sand, and water is available. Its application, in any given case, is a commercial question; the cost of digging, transporting, and damming the filling and of repumping the flushing water being set off against the net value of the recovered pillars and surface acreage.

CHAPTER XXIII

PRINCIPLES OF MINE EVALUATION

Mineral deposits, especially those of the rarer metals, are often irregular in size and uncertain in tenor and these peculiarities introduce the element of chance into their extraction. Though mining is thus naturally more or less speculative, it is by no means "just a gamble" as the average layman is apt to opine, for sound finance with the aid of applied science has in recent years gone far to place mining upon as legitimate an investment basis as agriculture or transportation.

In evaluating a mine we have two forms of property to consider, land-values and improvements. Of the two forms the first is paramount, for with a just sufficient quantity S of "ore" (payable metallic or non-metallic mineral) in the mine to return, with interest, the net expenditure on improvements necessary for its extraction, the property has no value at all as mineral land. Any excess of ore beyond S grants a land-value to the mine, while any deficit under S means a loss of liquid capital in the development.

It is thus essential to observe and study the physical features of the property, as the basis for evaluation, and these may be grouped under twelve topics as follows: (1) Quality of ore; (2) Quantity of ore; (3) Location for transportation; (4) Available markets; (5) Local topography; (6) Water conditions; (7) Climate; (8) Food supply; (9) Fuel supply; (10) Labor conditions; (11) Government; (12) Necessary investment.

Topic (1) is got by first sampling the ore body; and then making the chemical analyses and working tests on these samples necessary for determining the values in the ore and their commercial utilization. Topic (2) may be estimated from the development openings and from the geological nature of the deposit.

On this topic G. E. Collins has presented as a practical example, a section (Fig. 146) of a fissure-vein, the ore reserves of which he classifies as "positive" (proved on three sides), "probable" (proved on two sides), and "possible" (proved on one side). Upon the case thus supposed, he thus remarks.

"Most of us have known, many of us have experienced, cases where blocks of ore so exposed (even) on four sides have been found to enclose a large barren patch in the center. The amount of weight to be attached to exposure on four sides, on three, on two, or on one only, varies with the conditions of each particular case . . . Who, after examining the section (Fig. 146), and bearing in mind the phenomenal lateral extension of the ore-body shown, can doubt, that far more reliance can be placed on blocks marked in the section as "positive ore," even although some of them are exposed on three sides only, than on many blocks opened up on four sides in veins where the ore-bodies are of the bunchy and erratic type which we must recognize to be, after all, by far the most frequently encountered? Even the "possible ore" of Fig. 146—ground where the ore is proved to exist on one side only—can be depended on to a far greater extent than usual.

"My contention is that the amount of evidence to be required when making estimates of ore tonnage, and the number and nature of classes

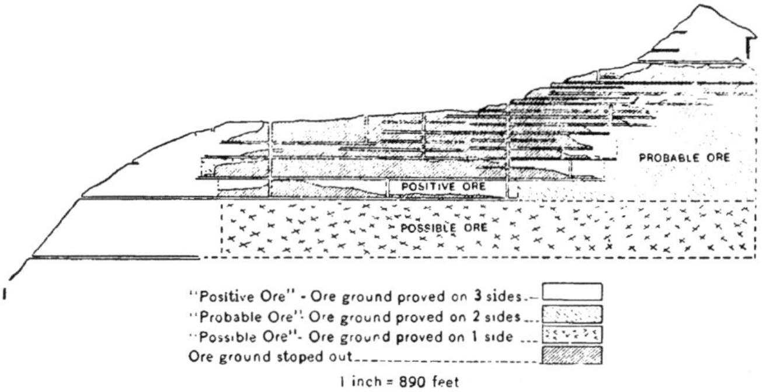

FIG. 146.

into which estimates should be divided, must depend on our general conclusions as to the nature and permanence of the ore-bodies in the mines under consideration, the distance between the workings, etc. I do not think that any rules can be made which will lessen the necessity for dependence on the examining engineer's individual judgment; and I distrust all cast-iron classifications, which do not allow for the infinite complexity of natural conditions. The only useful general rule is that no estimate of tonnage should be made unless accompanied by sketches indicating the basis on which it rests."

Topic (3) relates to the transportation facilities such as existing water or land-ways or the cost of constructing same to reach the nearest navigable water or railroad leading to market; also to the animal or other power available for transportation. This topic determines the cost of shipping the output to market and of importing those supplies not obtainable locally. The available markets of (4) may be fluctuating or steady in their demands and may accept in some cases the crude, in others only the finished mineral product. The local topography of (5) not only fixes the possible methods of mine development, but also affects the location of works, housing and transportation. Under (6), or water conditions, we consider water power, mine drainage, water supply for works and people, and the condition of roads and waterways.

The climate (7) of the past has affected the oxidation or enrichment of the ore-body and at present it affects wind power, health, food supply and the conditions of outdoor work. Topic (8) of food supply depends on the regions' soil and climate and on the skill, number, and industry of the present, or possible future, local farmers.

The fuel supply of (9) is a question of the existence near the mine of available forests, peat bogs, oil or gas wells, or coal seams. Topic (10) of labor hangs entirely on the local population in those many cases where induced immigration is impractical except for the official staff. The rate of wages, the adaption to mining work, the industry and the docility of the natives are all vital practical questions. The government of (11) is often the determining factor for an enterprise, not only does its system of taxation more or less affect costs but revolutions or predatory officials may sometimes render inadvisable any investment whatever.

The capital necessary for (12) can only be calculated after a summation and correlation of the previous eleven topics. The total investment will be apportioned between "fixed" and "working" capital; the former being the amount needful to expend in mine development and equipment and in the construction of works and roads, to make the property suitably productive, and the latter being the additional sum required during productive operations.

The following is an attempt to form a formula by which a mine can be quickly evaluated, after the above pertinent physical data have been collected from observations on the ground by a competent mining engineer.

Assumptions

Let G = price to be paid for the mining property.
Let M = cost of developing the mining property to yield y tons of ore daily.
Let P = cost of suitable plant to treat y tons of ore daily.
Let p = value of said plant when the mine has been exhausted.
Let C = total fixed capital investment.
Let W = working capital investment.
Then let $C+W$ = total capital investment.
Let y = required yield of ore daily in tons.
Let Y = yield of ore yearly in tons.
Let d = number of producing days per year.
Let u = average operating profit per ton of ore.
Let R = rate of interest to be earned on total investment of $(C+W)$.
Let r = rate of interest to be earned on sinking fund annuity.

Let v = tons of positive ore available in mine (see Fig. 146).
Let x = tons of probable ore available in mine (see Fig. 146).
Let z = tons of possible ore available in mine (see Fig. 146).
Let m = fractional factor to change probable to positive ore.
Let n = fractional factor to change possible to positive ore.
Let Q = tons of total ore available in mine.
Let t = time in years to exhaust mine at rate Y.
Let b = time in years for mine to reach production of y tons.
Let A = annuity to be paid to sinking fund to equal C at end of t years at r, compound interest.
Let a = annuity to be paid to sinking fund to equal C at end of t years at, r, simple interest.

Derivation of Formulæ

We have directly from the foregoing assumptions the following equations:

$$C = G + M + P \qquad (1).$$

$$Y = dy \qquad (2).$$

$$Q = v + mx + nz \qquad (3).$$

$$Q = tY \qquad (4).$$

Then since we must balance the two debits, of the interest to be paid during $(t+b)$ years and the sinking-fund annuity to be paid during t years, against the two credits, of the operating profit from the total available ore and the selling value of the abandoned plant, we have:

$$(t+b)(C+W)\,R + tA = tYu + p \qquad (5).$$

To determine A, we can substitute in that algebraic formula for the annuity which involves the rate of compound interest, the number of annuity payments and the capital to be refunded and get:

$$A = \frac{Cr}{(1+r)^t - 1} \qquad (6).$$

Substituting in (5) this value of A, we have:

$$(t+b)(C+W)R + \frac{tCr}{(1+r)^t - 1} = tYu + p \qquad (7).$$

With the observed data, and equations (1), (2), (3), and (7), we can proceed to evaluate the mine. Equation (7) includes nine factors, any one of which, except t, can be determined by solving a simple equation. When t is unknown (as is commonly the case), it is difficult to solve the equation by common algebraic methods. To obviate this difficulty it may be assumed that simple, instead of compound, interest is to be earned on the sinking-fund annuity; and the resulting difference in the result will act as a safeguard on the side of the mine-buyer. Additional safeguards for the buyer will be the assumption, in equation (7), that the factors C and W are invested at once, whereas their expenditure usually occupies a considerable period.

With simple interest, the sinking-fund annuity (a) becomes the first term of an arithmetical progression, of which the common difference is the annual interest (ar) gained on the annuity, the sum is the capital (C) to be repaid, and the number of terms is (t) the period of years. Then, by substitution in that algebraic formula for the sum which involves the first term, the common difference and the number of terms we have:

$$C=\frac{t}{2}\left(2a+(t-1)\,ar\right) \text{or } a=\frac{2C}{t(2+r(t-1))} \qquad (8).$$

Substituting this value of a for A in (5), we have:

$$(t+b)\ (C+W)\ R+\frac{2C}{2+r(t-1)}=tYu+p \qquad (9).$$

To determine from equation (9) the unknown factor t, will involve only the solution of a quadratic equation; and this is most conveniently performed after the substitution of the numerical values of the other factors; since the direct solution for t, of (9) as a literal equation, is very lengthy.

If no interest be earned on the annuity payments, the second term of equation (9) will become equal to C, and we shall have:

$$(t+b)\ (C+W)\ R+C=tYu+p \qquad (10).$$

Since equation (10) involves only first powers, it can be used as a quick check on the approximate accuracy of the solution of equation (9) for t.

Practical Examples

I

On a certain property in the southwest, examined by the author, it was required to find the minimum available ore that must be found by prospecting operations to warrant the capital expenditure required to inaugurate production on a given scale. The unknown factors were hence Q and t, and the known were estimated from the collected data to be:

G = \$40,000; M = \$56,000; P = \$39,000; p = \$10,000; W = \$15,000; y = 100 tons; d = 300 days; u = \$3; R = 0.15; r = 0.06; b = 2 years.

From equation (1), C = 40,000 + 56,000 + 39,000 = \$135,000.

From equation (2), Y = 300 × 100 = 30,000 tons.

From equation (9), $(t+2\ (135,000+15,000)\ 0.15+\frac{2\times 135,000}{2+0.06(t-1)}$

$=t(30,000\times 3)+10,000$; or

$$27t - 14 - \frac{108}{1.94 + 0.06t} = 0; \text{ or}$$

$$27t(1.94 + 0.06t) - 14(1.94 + 0.06t) - 108 = 0, \text{ or}$$
$$0.81t^2 + 25.77t - 67.58 = 0; \text{ whence}$$
$$t = 2.42 \text{ or} - 34.24.$$

It is evident that the positive value of 2.42 years is the one desired. Substituting the values of t and Y in equation (4), we have:

$$Q = 2.42 \times 30{,}000 = 72{,}600,$$

the number of tons of available ore, that should be found by prospecting, to satisfy the conditions of the case.

II

A consumer, using 5000 tons of a certain metal yearly, wishes to acquire a mine which would furnish his whole supply. He has found a mine which, by the expenditure, besides the purchase-price, of $200,000 for development and plant, and the provision of $50,000 working capital, would enable him to obtain annually the required supply of metal from 60,000 tons of its ore, at an operating profit of $4 per ton. A year will be required to develop and equip the property, and the available ore will last, at the required rate of production, for twenty years, at the end of which period the plant will be worthless. If interest on the total investment be reckoned at 6 per cent. and on the sinking-fund annuity at 5 per cent., and if, in addition, it is necessary, while working the mine, to make a net saving of 1 cent a pound on the whole metal supply of the consumer, what price could he afford to pay for the mine?

Here C and G are the unknown factors, and we have:

M + P = $200,000; W = $50,000; u = $4; p = 0; annual saving on metal, $100,000; Y = 60,000 tons; R = 0.06; r = 0.05; t = 20 years; and b = 1 year.

From equation (1), C = G + 200,000.

Substituting this value in equation (7), and remembering that the sinking-fund annuity is to be increased by the $100,000 of annual saving on metal supply, we have:

$$(20 + 1)(G + 200{,}000 + 50{,}000)\ 0.03 + \frac{20(G + 200{,}000)\ 0.05}{(1 + 0.05)^{20} - 1} + (20 \times 100{,}000)$$
$$= (20 \times 60{,}000 \times 4) + 0; \text{ or}$$

$$1.26(G + 250{,}000) + \frac{G + 200{,}000,}{2.654 - 1} + 2{,}000{,}000 = 4{,}800{,}000; \text{ or}$$

$$2.084(G + 250{,}000) + G + 200{,}000 + 1.654\ (2{,}000{,}000)$$
$$= 1.654\ (4{,}800{,}000); \text{ or}$$
$$G = \$1{,}264{,}500,$$

the maximum allowable price for the mine.

Conclusions

The practical value of such calculations as the foregoing may be plausibly questioned, at least as regards all mines other than collieries, for which the quantity of available reserves can be estimated with a degree of confidence and precision not usually attainable in mines of the metallic ores. In reply to this probable criticism the author begs to offer the following observations:

1. There are, in fact, besides collieries, more mines than we commonly realize, the actual reserves of which can be measured, and the probable or potential reserves estimated. Among such might be instanced many quarries, massive ore-bodies already explored by boring, etc. To all such cases, mathematical formulas of valuation are directly applicable.

2. With regard to the very large number of metal-mines, in which v, the certain ore-reserve, is relatively small, while x, the probable, and z, the possible reserve, are so uncertain as to make the determination of m and n, as the moduli representing reasonable expectation, merely a function of the temperament of the observer, yet even in such cases, the observer himself may, very likely, be steadied in his judgment, and aided to form prudent conclusions, by such calculations as will clearly show him to what extent his hopes, fears and guesses enter into his opinions. It may seem absurd to employ mathematical methods in the discussion of data so largely indefinite; but the quantitative determination, even of our ignorance, is a recognized application of mathematics; and "the probable error" has a value of its own, not less real, though it be less authoritative, than the rigorously demonstrated certainty. In other words, the discussion by exact methods, even of more or less uncertain data, is a valuable check upon the hasty, sentimental or temperamental general impressions which often claim the authority of conclusions. Such a check is the more important, because many mining investors are tempted to overlook the essential proposition that a mine is "a candle, burning as both ends"; that its value is constantly diminished by its product; and that its annual profits must cover, not only a satisfactory income from the investment which has been made in it, but also the progressive repayment of the investment itself. All this sounds very elementary; yet it is too often overlooked in the enthusiasm of speculation. Each investor, if he thinks of it at all, dismisses it from his mind with the reflection that he will have abundant opportunity to "unload" during the period of dazzling prosperity which he foresees for the mine. This is the chief reason why mining has not yet become universally a regular business. That desirable result will be greatly promoted when investors purchase a mine, either with the positive intention, not of selling it out, but of working it out, or else, at a price based upon the hypothesis of such an intention. Consequently, the more this consideration is emphasized by theoretical calculations of value like the preceding, the better.

APPENDIX I

Articles from which Much of the Subject-matter of the Book was Excerpted

Book			Published Article		
Chapter	Example	Title	Magazine	Date	Writer
1		Explosives	Mining & Eng. World	Apr. 29, May 13, 1911	The author
2		Blasting	Mining & Eng. World	Mar. 11, 1911	The author
3		Compressed Air	Eng & Min. Journal	Jan. 18, 1908	The author
4		Excavations	Mining & Eng. World	May 27, 1911	The author
5		Drainage	Mining & Eng. World	June 22 1911	The author
			Ditto.	Nov. 28, 1908	E. B. Kirby
6	1	Cuba	Trans Min Eng.	March, 1911	D. Woodbridge
	2	Mesabi	Mining Science	Dec. 24–31, 1908	The author
	3	Utah Copper	Mines & Methods	Sept., 1909	C. T. Rice
	4	Nevada-Con.	Mines & Methods	Sept., 1909	C. T. Rice
			Min. & Sc. Press	Mar. 4, 1911	E. E. Barker
	5	Danville	Mines & Minerals	Oct., 1907, Sept., 1910	Anonymous
7	6	Puertocitos	Mines & Methods	Feb., 1910	C. T. Rice
	7	Mesabi	Ditto, Example 2		
	8	Traders	Ditto, Example 2	Apr. 22, 1909	The author
	9	Alaska Treadwell	Trans. Min. Eng.	Vol. 34	R. A. Kinzie
8	10	Southeast Mo.	Mines & Minerals.	Nov.andDec.,1901	The author
			Eng. & Min. Journal	Feb. 12, 1910	Anonymous
	11	Southwest Mo.	Mines & Minerals	Nov., 1907	J. H. Polhemus
			Mining & Eng World	Sept. 26, 1908	Otto Ruhl
	12	Bisbee	Eng. & Min. Journal	July 23, 1910	M. J. Elsing
	13	Section 21	Mining Science	Jan. 7, 1909	The author
9	14	Wolverine	Mining & Eng. World	Mar. 26, 1910	The author
	15	Homestake	Eng. & Min. Journal	July 9, 1910	J. Tyssowski
	16	Gratz	Eng. & Min. Journal	Apr. 6, 1907	The author
	17	Alaska-Treadwell	Ditto, Example 9.		
	18	Veta Grande	Eng. & Min. Journal	Nov. 12, 1910	M. J. Elsing
10	19	South Range	Mining & Eng. World	Mar. 26, 1910	The author
	20	Minnesota	Mining Science	Dec. 17, 1908	The author
	21	Superior & Boston	Mines & Minerals.	Sept., 1910	R. L. Herrick
	22	Metcalf	Eng. & Min. Journal	July 16, 1910	P. B. Scotland
	23	Bisbee	Mines & Minerals	Feb. 1907	The author
11	24	Los Pilares	Mines & Minerals.	Sept., 1910	E. M. Robb, Jr.
	25	West Australia	Mines & Methods	Dec., 1910	R. Allen
	26	British	Mines & Minerals	May, 1907	A. J. Moore
	27	Proprietary	Mines & Minerals	May, 1907	A. J. Moore
12	28	Central	Mines & Minerals.	May, 1907	A. J. Moore
	29	King	Ditto, Example 22.		
	30	Coronado	Ditto, Example 22.		
	31	Los Pilares	Ditto, Example 24.		
13	32	Miami	Mines & Minerals	July, 1910	R. L. Herrick
	33	Boston-Con	Mines & Methods	Sept., 1910	C. T. Rice
	34	Duluth	Ditto, Example 18.		

Book			Published Articles		
Chapter	Example	Title	Magazine	Date	Writer
14	35	Hartford	Ditto, Example 13.		
	36	Pioneer	Mining Science	Dec. 10, 1908	The author
	37	Utah-Copper	Ditto, Example 3.		
15	38	Pewabic	Mining Science	Apr. 29, 1909	The author
	39	Mowry	Mines & Minerals	July, 1907	The author
	40	Detroit-Copper	Min. & Sci. Press	Dec. 24, 1910	W. L. Tovote
	41	Commercial	Mines & Minerals	Oct., 1907	The author
	42	Inspiration	Mines & Methods	June, 1909	C. T. Rice
16	43	Old Jordan	Mines & Minerals	Oct., 1907	The author
	44	Cumberland-Ely	Mines & Methods	Sept., 1909	C. T. Rice
	45	Oversight	Ditto, Exam le 6.		
17	46	Lake Superior	Mining Science	Feb. 18 and Apr. 22, 1909.	The author
	47	Mercur	Eng. & Min. Journal	June 18, 1910	R. H. Allen
	48	Kimberley	"Diamond Mines of S. A."		G. F. Williams
18		Roof-Pressure	Mines & Minerals	April, 1907	H. Briggs
19	50	Montour	Trans. Min. Eng.	October, 1891	H. H. Stoeck
	51	Providence	Mines & Minerals	April, 1907	Anonymous
	52	Vintondale	Coal Mining Inst.	June, 1907	J. I. Thomas
	53	Nova Scotia	Mines & Minerals	June, 1909	J. G MacKenzie
20	54	Gen. Pillar	Mines & Minerals	Jan., 1899	J T. Beard
	55	Retreating	E. & M. Journal	Dec. 26, 1908	H. J. Nelms
	56	Adv. Retreat	E. & M. Journal	Apr. 23, 1910	H. J. Nelms
	57	Connellsville	Mines & Minerals	July, 1907	G. S. Baton
	58	Pittsburg	Traps. Min. Eng.	March, 1910	F.Z.Schellenberg
21	60	Rand	E. & M. Journal	Feb. 5, 1910	Anonymous
23		Evaluation	E. & M. Journal.	March 14, 1903	G. E. Collins
			Trans. Min. Eng.	Apr. 1907	The author

INDEX

D

Spotted something that isn't quite right?

Let us know and we'll fix it for future reprints. And as a thank you for your diligence, we'll send you a free Sothis book of your choice.

publisher@sothis.press

This page intentionally left blank

Made in United States
Orlando, FL
05 November 2026

90231724R10192